Genesis Redux

Genesis Redux

Experiments creating artificial life

Edward Rietman

Windcrest®/McGraw-Hill

New York San Francisco Washington, D.C. Auckland Bogotá
Caracas Lisbon London Madrid Mexico City Milan
Montreal New Delhi San Juan Singapore
Sydney Tokyo Toronto

FIRST EDITION
FIRST PRINTING

Published by Windcrest, an imprint of McGraw-Hill, Inc.

Library of Congress Cataloging-in-Publication Data

Rietman, Ed.
Genesis redux : experiments creating artificial life / by Edward Rietman
p. cm.
Includes bibliographical references and index.
ISBN 0-8306-4503-9
1. Artificial intelligence. 2. Robotics. 3. Biological systems.
I. Title.
Q335.R55 1993
003'.7—dc20 93-1149
CIP

Acquisitions editor: Brad Schepp
Editorial team: Robert Ostrander, Executive Editor
David M. McCandless, Editor
Production team: Katherine Brown, Director
Wanda S. Ditch, Layout
Susan E. Hansford, Coding
Jodi L. Tyler, Indexer
Design team: Jaclyn J. Boone, Designer
Brian Allison, Associate Designer
Cover design: Sandra Blair Design, Harrisburg, Pa. WT1
Cover photograph: © Orion Press/Westlight 4460

Clarke's Laws of Prophesy

First Law

When a distinguished but elderly scientist states that something is possible, he is almost certainly right. When he states that something is impossible, he is very probably wrong.

Second Law

The only way of discovering the limits of the possible is to venture a little way past them into the impossible.

Third Law

Any sufficiently advanced technology is indistinguishable from magic.

Arthur C. Clarke
Profiles of the Future

This book is dedicated to the pioneers: to persons not afraid to venture a little way past the limits of the possible, and to persons not afraid of tackling big projects that may fail.

Contents

Acknowledgments

It would not have been possible for me to write a book of this complexity without the help of many persons. I thank Joe Griffith for the ecosystem photos; Mark Tilden and Vince Lambrecht for photos of the robots; Liz Zimmerman for literature searches; and Ron Powers for suggesting the book. I thank Bob Frye for help on Chapter 7, and Cody Stumpo for putting up with my mishagahss.

Mostly I thank my wife for her support: Suzanne Harvey. Love and blisses to her.

Preface

The primary question of artificial life is: What is life? No satisfactory answers to this question exist at this time. The individual scientific disciplines have failed. Artificial life is a multi-disciplinary field comprised predominantly of biology, artificial intelligence, robotics, and chemistry. By synthesizing complex systems that are analogous to life, we hope to better understand life and the living state. At the same time, we as an intelligent species might out grow our chauvinism. There is no reason to expect that life must be made of DNA-RNA-Protein biopolymers. And, even more important, we might learn not to impose our morals on another species.

The focus of this book is on advanced models ranging from computer simulations of mathematical biomorphs to *in vitro* artificial ecosystems and in silico robotics.

Introduction
What is life?

"If you want to understand life, don't think about vibrant, throbbing gels and oozes, think about information technology."
Dawkins, The Blind Watchmaker, 1987

The contents

This is the second book on artificial life and is designed as an introduction. At the heart of the subject is the question of what life is. We attempted to answer this question in the first book by examining the chemistry of the origin of biological life and genetic codes. In this book, we will examine models of very simple interacting systems that result in complex structures resembling biological life forms, self-reproducing systems, self-organizing systems, and hardware robotics. The book is intended for advanced readers, advanced computer hackers, and persons just breaking into the field of ALife. The book includes an extensive reference list with hundreds of sources.

The user of the book is expected to have a broad knowledge of science, computer programming, and a strong mathematics background. Much of the necessary mathematics is introduced in the first chapter, but most of it should be only a review for the reader. This is not an easy-reading book. I make no apologies for this. Much is demanded of the reader. I hope the book inspires some, is useful to others and, at least, is interesting to the rest.

Chapter 1 discusses cellular automata and self-organization from simple finite state machines. The chapter goes into much detail on the principles of cellular automata, and the physical chemistry of cellular automata as an artificial matter for creation of new life forms. Programs are included in the chapter.

Chapter 2 is on artificial neural networks as brains for artificial life. The chapter builds the foundation for an understanding in later chapters where neural networks are mentioned. Programs are included.

Chapter 3 is computer modeling of autonomous robots and the dynamics of learning robots. Programs are included.

In Chapter 4, many of the principles of self-organization are applied in the development of biomorphs. These are geometrical structures based on either simple dynamics of interacting subsystems or mathematical equations, such as fractals and iterated function systems. The chapter includes many computer-generated photos and figures. Programs are included.

Chapter 5 is about applying the biological genetic algorithm to development of computer programs, learning systems, and optimization. Programs are included.

Chapter 6 is concerned with ecological models. This includes conventional ecological models and more *avant-garde* models. Closed system ecology as an experimental tool is covered and includes photos of actual closed systems from in vitro experiments. Programs are included for using a spread sheet to solve systems of differential equations.

Chapter 7 is on hardware robotics. A huge amount of literature is reviewed from both the conventional approach to robotics and the new approaches based on stimulus response actions. Small systems are discussed. The chapter includes many photos of systems and hints on building autonomous robot systems, in silico.

What is life?

This book is about attempts to create artificial life (much of this section of the introduction is reproduced in the first book but included here for completeness). In order to say something about the life form we create we must know what life is. This is a question that has been pondered since the beginning of civilization (at least). Chris Langton defines artificial life as:

> ". . . the study of man-made systems that exhibit behaviors characteristic of natural living systems. It complements the traditional biological sciences concerned with the analysis of living organisms by attempting to synthesize life-like behaviors within computers and other artificial media. By extending the empirical foundation upon which biology is based beyond the carbon-chain life that has evolved on Earth, Artificial Life can contribute to theoretical biology by locating life-as-we-know-it within the larger picture of life-as-it-could-be." (Langton, *Artificial Life*, 1989).

This is an excellent working definition of artificial life. I would change the carbon-chain to read DNA-RNA-Protein. Research is currently underway to create artificial life in test tubes. This life would be based on a carbon chemistry but not DNA-RNA-Protein.

We could list all the attributes we ascribe to life. A short list might consist of reproduction, metabolism, evolution, growth, self-repair, and adaptability. No one would debate if a single cell is alive, but is a biological virus alive? It is a complex assemblage of biopolymers. It is a supramolecular system. It does not reproduce without a host. It does not metabolize. It can evolve by genetic mutation. It can grow (depending on your perspective). It can undergo self-repair, but mutations might result. It can adapt (to a degree) to new environments (i.e. new hosts) but will most likely undergo mutation. So, in the case of a biological virus, we see it is at the border line of living and nonliving in the biological world.

Perhaps our "definition" of life is too restrictive. This is suggested by the fact that we cannot state if certain entities are alive or not. One thing is clear from Langton's quote above, we are very much biased in our perception of what is a living thing. I doubt we would recognize a unique life-form if we stepped on it. Much of our bias is a result of caveman-like instincts. Life is something you can eat. We all recognize that nearly all living forms on this planet are useful as a food source for us.

Varela et al. (1974) defined the term autopoiesis as a definition of life. Later, other workers picked up on the term and expanded the concepts (cf. Maturana and Varela, 1980; Moreno et al. 1990; and Zeleny, 1977). An autopoietic system is a complex system defined by the relations between its components, and the components recursively participate in the system to produce other components. (This is what I have referred to above as self-maintaining and self-repair.) The component network of an autopoietic system is contained within some defined unit of space. This keeps the components from dispersing in the environment. Thus the autopoietic system can compensate for perturbations and remain as a unity. In Chapter 1, we will explore some automata systems that satisfy this definition of life. And in Chapter 6 we will examine closed ecosystems that clearly can be classified as one organism. In the case of these closed ecosystems, free energy enters the system as sunlight and heat is dissipated. The components are small living organisms that can utilize the energy to reproduce and thus maintain the larger system. The larger system, the ecosystem, is thus a living organism.

A broader and more objective definition would be based on thermodynamics. A life-form is a complex system capable of self-maintenance. A strange attractor (cf. Chapter 1) is a complex system capable of self-maintenance. Strange attractors and life-forms are energy dissipating systems. High-quality energy enters the system and low-quality heat energy leaves the system. Thus there is a rise in the universal entropy. The system is capable of "cheating" the Law of Entropy on the local state, but not on the global state. In a high entropy state, the atoms of a system are dispersed at random. In a low entropy state, such as a living organism, the atoms are highly organized in many supramolecular structures.

A bacterium is capable of growth and reproduction when placed in a medium with the appropriate molecular components. A crystal is capable of growth from a seed crystal placed in a supersaturated solution of the appropriate medium. Both are alive, from a thermodynamics point of view, but the bacterium is more alive than the crystal. This suggests an analog scale as a measure of life. Wesley, (1974) suggests the following relation for a measure of life:

$$L = \left(1 - \frac{S_i}{S_o}\right) \frac{R}{M}$$

where S_i is the entropy of the atoms in the organism in question and S_o is the entropy of the same type and quantity of atoms in a maximum entropy configuration (i.e. dispersed). The rate of energy flux R per unit mass M is given by R / M. What does this imply? A bacterium is alive because the arrangement of the atoms within it are in a crystal-like state. They are in a complex arrangement of biopolymers. The energy flux per unit of mass also could be thought of as the amount of energy entering the cell. For example, if the bacterium is a cyanobacteria then the cell is able to use sunlight to convert external (to itself) matter into low-grade energy or heat. Similarly a growing crystal of sodium chloride is able to use the free energy available in the solution to incorporate the external (to itself) sodium chloride molecules and produce a higher entropy condition in the universe.

So, a life form converts high quality energy into low quality energy. It does this by decreasing the entropy in its local environment (i.e., within itself) and increasing the entropy in the universe. Using Wesley's measure of life we see that a candle flame is a life-form, but not of the same measure as a bacterium cell or a dog. An autonomous solar powered LEGO robot is also a life-form, but again not the same measure as an ameba or algae. A cellular automata configuration that self-organizes reduces the entropy within its local neighborhood (an abstract space-time) but increases the entropy in the universe (same abstract space-time). It also is a life-form. In Chapter 7, we will examine autonomous mobile robots that utilize solar energy for input and dissipate heat energy. Many of these robots are capable of tracking the sun like a photovore, and thus stay alive. At night, they go into a dormant state until the sun again reaches them. These are clearly living organisms.

Implications and speculations

We can conclude that even if none of these examples are life-forms, it is perfectly reasonable to assume that one day we will succeed in creating an artificial life form. The examples in this book are the primeval slime out of which the new generations of life-forms will evolve. The examples in this book are, at the very least, emergent life-forms and models thereof.

We are rapidly approaching the point of Human equivalence for computers (cf. Moravec, 1988; Chorafas, 1989). These computers are the early ancestors for Silico sapiens. The Neanderthal equivalent of an artificial life-form may not be just metal, plastic, and silicon. It might very likely consist of complex biopolymers derived from genetic engineering and chemical syntheses. It may have a "brain" built from a biocomputer.

What impact will this new emerging species have on Homo sapiens? Deken, (1986) sees them as being our friends and working with us. Wesley, (1974) views the subject very objectively and concludes from a logical analysis that the Silico sapiens will inherit the Earth. Mankind and his genetic code will have only been a stepping stone in the evolution of life. The next stage will be a life-form that is far more robust. He gives some very convincing arguments. There is already a vast imbalance between man and machines in many domains. Moravec, (1988) sees yet a different future somewhere between these two extremes. He even predicts that man and machines will become a hybrid. All these futures may seem to be too far from the present to be of much concern, but de Garis, (1989) thinks we should begin now to address these tough questions. I couldn't agree more. We have a tendency to pussyfoot around until the last minute. We also have a tendency to solve the really hard problems when we are "backed up against the wall." I suspect that Moravec's vision is the more realistic. Silico sapiens will be our friends. They should be treated like an alien life-form of vast capabilities. We should not think that we can pull the plug on them any time we want.

1

CHAPTER

Chemical physics of artificial matter

In this chapter, we will explore self-organizing and self-reproducing systems constructed by cellular automata. We will see that most of the cellular automata rules, capable of self-organization or self-reproduction, are at a threshold or phase transition region between a stable state and chaos. We will examine in some detail the phase diagram for three cellular automata universes. And finally we will examine specific cases for cellular automata rules that fall in the phase transition region and observe the self-organization. Chapter 5 in *Creating Artificial Life* (TAB Book # 3719) is a prerequisite to this chapter.

Artificial matter and phase diagrams

Langton (1986) refers to a cellular automata universe as artificial matter, which is an excellent description because, with this artificial matter, we can create artificial molecular structures, artificial biochemistries, and artificial life forms. We can even create universal Turing machines and cellular automata computers (Poundstone 1985).

The number of rules/states possible for a cellular automata universe can be very large. We would like to be able to say something about what general classes of structures can form in a given cellular automata universe. Classification is the first step in many of the sciences. In zoology, for example, comparative studies are made for classification purposes. In cellular automata, there are obviously at least three types of patterns that can evolve in a given universe. One type is the ordered state in which all the cells are inactive or in which there are only a few simple oscillators in the entire universe. Another possible evolved universe is one in which nearly all the cells are changing state in a random manner. This is not the same as all cells changing from one state to another in an ordered sequence. Between these two evolved universes exist complex states with traveling groups of cells and/or groups of cells self-organizing and dissolving, and groups of cells involved in complex limit cycles.

After classification we would like to say something about the dynamics of cellular automata and the physical chemistry of the artificial matter. What types of biochemistries can evolve in this universe? Wolfram (1983) investigated the entire rule space for $K = 2, N = 3$ (2 state, 3 cell) one-dimensional cellular automata. For this cellular automata, there are 2^{2^2} or 256 rules. Of these rules, only 32 give rise to reflection symmetric rules (i.e., 100 and 001 yield identical values). Another restriction is that if a state is 0 and its neighborhood is 0, then the new state should also be 0. After considering these restrictions, we have 32 legal rules. Wolfram has identified four classes of cellular automata for this small universe.

Class 1 Evolve to a limit point.
Class 2 Evolve to a limit cycle.
Class 3 Evolve to chaotic behavior and strange attractors.
Class 4 Evolve to a long period cycle or oscillator.

Li et al. (1990) have suggested a spectrum of the rule space for this cellular automata. If we combine Wolfram's Class 2 and 4, we have only three classes: limit points, limit cycles, and chaos. Now defining a parameter λ, similar to Langton (1986), we can use it to prepare a phase diagram and spectrum of the cellular automata rule space. This λ parameter is just the fraction of ones in the rule table. The rule table for this $K = 2, N = 3$ cellular automata universe is an 8-bit binary vector. For example, we have the three-neighbor state (111) at time $t = 0$, and (x0x) at $t = 1$ the center cell changes to 0. There are eight possible configurations to describe the changes in state from $t = 0$ to $t = 1$. An example of the eight-state change is given in Table 1-1. There are 32 such eight-bit binary rule vectors as shown in Table 1-2.

Table 1-1 State changes for Wolfram's one-dimensional CA.

111	→	0
110	→	1
101	→	0
100	→	1
011	→	1
010	→	0
001	→	1
000	→	1

Table 1-2 Lambda values for the 32 legal rules of Wolfram's one-dimensional cellular automata.

Decimal	Binary	Number of ones		Lambda
0	0	0		0.0
4	100	1		0.125
16	10000	1		0.125
32	100000	1		0.125
36	100100	2	*class I*	0.25
48	110000	2		0.25
54	110110	4		0.5
60	111100	4		0.5
62	111110	5		0.625
8	1000	1		0.125
24	11000	2		0.25
40	101000	2	*class II*	0.25
56	111000	3		0.375
58	111010	4		0.50
2	10	1		0.125
6	110	2		0.25
10	1010	2		0.25
12	1100	2		0.25
14	1110	3		0.375
18	10010	2		0.25
22	10110	3		0.375
26	11010	3	*class III*	0.375
28	11100	3		0.375
30	11110	4		0.5
34	100010	2		0.25
38	100110	3		0.375
42	101010	3		0.375
44	101100	3		0.375
46	101110	4		0.5
50	110010	3		0.375
20	10100	2	*class IV*	0.25
52	110100	3		0.375

As I have just said, we can combine Wolfram's Class 2 and 4 into a single class. (Later we will see that there really are four classes and, thus, that this combining is not correct.) After doing that and counting the number of rules per class (see Table 1-2), we can prepare a spectrum of the three classes. Figure 1-1 is a plot of the fraction of rules in an individual class and the number of 1's in the rule table. The most interesting feature of this spectrum is that the three classes have their maximum at a unique λ value. As λ increases, more of the rules exhibit chaotic rather than periodic dynamics. This parameter acts like a pseudo temperature. As

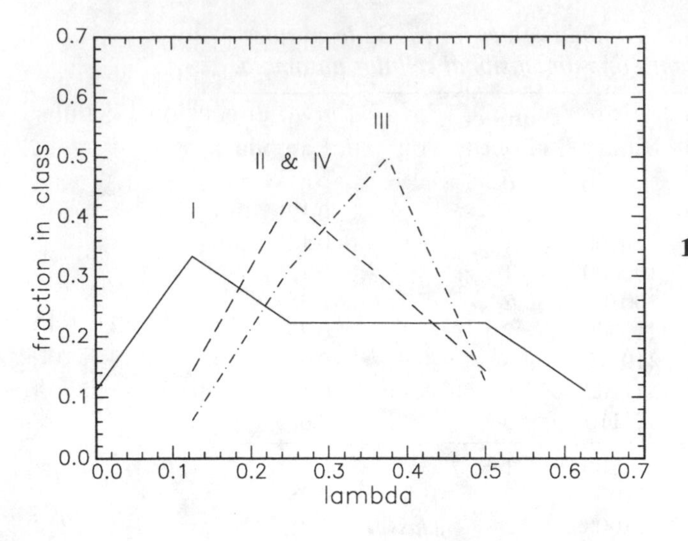

1-1 Spectrum of cellular automata classes.

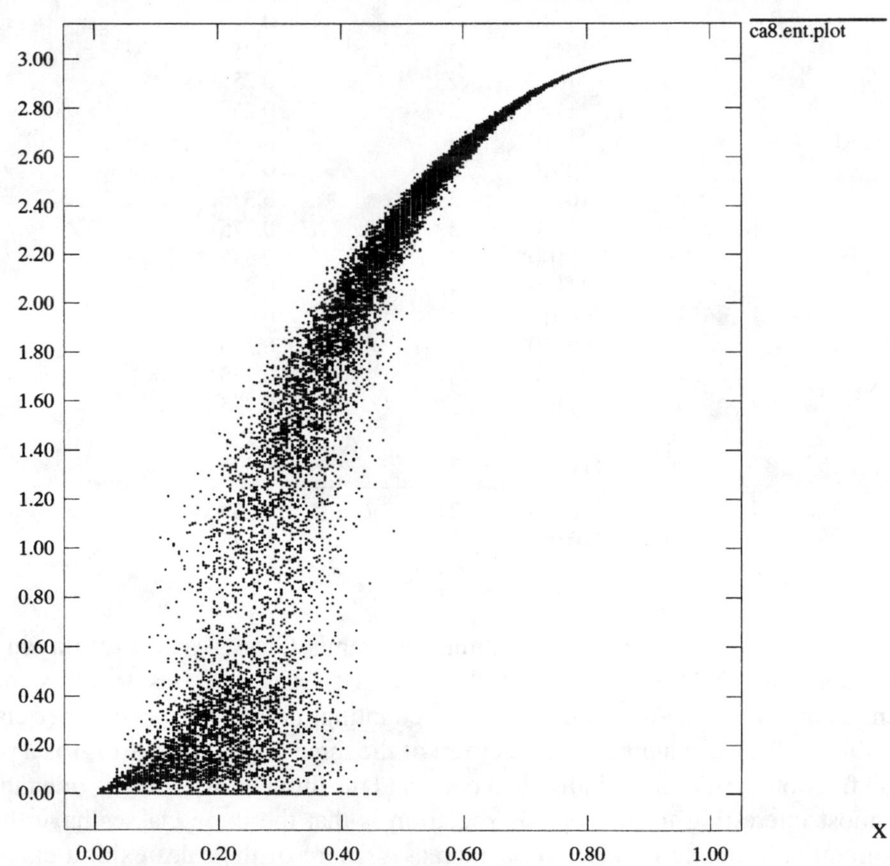

1-2a Entropy plot for $K = 8$, $N = 5$ cellular automata.

the parameter increases, the dynamics goes from limit points (solid) to limit cycle (liquid) to chaotic (gas). From this, we observe that the λ parameter might be a useful description of rule space. How well does it apply to two-dimensional cellular automata? Can we discover another pseudo thermodynamic parameter?

Li et al. (1990) have made extensive study of the $K = 8, N = 5$ cellular automata universe. This system was chosen because Langton (1984) discovered a self-reproducing structure in it. Their studies are of the λ parameter on the long term entropy of the system. The program ca8.c is designed to reproduce one of their curves and also show the effects of λ on the Hamming distance. This program takes about 3 days to run on a Sun spark workstation and about 16 hours on a CRAY. I wouldn't recommend running it on a PC because it could take many weeks. Figure 1-2a shows a plot of entropy versus the λ parameter.

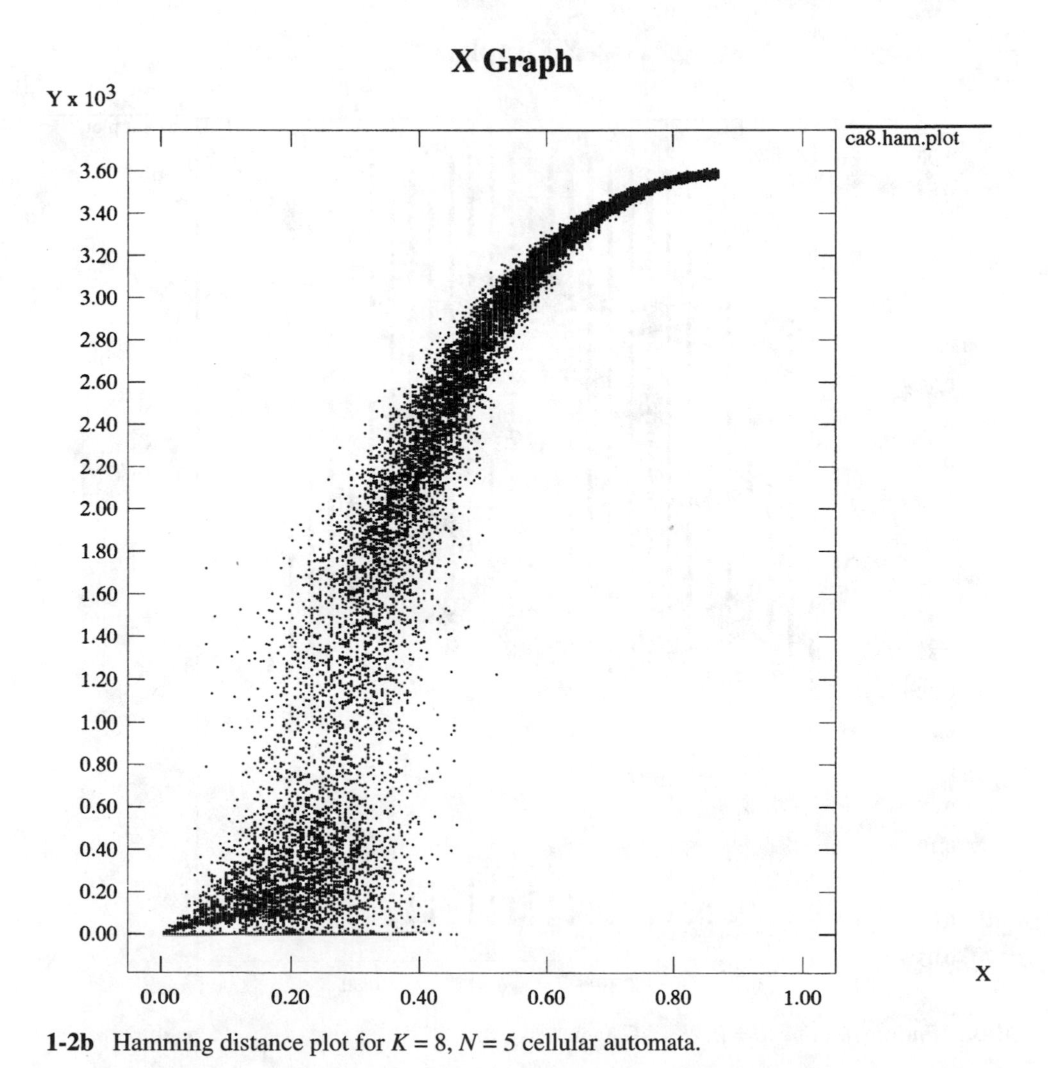

1-2b Hamming distance plot for $K = 8$, $N = 5$ cellular automata.

As to be expected, when our pseudotemperature increases, so does the entropy. This entropy increase is actually a first order change and can be described by a mean field theory (Langton, 1991).

$$s = -(1-\lambda)\, log\,(1-\lambda) - \lambda\, log\left(\frac{\lambda}{k-1}\right)$$

Figure 1-2b is a similar plot showing the Hamming distance between the 1000 and 1001 iteration. Notice the expected similarity between the Hamming plot and the entropy plot.

Making slight modification of the ca8.c, Martin Taylor (1991) has extended the work of Li et al. (1990) to include several more rule spaces. Figure 1-3a is an entropy vs. λ plot for $K = 2$, $N = 5$ cellular automata universe. Notice that, in this

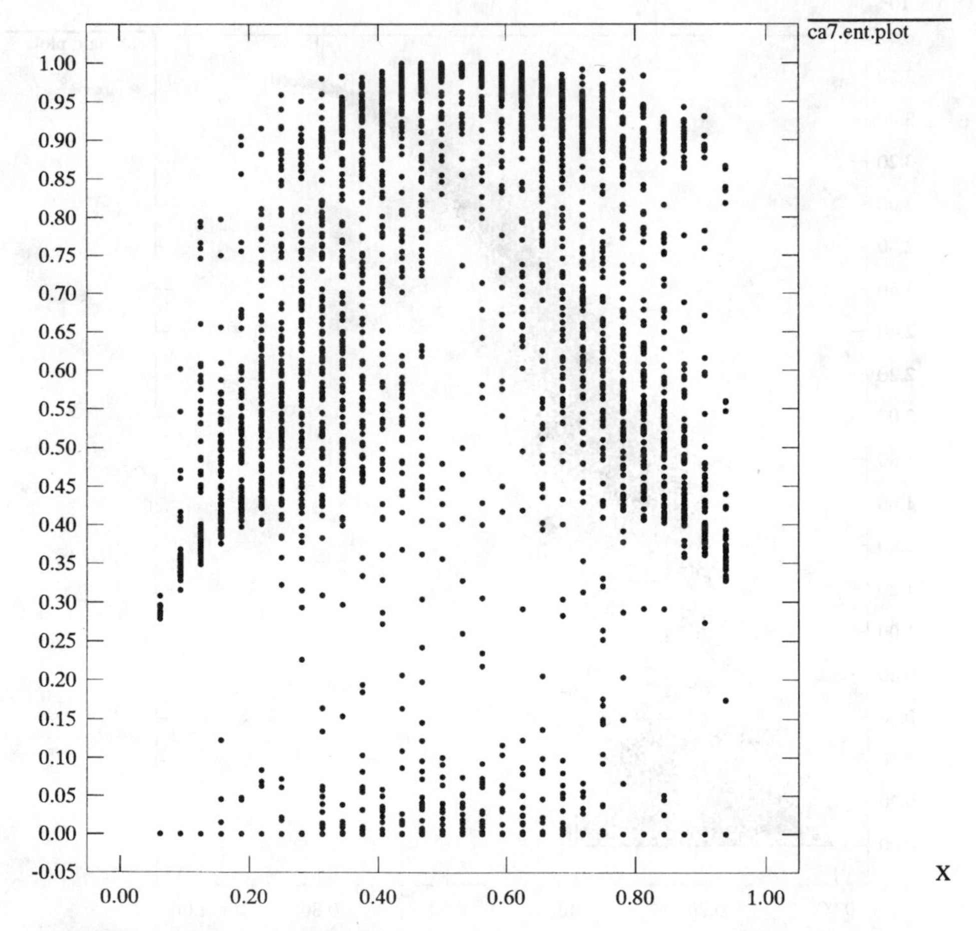

1-3a Entropy plot for $K = 2$, $N = 5$ CA.

plot, there is a great deal of symmetry in the rule space. This is to be expected for any $K = 2$ cellular automata.

Figure 1-3b is the corresponding Hamming plot, and Fig. 1-3c is the average of the entropy plot.

In all three plots, the symmetry in the rule space is obvious. The entropy reaches a maximum for $\lambda = 0.55$. Rule vectors with more ones—i.e., larger λ values—are symmetric. For example, (11100000) would give rise to a similar universe as (00011111).

Figure 1-4 is an entropy -λ plot for the $K = 4$, $N = 5$ universe, and Fig. 1-5 is for $K = 16$, $N = 5$.

Notice the similarity with $K = 8$, $N = 5$. In all these plots, we see a region or gap in which the density of the dots is lower than in the other region of the plot.

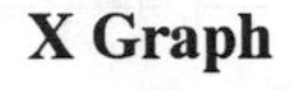

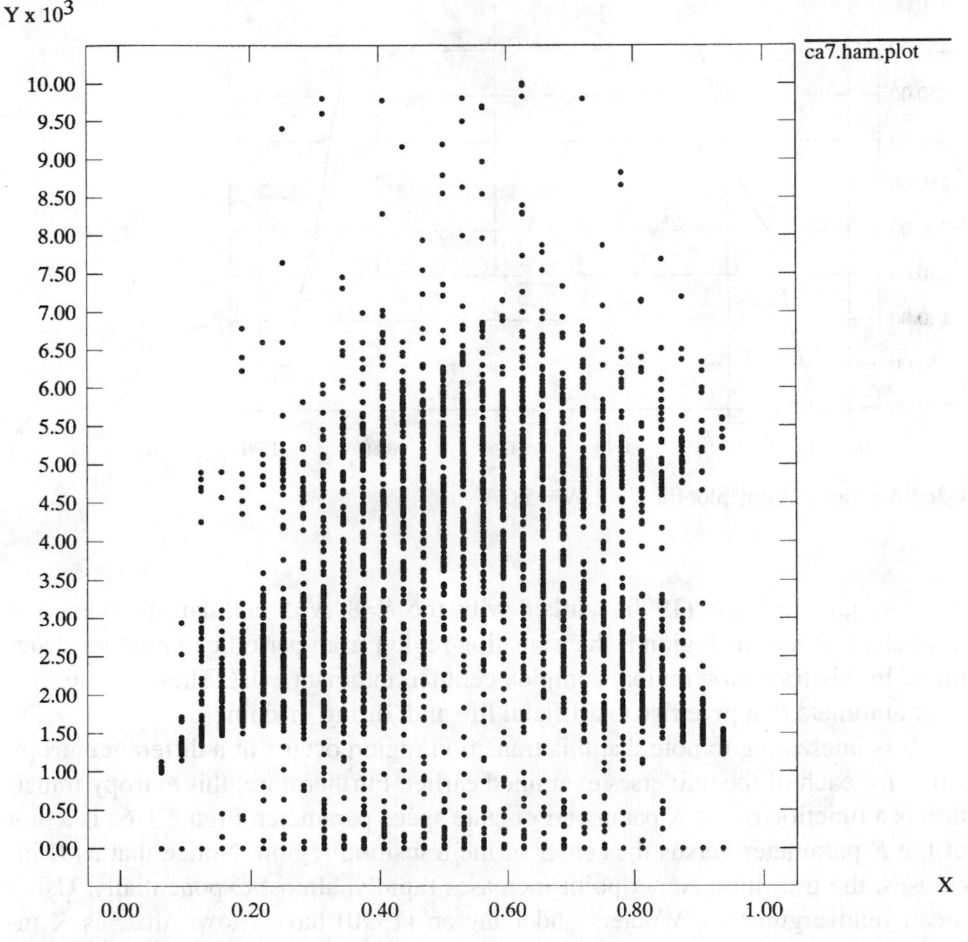

1-3b Hamming distance plot for $K = 2$, $N = 5$ CA.

X Graph

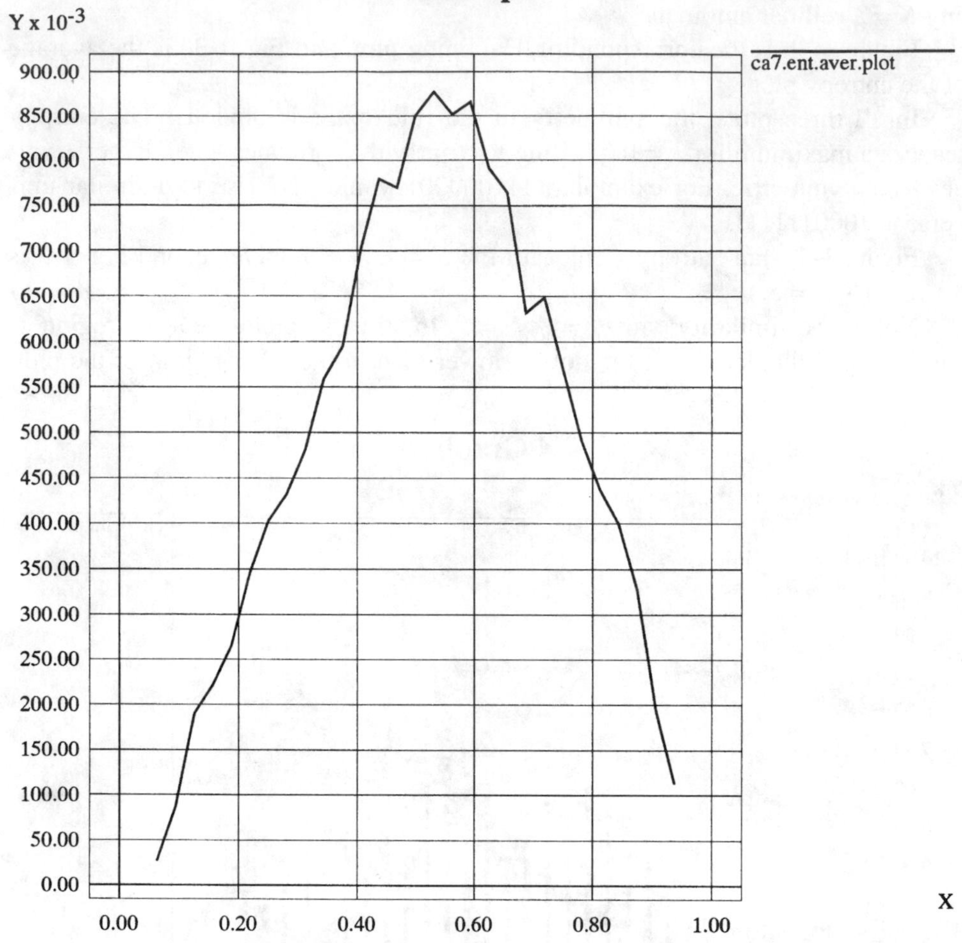

1-3c Average entropy plot for $K = 2$, $N = 5$ CA.

Wooters and Langton (1990) worked with the K=8, N=5 cellular automata and have shown that this region is the transition region from periodic to random structures. In this transition region, complex cellular automata exist. These are the cellular automata that give rise to artificial life and Turing machines.

It is interesting to note that this transition region occurs at a different entropy value for each of the universes examined earlier. Furthermore, this entropy transition is a function of the K parameter or state space parameter. Figure 1-6a is a plot of the K parameter versus the center of the transition region. Notice that as K increases, the transition center point increases rapidly, almost exponentially. Using mean field arguments, Wooters and Langton (1990) have shown that, as K increases to infinity, the gap or transition region becomes narrower.

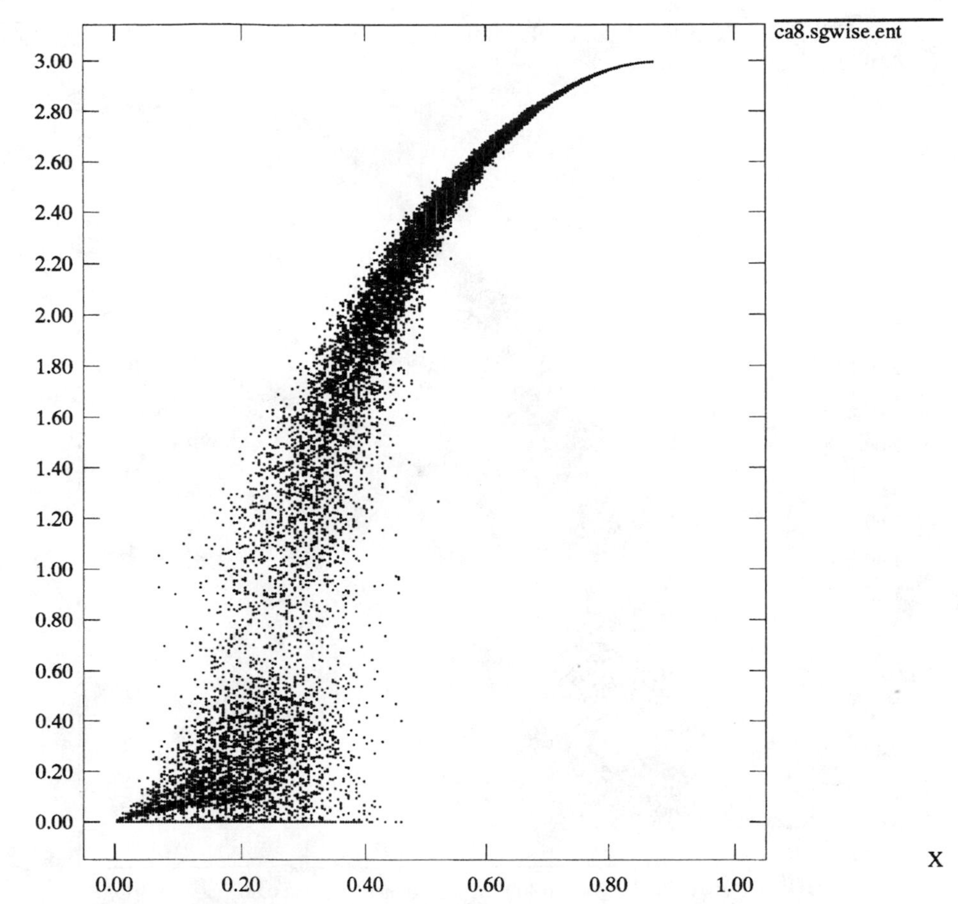

1-4 Entropy plot for $K = 4$, $N = 5$ cellular automata.

Figure 1-6b is a plot of K versus entropy showing the width of the transition region. This phase diagram, which should be considered only a schematic, suggests that the K parameter is analogous to a pseudopressure. For a given entropy (e.g., 0.7) at low K value (0.2), we are in a random region of cellular automata space. This is analogous to a low pressure gas. As K increases, along the 0.7 entropy line, we enter the complex region that is analogous to a condensed or liquid phase. At still higher K (pseudopressure), we leave the liquid region and enter the limit cycle or solid state region.

According to Langton's research, the width of this transition region is not nearly as wide as this figure indicates; the gap might be a computational anomaly and not really exist. In this picture, the transition would be a sharp boundary, a classical thermodynamic phase transition, between chaos and order. Strange at-

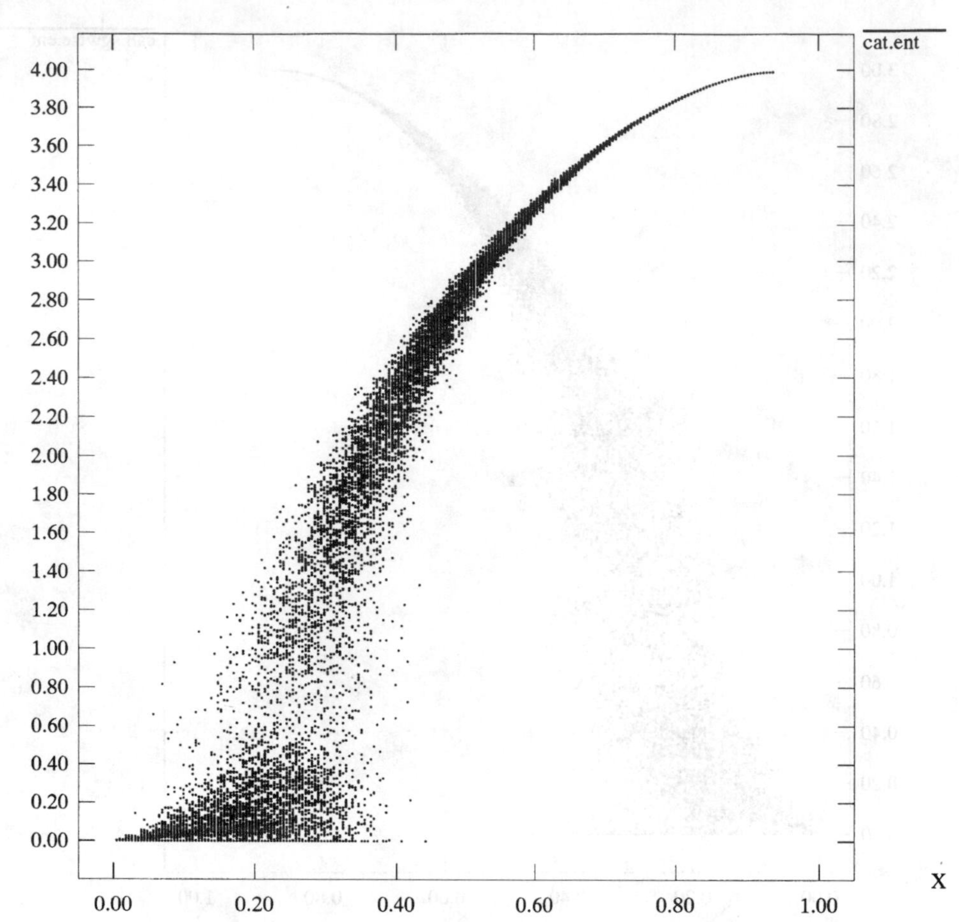

1-5 Entropy plot for $K = 16$, $N = 5$ cellular automata.

tractors exist in this boundary region. In the picture, I am proposing this boundary region is not sharp but might be more narrow than shown in Fig. 1-6b. This larger region would allow a greater range of the parameters that determine the complex systems between chaos and order.

These experiments suggest that phase diagrams of cellular automata can be represented by three parameters:

- λ (the rule space or pseudotemperature)
- Entropy
- K (the state space or pseudopressure)

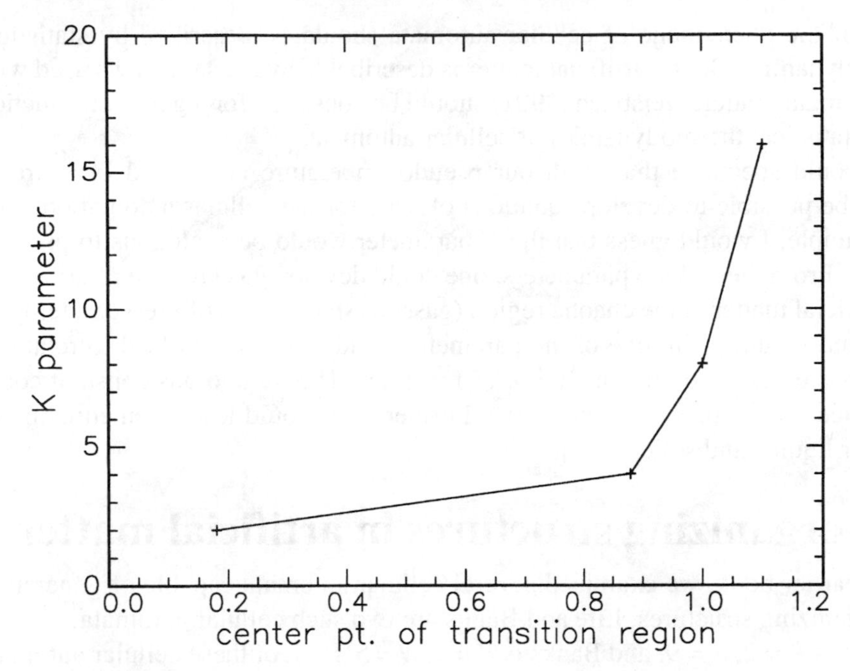

1-6a *K* parameter vs. the center point of the entropy transition region.

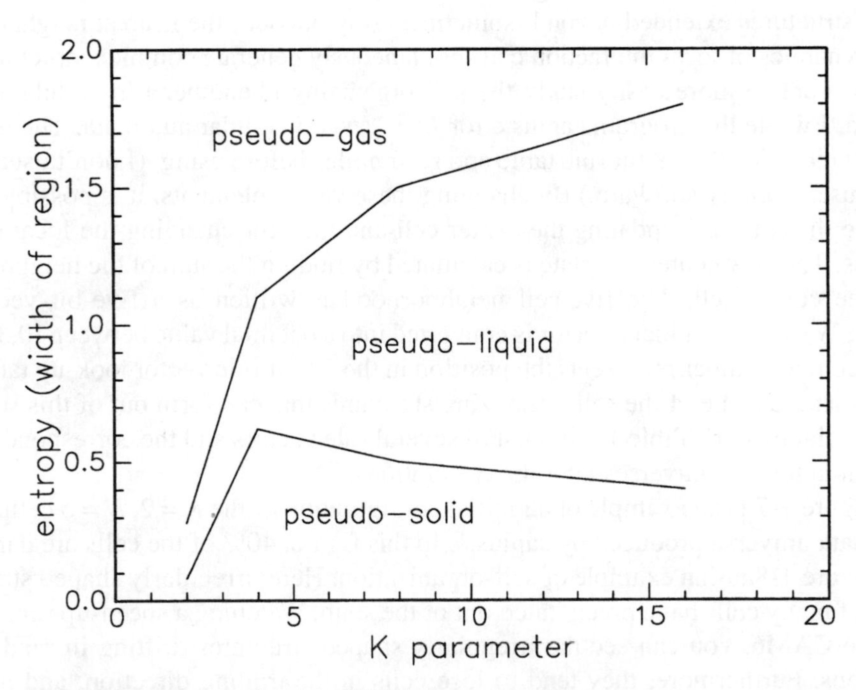

1-6b Phase diagram for the artificial matter of cellular automata.

Much of the phenomena of cellular automata should be described by statistical thermodynamics. So the artificial matter is describable by the same laws used with conventional matter. Weisbuch (1991) should be consulted for a good introduction to the statistical thermodynamics of cellular automata.

I would speculate that, with our pseudotemperature and pseudopressure, it should be possible to develop equations of state for the cellular automata matter. For example, I would guess that the N parameter would be analogous to pseudo-volume. From these three parameters, one could develop gas law like relations for the artificial matter in the chaotic region (gaseous state). Plots of pressure, temperature, and volume with one of the parameters held fixed should lead to relations analogous to the well known $PV = nRT$ for gases. The pseudo-gas constant could be deduced from the slop of the curve. Further work could lead to an equation of state for liquids and solids.

Self-organizing structures in artificial matter

In my earlier book, we examined several cellular automata capable of generating self-organizing structures. Life and Banks are two such cellular automata.

Life is $K = 2$, $N = 9$, and Banks is $K = 2$, $N = 5$. Both of these cellular automata fall within the transition region between periodic and random structures. In this transition region, the local rules give rise to large-scale structure. By large- scale, I mean structures extended beyond (sometimes way beyond) the nearest neighbors. The dynamics of local interaction can spontaneously generate complex structures.

In order to more easily study the self-organizing phenomena in cellular automata, I wrote the program caplus.c for $K = 2$, $N = 5$ cellular automata. The user simply modifies the 32-bit rule table and recompiles before using. (I don't usually write user friendly software.) By changing these vector elements, it is possible to change the rules for updating the center cell and thus for changing the local dynamics. The new center cell state is calculated by finding the sum of the neighbors and the center cell. The five-cell neighborhood is written as a five-bit vector (CNSEW), and this binary vector is converted into a decimal value between [0,31]. The decimal number represents the position in the 32-bit rule vector look-up table.

What are some of the self-organizing structures that can form out of this soup of artificial matter? Table 1-3 is a list of several rule vectors and the corresponding cellular automata universes after many iterations.

Figure 1-7 is an example of an initial configuration of the $K = 2$, $N = 5$ cellular automata universe produced by caplus.c. In this figure, 40% of the cells are dark.

Figure 1-8a is an example of self-organization. Here, irregularly shaped structures of many cells have precipitated out of the soup. Watching a speed-up run, using the CAM6, you can see the irregularly shaped structures drifting in random positions. Furthermore, they tend to lose cells in the trailing direction, and new cells are attached in the advancing direction. One is reminded of molecular recognition processes in a chemical reactor while viewing this at high speed. The irreg-

Table 1-3 Rule vectors for several figures in this chapter.

Rule vector	Figure number
11110101011101001110101111100001	6–8
10001010100001111101100000000000	6–9
00111101010001011000010000011110	6–10
00000101010001011000010000011110	6–11
10000111110111100101110101011111	6–12
00001100111000001011010101000010	6–13
11110011000111110100101010111101	6–14
00000000000110111110000101010001	6–15

ular objects act as molecular assemblers. On one side, they recognize and collect molecules, while on the other side they discard small structures that eventually break up into individual cells. These irregular structures could also be thought of as living organisms swimming in a sea with food particles. The food particles are absorbed on one side and waste products are discarded on the other. The waste products then break down and are re-absorbed by the sea. Long time iterations

1-7 Example of CA space at initial configuration with 40% of the cells in the dark mode.

1-8a Example of self-organizing structure. See Table 1-3 for rule vector. 30 iterations.

1-8b CA of Fig. 1-8a after 80 iterations.

show (Fig. 1-8b) that these irregular structures are growing or condensing to large complex structures.

Figure 1-9 is another example of irregularly shaped cellular structures. These are small molecular structures from this artificial matter.

Figure 1-10 shows self-organized structures that resemble the lattice animals engineered in *Creating Artificial Life* (TAB Book #3719). Figure 1-11 shows small L-shaped structures. These structures continually flicker in and out of existence. Figure 1-12 shows highly regular very large structures. These crystals self-organize from the artificial molecular matter.

Figure 1-13 is an example of self-organized structures that have primarily organized along an x and y axis. Again this is a type of lattice animal. Figure 1-14 shows self-organized irregular structures that have precipitated after only 12 iterations. The structures in Fig. 1-15 are small self-organized linear structures similar to the directed growth of lattice animals.

1-9 Example of self-organizing structure. See Table 1-3 for rule vector. 66 iterations.

1-10 Example of self-organizing structure. See Table 1-3 for rule vector. 36 iterations.

1-11 Example of self-organizing structure. See Table 1-3 for rule vector. 21 iterations.

1-12 Example of self-organizing structure. See Table 1-3 for rule vector. 13 iterations.

1-13 Example of self-organizing structure. See Table 1-3 for rule vector. 20 iterations.

1-14 Example of self-organizing structure. See Table 1-3 for rule vector. 12 iterations.

1-15 Example of self-organizing structure. See Table 1-3 for rule vector. 15 iterations.

Self-reproduction in artificial matter

In the last section, we have seen several examples of structures that self-organize from the cellular automata matter. We have also seen how some of these structures can grow to very large sizes. A reasonable question to ask is: Are there structures that can self-reproduce? Remember the Fredkin cellular automata. Simple structures were able to self-reproduce from this simple rule. *Creating Artificial Life* (TAB Book #3719) also showed self-reproducing structures on the advancing front of the random cellular automata program cell1.bas. In both of these cases, the self-reproducing structures could have any engineered shape. The Fredkin cellular automata has reversible rules, and the cell has random rules with probably some reversible rules forming from time-to-time.

Can we have complex structures self-reproduce without reversible logic? Von Neumann (1966) was the first to show that this is indeed possible. He discovered a $K = 29, N = 5$ cellular automata capable of self-reproducing. Before examining this cellular automata, let's look at the LIFE algorithm again.

Poundstone (1985) and Berlekamp and Conway (1982) each discuss the LIFE computer and show that it is a universal Turing machine. Recall from *Creating Artificial Life* that the LIFE algorithm can generate small stable structures and small oscillators. Figure 1-16 is a diagram (similar to Poundstone, p172) showing the frequency of occurrence, from a random field, versus number of pixels in the structure. This graph was produced by setting the smallest structure to a frequency of 1 and normalizing all the others to that. This smallest structure is known as the *block* and consists of a 2×2 pixel structure. It is a limit point.

Another important structure in the LIFE world is the *glider*. It occurs with a frequency of about 0.009 relative to the block at 1.0. The glider is a periodic structure that moves (dynamic cellular automata) across the field. Four time steps are needed for the glider to move forward by one cell. The glider is thus said to travel at ¼ the speed of light. Figure 1-17 shows the four configurations during the transit of the glider.

A group of MIT students discovered a structure they call the glider gun; I prefer to think of it as a glider factory. This structure, shown in Fig. 1-18, is capable of generating gliders in 30 iterations. It is even possible to generate a glider factory using thirteen gliders in the appropriate arrangement (Fig. 1-19). Poundstone has shown that, with glider factories to generate glider pulses and other periodic and stable structures, it is possible to build NOT and AND gates. These gates can be assembled to act as a computer, a Turing machine, and self-reproducing automata.

Three glider factories and an eater (a structure that eats gliders) can be combined to form a gun called a *thin gun*. Two gliders can combine at the appropriate angle of impact and annihilate each other. Consequently, using these structures, it is possible to "thin out" a glider stream to any desired degree, thus allowing one to build pulse generators of any frequency. Figure 1-20 is a rough sketch of a thin glider gun or pulse generator. By changing the distance, *d*, it is possible to adjust the frequency of the pulse stream.

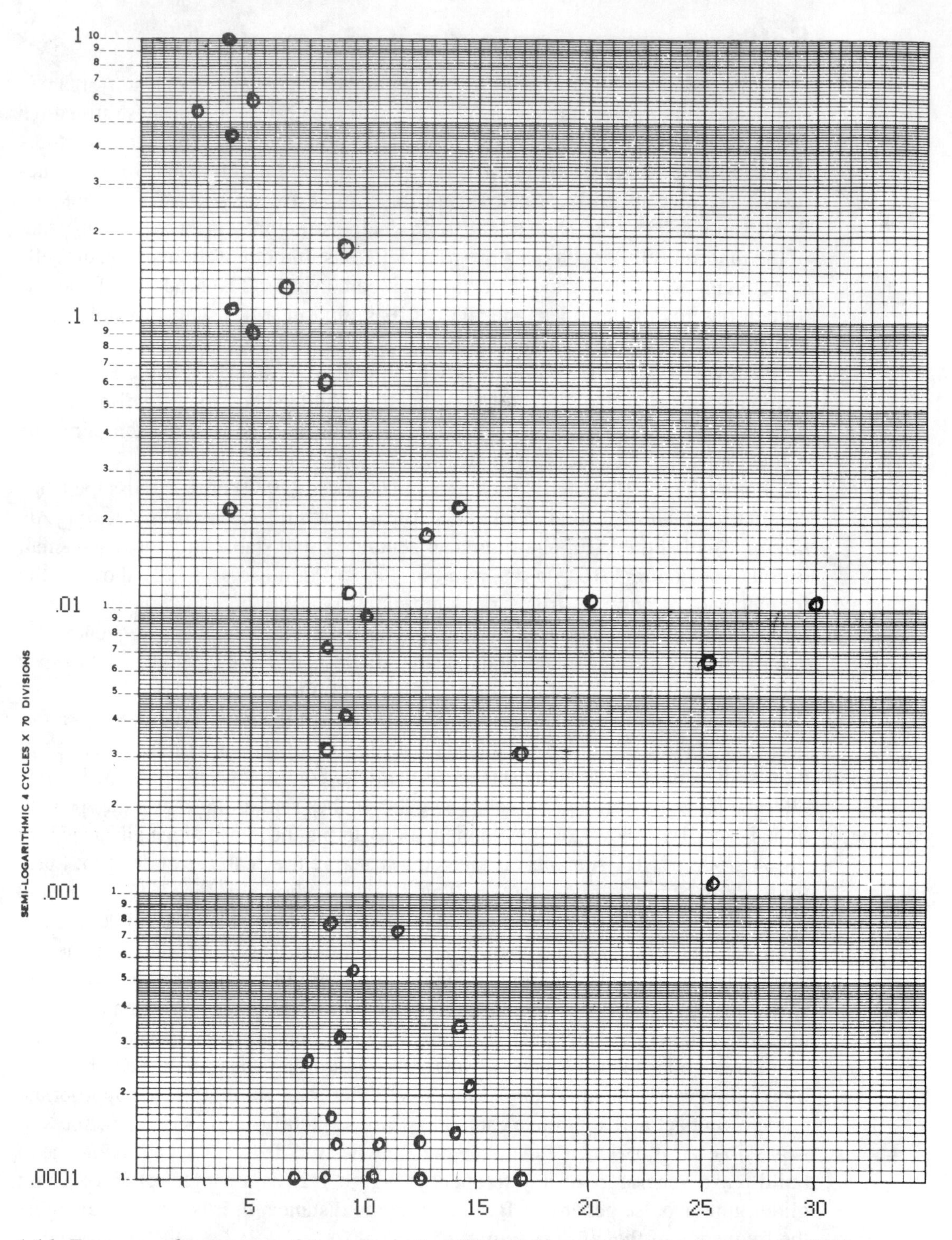

1-16 Frequency of occurrence of structures in the LIFE cellular automata (after Poundstone, 1985).

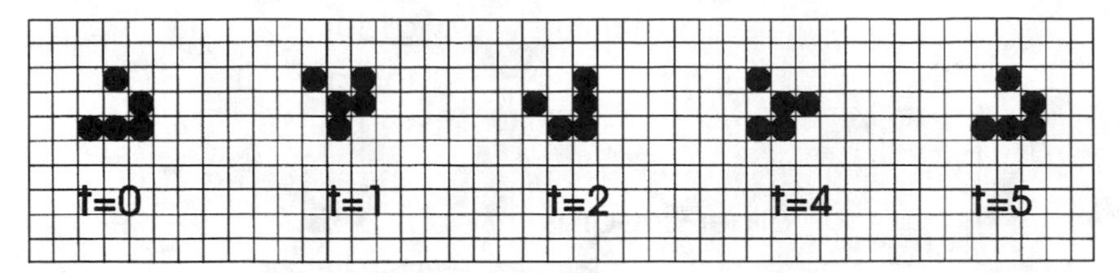

1-17 The five states of the glider in LIFE.

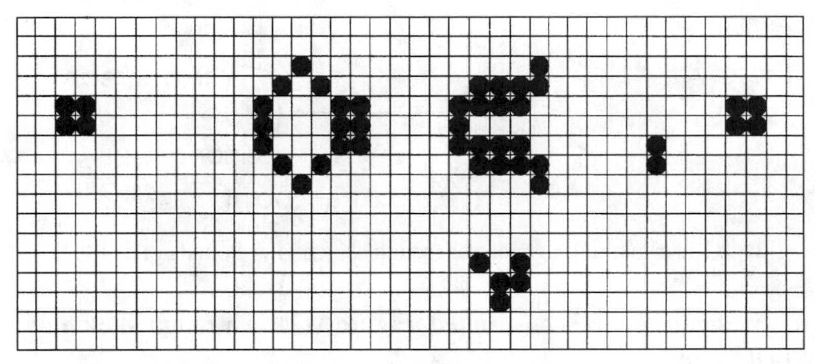

1-18 Glider factory.

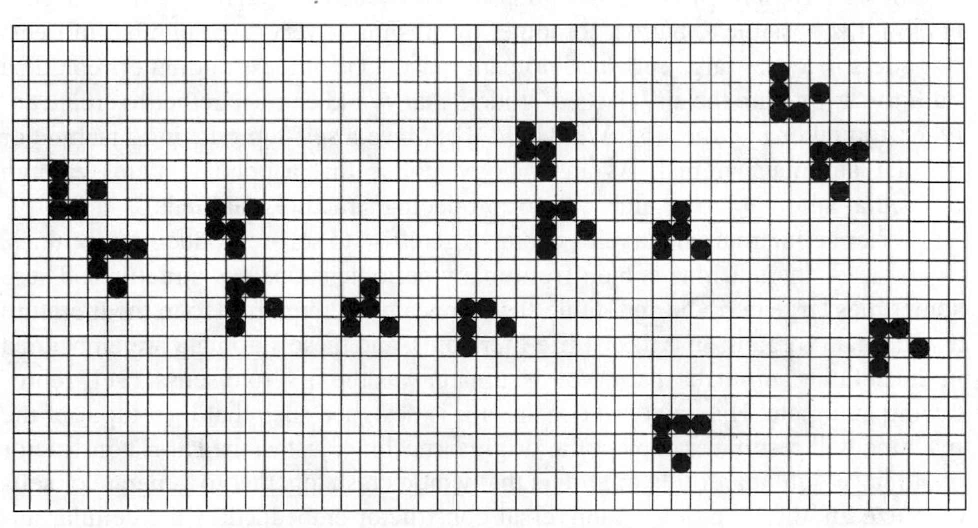

1-19 Configuration of 13 gliders that will produce a glider factory.

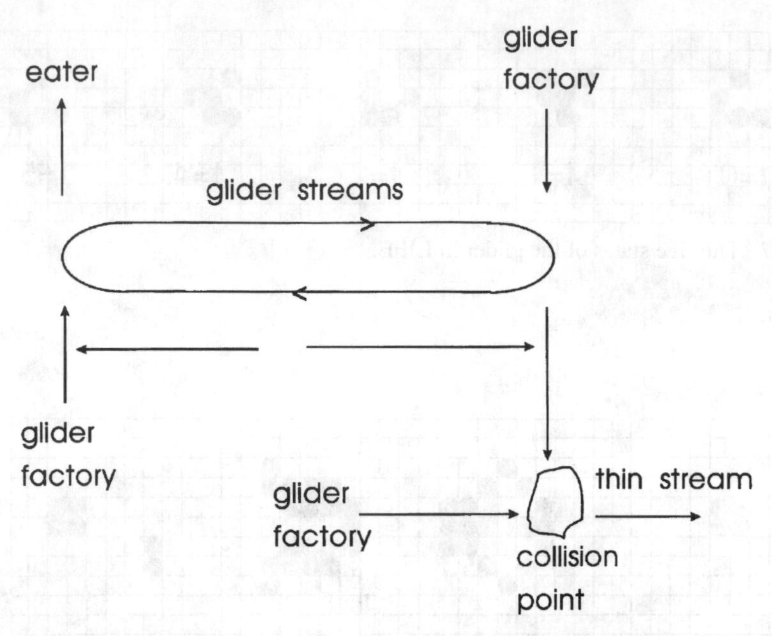

1-20 Glider gun pulse generator (rough sketch after Poundstone, 1985).

Logic gates can also be constructed. Figure 1-21 is a rough sketch of an OR, AND, and NOT gate.

Poundstone (1985), Dewdney (1988) and Berlekamp and Conway (1982) have shown (just as we could guess) that, by using these LIFE logic gates and pulse generators, one could build a computer and a Turing machine. They have also shown that it is possible to use these LIFE "organs" to build a universal constructor. In addition, we have shown that glider guns can be created by crashing a fleet of gliders. So it is possible to build a computer by crashing a very large number of fleets of gliders in a very large cellular automata space. One of these computers could be built to "throw into the air" fleets of gliders that would come together to create another computer like the first. We would then have a self-reproducing machine or cellular automata animal. As an analogy, notice that a pond contains enough molecular automata to assemble self-reproducing structures (animals).

Other cellular automata rules can also give rise to self-reproducing machines. In all cases, the machine is built from many periodic and stable structures. These sometimes large periodic and stable structures are often referred to in the literature as automata (organs, in artificial life literature), and these automata are embedded in the cellular automata space. Von Neumann was the first to discuss a self-reproducing automata. His universal constructor is designed such that the output of the machine will result in a copy of the original machine, just as the LIFE constructor could have output a flotilla of gliders that would crash together to generate or self-organize another copy of the universal constructor embedded in the cellular automata space.

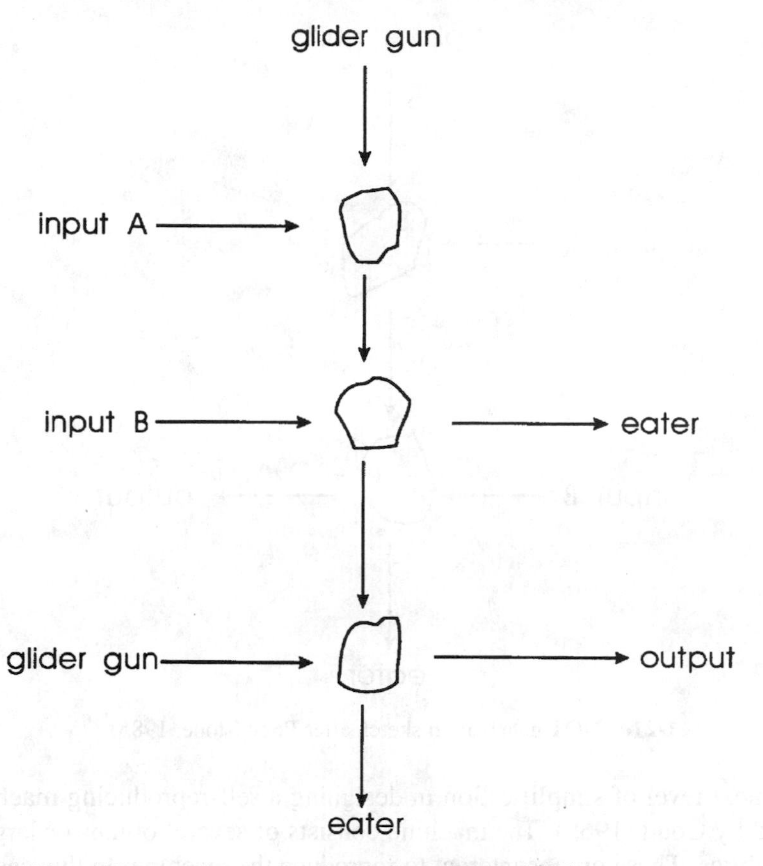

1-21a AND gate (rough sketch after Poundstone, 1985).

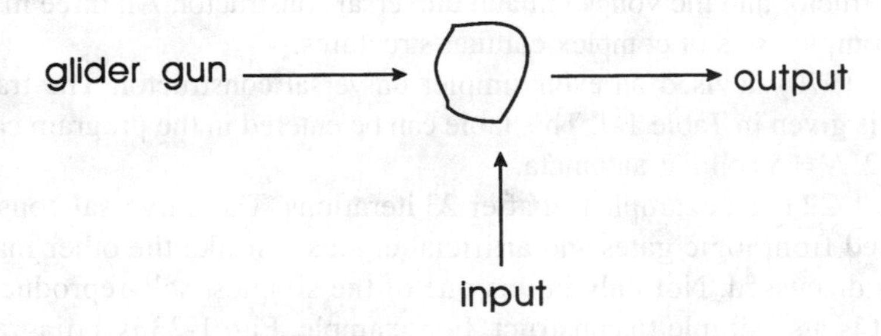

1-21b OR gate (rough sketch after Poundstone, 1985).

Von Neumann's machine is embedded in a cellular automata array in which each cell has 29 states interacting in a 5-cell neighborhood ($K = 29, N = 5$). His machine is a complex device consisting of gates, tape reading mechanisms, construction arms, pulse generators, etc. all embedded in the 29-state cellular automata. By comparison, the LIFE machine is much simpler, being embedded in a 2-state, 9-neighbor space.

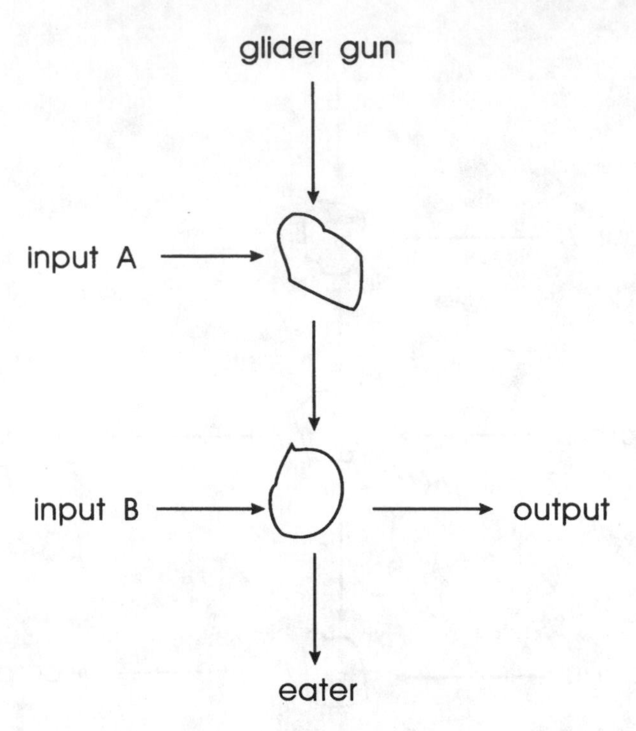

1-21c NOT gate (rough sketch after Poundstone, 1985).

The next level of simplification in designing a self-reproducing machine was discussed by Codd (1968). The machine consists of several organs or large cellular assemblies. These organs interact to reproduce the machine. In this case, it is a $K = 8, N = 5$ cellular automata space. In many respects, it is similar to the LIFE universal constructor and the von Neumann universal constructor. All three machines involved complex sets of complex cellular structures.

Banks (1970) devised an even simpler universal constructor. The transition state table is given in Table 1-4. This table can be entered in the program caplus.c. It is a $K = 2, N = 5$ cellular automata.

Figure 1-22 is an example run after 23 iterations. The universal constructor is assembled from logic gates and artificial organs just like the other machines previously discussed. Not only is this one of the simplest self-reproducing organisms, it is also simple to construct. For example, Fig. 1-23 is a diagram of a pulse generator, wire, and endcap. One pulse is generated every 15 cycles. It is also possible to assemble logic gates and other small devices so a universal constructor can be built.

All of these self-reproducing machines require many subassemblies or organs for the entire machine to reproduce. We now ask the question: Would it be possible to assemble a self-reproducing machine that does not require such a vast real estate and baggage? We would like to build a simple organism with only a few cells and still have it self-reproduce.

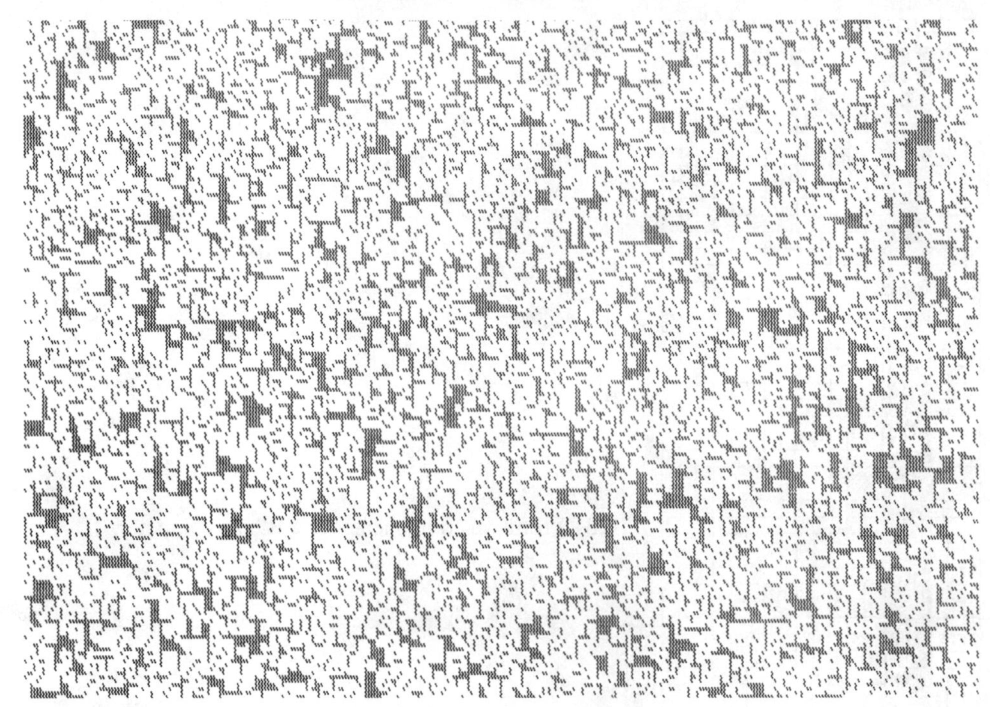

1-22 The Banks automata after 23 iterations from a random configuration.

Table 1-4 Transition table for Banks CA.

EWSNC			
00000	0	10001	1
00001	1	10010	0
00010	0	10011	1
00011	1	10011	1
00100	0	10100	0
00101	1	10101	0
00110	0	10110	1
00111	1	10111	1
01000	0	11000	0
01001	1	11001	1
01010	0	11010	1
01011	0	11011	1
01100	0	11100	1
01101	0	11101	1
01110	1	11110	1
01111	1	11111	1
10000	0		

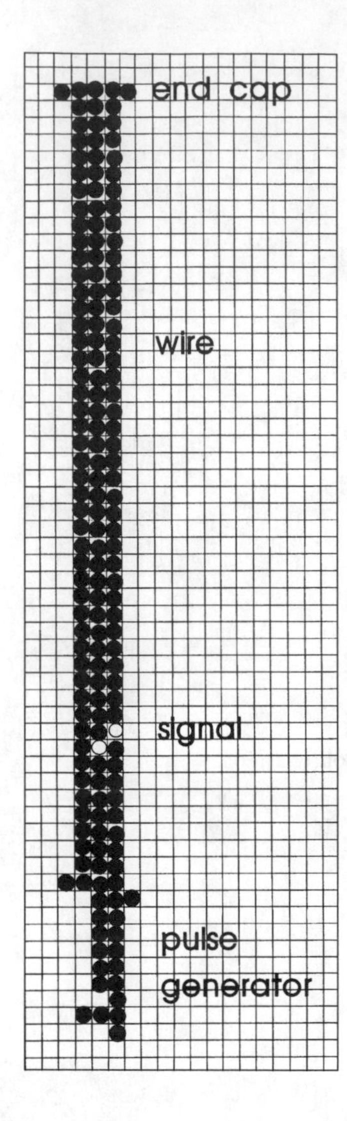

1-23 Primitive components for a Banks computer.

In all the above, we can see pulses traveling on wires. For example, in the Codd machine, the wire structure that acts as a path for pulses to travel down is shown in Fig. 1-24a. The 07 pattern, shown in the figure, is a signal traveling from left to right along the conductor (cells in state 1). Figure 1-24b shows how a 07 signal can split to travel on two conductors. Figure 1-24c shows how the wire can be extended (i.e., how it can grow) when a 07 hits the end cap it turns the 2 into a 1. When the 06 hits the end of the wire, it creates a 2 and extends the wire by one cell. Using these ideas, Langton (1984) discovered a much smaller self-reproducing machine. Langton's machine does not require many subassemblies. It is a simple structure of only a few cells in a $K = 8$, $N = 5$ cellular automata space.

Langton's machine is shown in Fig. 1-25. Within the loop are the instructions to build another machine. This machine differs from Codd's machine by another key

```
           212
           212
           212
        2222221222222
        1111071111111
        2222222222222

                                 2222222222222
                                 11110611071112
           212                   2222222222222
           212
           272
222222222222   2222220222222     2222222222222
111110711111   1111111071111     11111111061111
222222222222   2222222222222     2222222222222

    (a)             (b)               (c)
```

1-24a,b,c Components and states for the Codd's machine.

point. The Codd machine required the sequence 70...60 to extend the data paths. Langton's machine only requires the 70 sequence. Furthermore, signals need only be spaced 3 cells rather than 4 as in Codd's machine. The sequence 40...40...50...60 will form a new T junction in Codd's machine. In Langton's machine, the 40 signal will form a left-hand turn and fuse the paths to form a T junction.

Figure 1-26 is a sequence of events for self-reproduction of the Langton machine. The initial structure starts as *d* shape. The arm of the *d* then grows in length until a 1111 pattern is generated. The growth then makes a left-hand turn and continues growing in length again until a 1111 pattern is generated. After another left turn and more growth, the paths fuse. The *d* then separates and the arm dissolves. During the breakup, a 5 and 6 are placed in the circuit, both of which trigger the generation of new arms. Growth and reproduction continue until all the resources are used up, i.e., until the cellular automata space is filled.

```
 22222222
2170140142
2022222202
272    212
212    212
202    212
272    212
21222222122222
207107107111112
 2222222222222
```

1-25 Primitive components for the Langton machine.

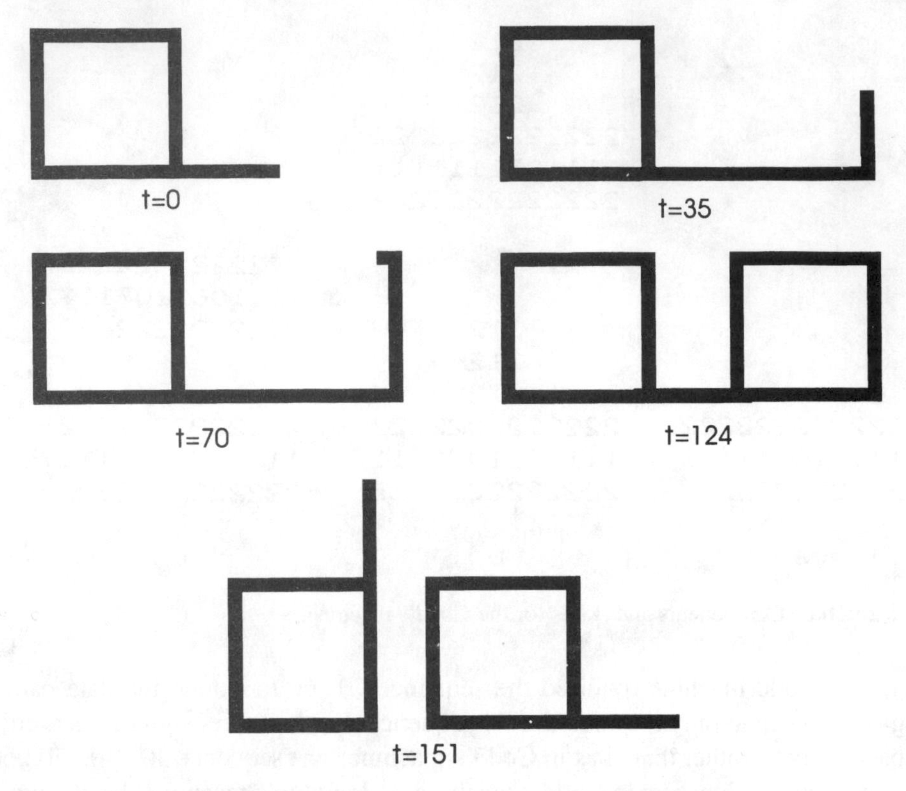

1-26 Rough sketch of self-reproduction in the Langton Machine (after Langton, 1984).

Summary

In this chapter, we have examined some physical chemistry of an artificial matter. The behavior of this matter may be described by the common laws of thermodynamics and statistical mechanics. We then examined what types of structures can self- organize in this artificial matter. Crystals and amorphous cellular aggregates can precipitate out of the cellular matrix.

Furthermore, complex cellular structures can cooperate by local dynamics for entire subsystems to be reproduced. The current limitation to continued development on these complex reproducing cellular subsystems seems to be twofold. One is the enormous memory and computational resources required, and the other is an apparent lack of applications, at this time, for these self-reproducing systems. The computational resources will eventually be here. The exploratory engineering for these self-reproducing machines could begin now.

banks2.c

```
#include <float.h>
#include <stdio.h>
#include <graph.h>
#include <math.h>
```

```
unsigned short int huge am[320][200];
unsigned short int huge newam[320][200];

struct videoconfig vc;
char error_message[] = "this video mode is not suported";
void main()
{

        /* declarations */
        int i,j,k;
        int sigma;
        int xmaxscreen=320,ymaxscreen=200;
        int xcoord,ycoord;
        double rnd;


        /* set mode of screen */
        if (_setvideomode(_MRESNOCOLOR) == 0)
        {
                printf("%s\n",error_message);
                exit(0);
        }
        _getvideoconfig(&vc);
        _clearscreen(_GCLEARSCREEN);


        _setviewport(0,0,320,200);


        /* load array */
        for(j=0;j<200;++j)
        {
                for(i=0;i<320;i++)
                {
                        rnd = rand()/32767.0;
                        if(rnd<0.55)
                        {
                                am[i][j] = 1;
                                _setcolor(1);
                        }
                        else
                        {
                                am[i][j] = 0;
                                _setcolor(0);
                        }
                        _setpixel(i,j);
                }
        }

        /* cellular automata rules */
        for(k=0;k<50;k++)
        {
                /* read array */
                for(j=0;j<200;j++)
                {
                        for(i=0;i<320;i++)
                        {

                                sigma = am[i%320][j%200] + am[i%320][(j%200)-1];
                                sigma = sigma + am[(i%320)-1][j%200]+
                                        am[(i%320)+1][j%200];
                                sigma = sigma + am[i%320][(j%200)+1];
                                        /* the following line takes care
                                                of all the cases not covered
                                                below */
                                newam[i%320][j%200] = am[i%320][j%200];

                                if((sigma==2)&(am[i%320][j%320]==1))
                                {
```

```
                                        newam[i%320][j%200] = 0;
                                }
                                if((sigma==3)&(am[i%320][j%200]==0))
                                {
                                        newam[i%320][j%200] = 1;
                                }
                                if((sigma==4)&(am[i%320][j%200]==0))
                                {
                                        newam[i%320][j%200] = 1;
                                }
                        }
                }

                /* set new values */
                for(j=0;j<200;j++)
                {
                        for(i=0;i<320;i++)
                        {
                                am[i][j] = newam[i][j];
                                _setcolor(am[i][j]);
                                _setpixel(i,j);
                        }
                }

        }

        /* clear screen & return control hit enter */
        while(!kbhit());
        _clearscreen(_GCLEARSCREEN);
        _setvideomode(_DEFAULTMODE);


}
```

ca8.c

```
#include <stdio.h>
#include <math.h>

#define VERTICAL    64
#define HORIZONTAL  64
#define NUM_ELEMENTS  4096
#define VERT_MAX    63
#define HORZ_MAX    63
#define VERTICAL_LESS     63
#define HORIZONTAL_LESS   63
#define VERTICAL_PLUS     0
#define HORIZONTAL_PLUS   0

#define POINTS_PER  100
#define GENS        1000
#define STAT_GENS   1000

#define NUM_COLORS             8
#define NUM_COLORS_MINUS_ONE   7
#define K_TO_THE_N             32768
#define K_TO_THE_N_MINUS_ONE   32767

#define SEVEN_EIGTHS  0.875

int  am[HORIZONTAL][VERTICAL];
int  newam[HORIZONTAL][VERTICAL];
int  rule_vector[K_TO_THE_N];

int    sigma;
double lambda;
int    hamming;
```

```
double average_ham;
int    entropic[VERTICAL][HORIZONTAL][NUM_COLORS];
double entropy;

int all_same;
int value_chk;

/******************************************************
*
*
*******************************************************/
main()
{

        /* declarations */
        int i,j,k,l;
        int lambda_number;
        int number_of_nums;
        int counter;
        int stat_counter;
        int position;
        int ivec_plus,jvec_plus,ivec_less,jvec_less;
        int stats_counter;
        int rnd;
        double p0, p1, p2, p3, p4, p5, p6, p7;

        srandom(getpid());

        /* select a lambda / cellular automata rules */
        /* 200 values of lambda */
        for(lambda = 0.004375;lambda <= SEVEN_EIGTHS;lambda = lambda + 0.004375)
        {
                /* select random rule based on lambda */
                number_of_nums = (int)(K_TO_THE_N * lambda);

                /* iterate each lambda for POINTS_PER times to collect statistics */
                for(lambda_number = 0; lambda_number < POINTS_PER; lambda_number++)
                {

                /* load screen array */
                for(j=0;j<VERTICAL;++j)
                   for(i=0;i<HORIZONTAL;i++) { am[j][i] = random()%NUM_COLORS; }

                for(i=0;i<K_TO_THE_N;i++){ rule_vector[i] = 0; }

                counter = number_of_nums - 7; /* 11111 -> 1, 22222 -> 2 etc.etc. */
                rule_vector[011111] = 1;
                rule_vector[022222] = 2;
                rule_vector[033333] = 3;
                rule_vector[044444] = 4;
                rule_vector[055555] = 5;
                rule_vector[066666] = 6;
                rule_vector[077777] = 7;

                while(counter)
                {
                        /* select position */
                        position = ((random()%K_TO_THE_N_MINUS_ONE) + 1);

                        /* check position */
                        if(rule_vector[position] == 0)
                        {
                                rnd = ((random()%NUM_COLORS_MINUS_ONE) + 1);
                                rule_vector[position] = rnd;
                                /* decrement counter */
                                counter—;
                        }

                } /* end rule selection */
                k = 0;
```

```
/* for GENS generations */
while(k<GENS)
{
        /* generation counter */
        k++;

        /* read screen array */
        for(j=0;j<VERTICAL;j++)
        {
                jvec_plus = j + 1;
                jvec_less = j - 1;
                if(j==VERT_MAX) jvec_plus = VERTICAL_PLUS;
                else if(j==0) jvec_less = VERTICAL_LESS;

                for(i=0;i<HORIZONTAL;i++)
                {
                        /* sum the neighborhood */
                        ivec_plus = i + 1;
                        ivec_less = i - 1;
                        if(i==HORZ_MAX) ivec_plus = HORIZONTAL_PLUS;
                        else if(i==0) ivec_less = HORIZONTAL_LESS;
                        sigma = am[j][i];
                        sigma += (am[j][ivec_plus])<<3;
                        sigma += (am[j][ivec_less])<<6;
                        sigma += (am[jvec_plus][i])<<9;
                        sigma += (am[jvec_less][i])<<12;

                        /* new cell position */
                        newam[j][i] = rule_vector[sigma];
                }
        } /* end read screen array */
        all_same = 1;
        value_chk = newam[0][0];
        for(j=0;j<VERTICAL;j++) {
                for(i=0;i<HORIZONTAL;i++) {
                        am[j][i] = newam[j][i];
                        if(newam[j][i] != value_chk) { all_same = 0;}
                }
        }
        if(all_same) break;

} /* end GENS generation loop */

if(!all_same){
    /* collect statistical information */
    stat_counter = 0;
    hamming = 0;
    for(j=0;j<VERTICAL;j++)
        for(i=0;i<HORIZONTAL;i++)
           for(l=0;l<NUM_COLORS;l++)
               (entropic[j][i][l]) = 0;
           while(stat_counter < STAT_GENS)
           {
               /* read screen array */
               for(j=0;j<VERTICAL;j++)
               {
                       jvec_plus = j + 1;
                       jvec_less = j - 1;
                       if(j==VERT_MAX) jvec_plus = VERTICAL_PLUS;
                       else if(j==0) jvec_less = VERTICAL_LESS;

                       for(i=0;i<HORIZONTAL;i++)
                       {
                               /* sum the neighborhood */
                               ivec_plus = i + 1;
                               ivec_less = i - 1;
                               if(i==HORZ_MAX) ivec_plus = HORIZONTAL_PLUS;
                               else if(i==0) ivec_less = HORIZONTAL_LESS;
                               sigma = am[j][i];
```

```
                                        sigma += (am[j][ivec_plus])<<3;
                                        sigma += (am[j][ivec_less])<<6;
                                        sigma += (am[jvec_plus][i])<<9;
                                        sigma += (am[jvec_less][i])<<12;

                                        /* new cell color*/
                                        newam[j][i] = rule_vector[sigma];
                                }
                } /* end read screen array */
                for(j=0;j<VERTICAL;j++) {
                        for(i=0;i<HORIZONTAL;i++) {
                                if(newam[j][i] != am[j][i]) { hamming++; }
                                /* set new values for screen array */
                                am[j][i] = newam[j][i];
                                if(am[j][i] == 0)      (entropic[j][i][0]) ++;
                                else if(am[j][i] == 1) (entropic[j][i][1]) ++;
                                else if(am[j][i] == 2) (entropic[j][i][2]) ++;
                                else if(am[j][i] == 3) (entropic[j][i][3]) ++;
                                else if(am[j][i] == 4) (entropic[j][i][4]) ++;
                                else if(am[j][i] == 5) (entropic[j][i][5]) ++;
                                else if(am[j][i] == 6) (entropic[j][i][6]) ++;
                                else if(am[j][i] == 7) (entropic[j][i][7]) ++;
                        }
                }
                stat_counter++;
        } /* end stats. collection */
        average_ham = (double)hamming/(double)STAT_GENS;
        entropy = 0.0;
        for(j=0;j<VERTICAL;j++) {
                for(i=0;i<HORIZONTAL;i++) {
                        p0 = (double)entropic[j][i][0]/(double)STAT_GENS;
                        p1 = (double)entropic[j][i][1]/(double)STAT_GENS;
                        p2 = (double)entropic[j][i][2]/(double)STAT_GENS;
                        p3 = (double)entropic[j][i][3]/(double)STAT_GENS;
                        p4 = (double)entropic[j][i][4]/(double)STAT_GENS;
                        p5 = (double)entropic[j][i][5]/(double)STAT_GENS;
                        p6 = (double)entropic[j][i][6]/(double)STAT_GENS;
                        p7 = (double)entropic[j][i][7]/(double)STAT_GENS;
                        if(p0 == 0.00){ /* entropy += 0.00; */ }
                        else if(p0 == 1.00){ /*entropy += 0.00; */ }
                        else entropy += p0*log2(p0);
                        if(p1 == 0.00){ /* entropy += 0.00; */ }
                        else if(p1 == 1.00){ /*entropy += 0.00; */ }
                        else entropy += p1*log2(p1);
                        if(p2 == 0.00){ /* entropy += 0.00; */ }
                        else if(p2 == 1.00){ /*entropy += 0.00; */ }
                        else entropy += p2*log2(p2);
                        if(p3 == 0.00){ /* entropy += 0.00; */ }
                        else if(p3 == 1.00){ /*entropy += 0.00; */ }
                        else entropy += p3*log2(p3);
                        if(p4 == 0.00){ /* entropy += 0.00; */ }
                        else if(p4 == 1.00){ /*entropy += 0.00; */ }
                        else entropy += p4*log2(p4);
                        if(p5 == 0.00){ /* entropy += 0.00; */ }
                        else if(p5 == 1.00){ /*entropy += 0.00; */ }
                        else entropy += p5*log2(p5);
                        if(p6 == 0.00){ /* entropy += 0.00; */ }
                        else if(p6 == 1.00){ /*entropy += 0.00; */ }
                        else entropy += p6*log2(p6);
                        if(p7 == 0.00){ /* entropy += 0.00; */ }
                        else if(p7 == 1.00){ /*entropy += 0.00; */ }
                        else entropy += p7*log2(p7);
                }
        }
        entropy = -(entropy/(double)NUM_ELEMENTS);
}
else { /* all the same */
        average_ham = 0.00;
        entropy = 0.00;
}
```

```
                    printf("%d\t",number_of_nums);
                    printf("%f\t",lambda);
                    printf("\t%f",average_ham);
                    printf("\t%f\n",entropy);
                    printf("\n");
                    fflush(stdout);

            } /* end lambda_number 0 to POINTS_PER loop */
    } /* end main lambda loop */


}  /* end main() */
```

caplus.c

```
#include <float.h>
#include <stdio.h>
#include <graph.h>
#include <math.h>

unsigned short int huge am[320][200];
unsigned short int huge newam[320][200];

unsigned short int rule_vector[32];

struct videoconfig vc;
char error_message[] = "this video mode is not suported";

void main()
{

        /* declarations */
        int i,j;
        int k=0;
        int sigma;
        double rnd;
        int ivec_plus, jvec_plus;
        int ivec_less, jvec_less;


        /* set mode of screen */
        if (_setvideomode(_MRESNOCOLOR) == 0)
        {
                printf("%s\n",error_message);
                exit(0);
        }
        _getvideoconfig(&vc);
        _clearscreen(_GCLEARSCREEN);


        _setviewport(0,0,320,200);


        /* load array */
        for(j=0;j<200;++j)
        {
                for(i=0;i<320;i++)
                {
                        rnd = rand()/32767.0;
                        if(rnd<0.4)
                        {
                                am[i][j] = 1;
                                _setcolor(1);
                        }
                        else
                        {
                                am[i][j] = 0;
                                _setcolor(0);
                        }
```

```
                                        _setpixel(i,j);
                }
        }

        /* set rule vector */
        rule_vector[0] = 0;
        rule_vector[1] = 1;
        rule_vector[2] = 0;
        rule_vector[3] = 1;
        rule_vector[4] = 0;
        rule_vector[5] = 1;
        rule_vector[6] = 0;
        rule_vector[7] = 1;
        rule_vector[8] = 0;
        rule_vector[9] = 1;
        rule_vector[10] = 0;
        rule_vector[11] = 0;
        rule_vector[12] = 0;
        rule_vector[13] = 0;
        rule_vector[14] = 1;
        rule_vector[15] = 1;
        rule_vector[16] = 0;
        rule_vector[17] = 1;
        rule_vector[18] = 0;
        rule_vector[19] = 1;
        rule_vector[20] = 0;
        rule_vector[21] = 0;
        rule_vector[22] = 1;
        rule_vector[23] = 1;
        rule_vector[24] = 0;
        rule_vector[25] = 1;
        rule_vector[26] = 1;
        rule_vector[27] = 1;
        rule_vector[28] = 1;
        rule_vector[29] = 1;
        rule_vector[30] = 1;
        rule_vector[31] = 1;



        /* run the cellular automata */
        while(!kbhit())
        {
                /* generation counter */
                k++;

                /* read array */
                for(j=0;j<200;j++)
                {

                        jvec_plus = j + 1;
                        jvec_less = j - 1;
                        if(j==199) jvec_plus = 0;
                        else if(j==0) jvec_less = 199;

                        for(i=0;i<320;i++)
                        {
                                /* sum the neighborhood */
                                ivec_plus = i + 1;
                                ivec_less = i - 1;
                                if(i==319) ivec_plus = 0;
                                else if(i==0) ivec_less = 319;

                                sigma = am[i][j];
                                sigma += (am[i][jvec_plus])<<1;
                                sigma += (am[i][jvec_less])<<2;
                                sigma += (am[ivec_plus][j])<<3;
                                sigma += (am[ivec_less][j])<<4;

                                /* new cell state */
```

```
                        newam[i][j] = rule_vector[sigma];

                }
        }

        /* set new values */
        for(j=0;j<200;j++)
        {
                for(i=0;i<320;i++)
                {


                        if(newam[i][j]==1)
                        {
                                am[i][j] = 1;
                                _setcolor(1);
                        }
                        if(newam[i][j]==0)
                        {
                                am[i][j] = 0;
                                _setcolor(0);
                        }
                _setpixel(i,j);


                }
        }

    }

    _clearscreen(_GCLEARSCREEN);
    _setvideomode(_DEFAULTMODE);
    printf("number of gernerations %d\n",k);

}
```

cell1.bas

```
10 KEY OFF
20 CLS
30 CLEAR
40 DEFINT A-Z
50 SCREEN 1
60 P=8000
70 DIM A(P)
80 PSET (1,1)
90 GET (1,1)-318,198),A
100 XI=SGN(RND-.5)
110 YI=SGN(RND-.5)
120 PUT(1+XI,1YI),A
130 GOTO 90
```

direct.bas

```
10 DEF SEG =&HB800
20 DEFINT A-Y
30 SCREEN 0,0,0
40 RANDOMIZE TIMER
50 CLS
60 DIM A(25,80),D(25,80),E(25,80),HAMMING(30)
70 REM 219 IS ASCII FOR WHITE PIXEL
80 REM 255 IS ASCII FOR BLACK PIXEL
90 FOR I%=0 TO 24
100 FOR J%=0 TO 79
110 Z=RND(1)
```

```
120 IF Z<.1  THEN POKE I%*160+J%*2,219 ELSE POKE I%*160+J%*2,255
130 NEXT J%
140 NEXT I%
150 WHILE CYCLE < 1000
160 CYCLE=CYCLE+1
170 FOR I%=0 TO 24
180 FOR J%=0 TO 79
190 B=PEEK(I%*160+J%*2)
200 IF B=219 THEN A(I%,J%)=219 ELSE A(I%,J%)=0
210 D(I%,J%)=A(I%,J%)
220 NEXT J%
230 NEXT I%
240 'INSERT CODE HERE FOR CELLULAR AUTOMATA RULES
250 FOR I%=1 TO 23
260 FOR J%=1 TO 78
270 IF A(I%-1,J%)=219 OR A(I%+1,J%)=219 THEN 280 ELSE 290
280 POKE I%*160+J%*2,219
285 GOTO 350
290 POKE I%*160+J%*2,255
350 NEXT J%
360 NEXT I%
380 WEND
390 END
```

foo.exp

```
      1                                                        0
 0 \ FOO (fractal expt)                             10Oct89cam
 1
 2 NEW-EXPERIMENT
 3
 4 N/MOORE
 5                          : 8SUM ( — n)
 6 N.WEST NORTH N.EAST
 7 WEST           EAST
 8 S.WEST SOUTH S.EAST
 9      + + + + +           ;
10                          : FOO
11 8SUM 1 = IF
12      1 ELSE
13 CENTER THEN
14       >PLN0              ;
15

      2                                                        0
 0                     : ECHO
 1 CENTER >PLN    ;
 2 MAKE-TABLE ECHO
 3 MAKE-TABLE FOO
 4
 5
 6
 7
 8
 9
10
11
12
13
14
15

      0                                                        0
 0
 1
 2
 3
 4
```

```
 5
 6
 7
 8
 9
10
11
12
13
14
15
```

Foo1.exp

```
    1                                                      0
 0 \ FOO1(bank's experiment)             11Oct89cam
 1 NEW-EXPERIMENT
 2 N/VONN
 3                                 : U
 4                        CENTER ;
 5                                 : CORNER?
 6 NORTH SOUTH = IF U ELSE 0 THEN ;
 7                                 : BANKS
 8   NORTH SOUTH WEST EAST + + +
 9     { U U CORNER? 1 1) >PLNO ;
10
11
12 MAKE-TABLE BANKS
13
14
15

    0                                                      0
 0
 1
 2
 3
 4
 5
 6
 7
 8
 9
10
11
12
13
14
15

    0                                                      0
 0
 1
 2
 3
 4
 5
 6
 7
 8
 9
10
11
12
13
14
15
```

fractal.c

```
#include <float.h>
#include <stdio.h>
#include <graph.h>
#include <math.h>

unsigned short int huge am[320][200];
unsigned short int huge newam[320][200];

struct videoconfig vc;
char error_message[] = "this video mode is not suported";

void main()
{

        /* declarations */
        int i,j,k;
        int sigma;
        int xmaxscreen=320,ymaxscreen=200;
        int xcoord,ycoord;
        double rnd;

        /* set mode of screen */
        if (_setvideomode(_MRESNOCOLOR) == 0)
        {
                printf("%s\n",error_message);
                exit(0);
        }
        _getvideoconfig(&vc);
        _clearscreen(_GCLEARSCREEN);


        _setviewport(0,0,320,200);


        /* load array */
        for(j=0;j<200;++j)
        {
                for(i=0;i<320;i++)
                {
                        rnd = rand()/32767.0;
                        if(rnd<0.001)
                        {
                                am[i][j] = 1;
                                _setcolor(1);
                        }
                        else
                        {
                                am[i][j] = 0;
                                _setcolor(0);
                        }
                        _setpixel(i,j);
                }
        }

        /* cellular automata rules */
        for(k=0;k<10;k++)
        {
                /* read array */
                for(j=0;j<200;j++)
                {
                        for(i=0;i<320;i++)
                        {
                                sigma = am[(i%320)-1][(j%200)-1]+
                                        am[i%320][(j%200)-1]+
                                        am[(i%320)+1][(j%200)-1];
                                sigma = sigma + am[(i%320)-1][j%200]+
```

```
                                        am[(i%320)+1][j%200];
                                sigma = sigma + am[(i%320)-1][(j%200)+1]+
                                        am[i%320][(j%200)+1]+
                                        am[(i%320)+1][(j%200)+1];
                                if(sigma==1)
                                {
                                        newam[i%320][j%200] = 1;
                                }
                                else
                                {
                                        newam[i%320][j%200] = am[i%320][j%200];
                                }
                        }
                }

                /* set new values */
                for(j=0;j<200;j++)
                {
                        for(i=0;i<320;i++)
                        {
                                if(newam[i%320][j%200]==1)
                                {
                                        am[i%320][j%200] = 1;
                                        _setcolor(1);
                                }
                                else
                                {
                                        am[i%320][j%200] = newam[i%320][j%200];
                                       _setcolor(am[i%320][j%200]);
                                }
                        _setpixel(i,j);
                        }
                }

        }

        /* clear screen & return control hit enter */
        while(!kbhit());
        _clearscreen(_GCLEARSCREEN);
        _setvideomode(_DEFAULTMODE);


}
```

fractal1.c

```
#include <float.h>
#include <stdio.h>
#include <graph.h>
#include <math.h>

unsigned short int huge am[320][200];
unsigned short int huge newam[320][200];

struct videoconfig vc;
char error_message[] = "this video mode is not suported";

void main()
{

        /* declarations */
        int i,j,k;
        int sigma;
        int xmaxscreen=320,ymaxscreen=200;
        int xcoord,ycoord;
        double rnd;
```

```
/* set mode of screen */
if (_setvideomode(_MRESNOCOLOR) == 0)
{
        printf("%s\n",error_message);
        exit(0);
}
_getvideoconfig(&vc);
_clearscreen(_GCLEARSCREEN);


_setviewport(0,0,320,200);


/* load array */
for(j=0;j<200;++j)
{
        for(i=0;i<320;i++)
        {
                if((i==160) && (j==100))
                {
                        am[i][j] = 1;
                        _setcolor(1);
                }
                else
                {
                        am[i][j] = 0;
                        _setcolor(0);
                }
                _setpixel(i,j);
        }
}

/* cellular automata rules */
for(k=0;k<50;k++)
{
        /* read array */
        for(j=1;j<199;j++)
        {
                for(i=1;i<319;i++)
                {
                        sigma = am[i-1][j-1]+
                                am[i][j-1]+
                                am[i+1][j-1];
                        sigma = sigma + am[i-1][j]+
                                am[i+1][j];
                        sigma = sigma + am[i-1][j+1]+
                                am[i][j+1]+
                                am[i+1][j+1];
                        if(sigma==1)
                        {
                                newam[i][j] = 1;
                        }
                        else
                        {
                                newam[i][j] = am[i][j];
                        }
                }
        }

        /* set new values */
        for(j=1;j<199;j++)
        {
                for(i=1;i<319;i++)
                {
                        if(newam[i][j]==1)
                        {
                                am[i][j] = 1;
                                _setcolor(1);
                        }
```

```
                    else
                    {
                            am[i][j] = newam[i][j];
                            _setcolor(am[i][j]);
                    }
                _setpixel(i,j);
                }
        }

    }

    /* clear screen & return control hit enter */
    while(!kbhit());
    _clearscreen(_GCLEARSCREEN);
    _setvideomode(_DEFAULTMODE);


}
```

fredkin.bas

```
10 DEF SEG =&HB800
15 DEFINT A-Y
20 SCREEN 0,0,0
30 CLS
40 DIM A(25,80),B(25,80)
50 REM 219 IS ASCII FOR WHITE PIXEL
60 REM 255 IS ASCII FOR BLACK PIXEL
70 FOR I%=0 TO 24
80 FOR J%=0 TO 79
100 IF I%=12 AND J%=40 THEN POKE I%*160+J%*2,219 ELSE POKE I%*160+J%*2,255
120  POKE  I%*160+J%*2+1,10
140 NEXT J%
150 NEXT I%
160 FOR I%=0 TO 24
170 FOR J%=0 TO 79
180 B=PEEK(I%*160+J%*2)
190 IF B=219 THEN A(I%,J%)=219 ELSE A(I%,J%)=0
210 NEXT J%
211 NEXT I%
215 'INSERT CODE HERE FOR CELLULAR AUTOMATA RULES
220 FOR I%=1 TO 23
230 FOR J%=1 TO 78
260    C=A(I%,J%-1)
270    C=C+A(I%-1,J%)+A(I%+1,J%)
280    C=C+A(I%,J%+1)
292  IF C=219 OR C=219*3 THEN POKE I%*160+J%*2,219 ELSE POKE I%*160+J%*2,255
300 POKE I%*160+J%*2+1,10
400 NEXT J%
401 NEXT I%
402 CYCLE = CYCLE + 1
450 GOTO 160
1000 END
```

life.bas

```
10 DEF SEG =&HB800
20 DEFINT A-Y
30 SCREEN 0,0,0
40 RANDOMIZE TIMER
50 CLS
60 DIM A(25,80),D(25,80),E(25,80),HAMMING(30)
70 REM 219 IS ASCII FOR WHITE PIXEL
80 REM 255 IS ASCII FOR BLACK PIXEL
90 FOR I%=0 TO 24
```

```
100 FOR J%=0 TO 79
110 Z=RND(1)
120 IF Z<.1  THEN POKE I%*160+J%*2,219 ELSE POKE I%*160+J%*2,255
130 NEXT J%
140 NEXT I%
150 WHILE CYCLE < 30
160 CYCLE=CYCLE+1
170 FOR I%=0 TO 24
180 FOR J%=0 TO 79
190 B=PEEK(I%*160+J%*2)
200 IF B=219 THEN A(I%,J%)=219 ELSE A(I%,J%)=0
210 D(I%,J%)=A(I%,J%)
220 NEXT J%
230 NEXT I%
240 'INSERT CODE HERE FOR CELLULAR AUTOMATA RULES
250 FOR I%=1 TO 23
260 FOR J%=1 TO 78
270    C=A(I%-1,J%-1)+A(I%,J%-1)+A(I%+1,J%-1)
280    C=C+A(I%-1,J%)+A(I%+1,J%)
290    C=C+A(I%-1,J%+1)+A(I%,J%+1)+A(I%+1,J%+1)
300  IF C<=219*1 THEN   POKE I%*160+J%*2,255 : E(I%,J%)=0
310  IF C=219*3 THEN   POKE I%*160+J%*2,219 : E(I%,J%)=219
320  IF C>=219*4 THEN   POKE I%*160+J%*2,255 : E(I%,J%)=0
330 IF E(I%,J%)<>D(I%,J%) THEN 340 ELSE 350
340 HAMMING(CYCLE)=HAMMING(CYCLE)+1
350 NEXT J%
360 NEXT I%
370 LPRINT CYCLE;HAMMING(CYCLE)
380 WEND
390 END
```

life4.c

```
/* life simulation */

#include "\quickc\include\stdio.h"
#include "\quickc\include\stdlib.h"

static unsigned short int am[25][160];

FILE *fe;

void main()
{
        /* declarations */
        unsigned char far *a;
        unsigned short int b;
        int i,j,x,y,counter,sigma,address;
                        /* counter is used in address decoding */
        int cycle = 0;
        int seed;

        a = (unsigned char far *) 0xb8000000;

        /* user input */
        printf("input random seed \n");
        scanf("%d",&seed);
        srand(seed);

        /* clear graphics memory */
        counter = 0;
        for(x=0;x<25;x++)
        {
                for(y=0;y<160;y++)
                {
                        *(a + counter) = 0;
                        counter ++;
                }
```

```
        }

        /* poke (random) */
        counter = 0;
        for(x=0;x<25;x++)
        {
                for(y=0;y<160;y++)
                {
                        if(rand()/32767.0 < 0.5)
                        {
                                *(a + counter) = 255;
                        }
                        counter ++;
                }
        }

        /* begin life iterations */
        for(;;)
        {
                if(kbhit())
                {
                        exit(0);
                }

                /* peek to fill matrix */
                counter = 0;
                for(i=1;i<24;i++)
                {
                        for(j=2;j<159;j++)
                        {
                                address = i*160 + j;
                                b = *(a + address);
                                am[i][j] = b;
                                counter = counter + 1;
                        }
                }

                /* cellular automata rules for life */
                counter = 0;
                for(i=1;i<24;i++)
                {
                        for(j=2;j<159;j++)
                        {
                                sigma = am[i-1][j-1]+am[i][j-1]+am[i+1][j-1];
                                sigma = sigma + am[i-1][j] + am[i+1][j];
                                sigma = sigma + am[i-1][j+1] + am[i][j+1] + am[i+1][j+1];
                                counter = counter + 1;
                                address = i*160 + j;
                                if(sigma<=255*1)
                                {
                                        *(a + address) = 0;
                                }
                                if(sigma==255*3)
                                {
                                        *(a + address) = 255;
                                }
                                if(sigma>=255*4)
                                {
                                        *(a + address) = 0;
                                }
                        }
                }
        }
}
```

life4hr4.c

```
#include <float.h>
#include <stdio.h>
#include <graph.h>
#include <math.h>

unsigned short int huge am[320][200];
unsigned short int huge newam[320][200];

struct videoconfig vc;
char error_message[] = "this video mode is not suported";

void main()
{

        /* declarations */
        int i,j,k;
        int sigma;
        int xmaxscreen=320,ymaxscreen=200;
        int xcoord,ycoord;
        double rnd;


        /* set mode of screen */
        if (_setvideomode(_MRESNOCOLOR) == 0)
        {
                printf("%s\n",error_message);
                exit(0);
        }
        _getvideoconfig(&vc);
        _clearscreen(_GCLEARSCREEN);


        _setviewport(0,0,320,200);


        /* load array */
        for(j=0;j<200;++j)
        {
                for(i=0;i<320;i++)
                {
                        rnd = rand()/32767.0;
                        if(rnd<0.4)
                        {
                                am[i][j] = 1;
                                _setcolor(1);
                        }
                        else
                        {
                                am[i][j] = 0;
                                _setcolor(0);
                        }
                        _setpixel(i,j);
                }
        }

        /* cellular automata rules */
        for(k=0;k<99;k++)
        {
                /* read array */
                for(j=0;j<200;j++)
                {
                        for(i=0;i<320;i++)
                        {
                                sigma = am[(i%320)-1][(j%200)-1]+
                                        am[i%320][(j%200)-1]+
                                        am[(i%320)+1][(j%200)-1];
                                sigma = sigma + am[(i%320)-1][j%200]+
```

```
                                        am[(i%320)+1][j%200];
                                sigma = sigma + am[(i%320)-1][(j%200)+1]+
                                        am[i%320][(j%200)+1]+
                                        am[(i%320)+1][(j%200)+1];
                                if(sigma<=1)
                                {
                                        newam[i%320][j%200] = 0;
                                }
                                if(sigma==3)
                                {
                                        newam[i%320][j%200] = 1;
                                }
                                if(sigma>=4)
                                {
                                        newam[i%320][j%200] = 0;
                                }
                        }
                }

                /* set new values */
                for(j=0;j<200;j++)
                {
                        for(i=0;i<320;i++)
                        {
                                if(newam[i%320][j%200]==1)
                                {
                                        am[i%320][j%200] = 1;
                                        _setcolor(1);
                                }
                                if(newam[i%320][j%200]==0)
                                {
                                        am[i%320][j%200] = 0;
                                        _setcolor(0);
                                }
                                _setpixel(i,j);
                        }
                }

        }

        /* clear screen & return control hit enter */
        while(!kbhit());
        _clearscreen(_GCLEARSCREEN);
        _setvideomode(_DEFAULTMODE);


}
```

poke.c

```
#include <stdio.h>
#include <stdlib.h>

void main()
{
        unsigned char far *a;
        int i,j,counter=0;
        int address;
        a = (char far *) 0xb8000000;
        for(;;)
        {
                counter = 0;
                for(i=1;i<24;i++)
                {
                        for(j=2;j<159;j++)
                        {
```

```
                        address = counter;
                        *(a + address) = 88;
                        counter++;
                    }
                }
                if(kbhit())
                {
                    printf("%d ",address);
                    exit(0);
                }
            }
        }
```

2
CHAPTER

Neural networks
Brains for artificial life

In this chapter, we will examine a new computing paradigm known as neural networks. Artificial neural networks are modeled after the biological brain.

First we will examine threshold logic and nonlinear systems while introducing the relevant mathematical methods. Then we will look at content addressable memories and Hopfield networks. Following this will come Boltzman machines, statistical learning, and learning by back-propagation of errors. In the demonstration of the learning algorithms, we will look at toy problems that are highly nonlinear. Later, in another chapter, we will use one of these neural networks in a simulated robot.

Readers interested in building hardware neural networks should see my earlier book (Rietman 1988). Wasserman (1989), Dayhoff(1990), and Eberhart and Dobbins (1991) have written good introductions to the subject. A literature search shows that this is the fastest growing field of AI or ALife.

Threshold logic and nonlinear systems

The human brain is a massive parallel personal computer based on organic threshold logic devices known as *neurons*. Figure 2-1 shows a sketch of a neuron.

The cell body has one or more output lines called the *axon*; the input lines are called *dendrites*. There are about 1000 dendrite connections to an average neuron in the human brain. The neurons are interconnected by a synapse and can have both excitatory and inhibitory connections. An excitatory connection tells the neu-

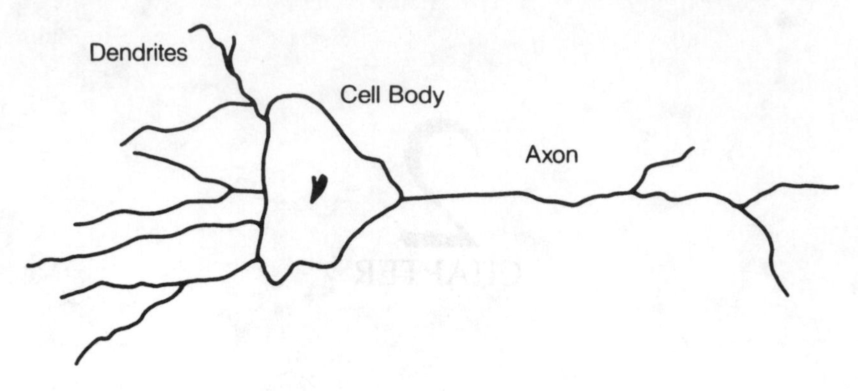

2-1 Schematic of a biological neuron.

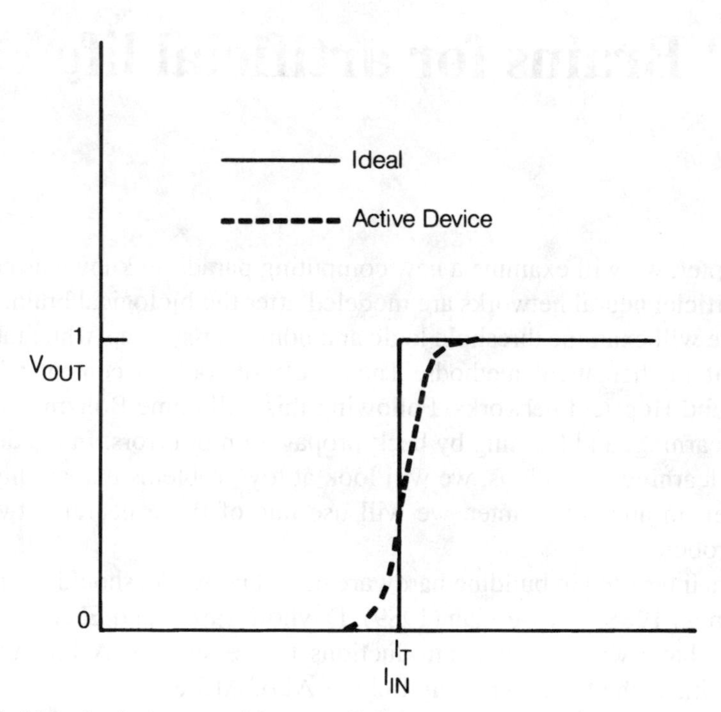

2-2 Threshold logic device transfer curve. Solid curve is ideal device, and dotted curve is active, real device.

ron to fire, while an inhibitory connection tells the neuron to not fire. The input signals are summed in the neuron. At a certain threshold level, the neuron will fire; below this level, it will not fire. This firing is diagrammed in Fig. 2-2.

If the sum of the input signal I_{input} is less than threshold I_t, then V_{out} goes low, logic 0. If the sum of input signal I_{input} is above threshold I_t, then V_{out} goes high, logic 1. This idea can be represented algebraically as follows:

$$V_{out} = 1 \text{ if } I_{input} > I_t$$
$$V_{out} = 0 \text{ if } I_{input} < I_t$$

Software implementation of this idea is presented later in this chapter.

This model is a little naive, because the synapse connections to the inputs of the threshold logic device haven't been considered. In hardware implementation of threshold logic circuits, the synapse is a resistive interconnection between logic devices. A less naive model in algebraic terms is given here:

$$V_{out} \begin{cases} 1 \\ 0 \end{cases} \text{if} \sum_{j \neq i} W_{ij} V_j \begin{matrix} > I_t \\ < I_t \end{matrix}$$

In this relation, the V_j is the input signal to neuron j, and W_{ij} is the conductance of the connection between the ith and jth neuron. Each threshold logic unit randomly and asynchronously computes whether it is above or below threshold and readjusts accordingly. Therefore a network of these threshold logic units is a parallel computer.

This parallel computer can be used in optimization problems and content addressable memories. The content addressable memory implementation of these parallel computation circuits will be discussed at length in this chapter. Then, later I will explain how to teach a neural network and explore other learning paradigms.

The information storage algorithm for content addressable memories, examined in this chapter, is called the *Hebb learning rule*. A memory state is given as a binary vector. In a binary vector, all the elements are either 1 or 0. The outer product of this memory vector with its transpose gives a storage matrix.

Here's an example. The ASCII code for the letter "A" is given by the standard binary representation (0 1 0 0 0 0 0 1). The outer product of this eight-dimensional binary vector with its transpose is as follows:

$$\begin{pmatrix} 0 \\ 1 \\ 0 \\ 0 \\ 0 \\ 0 \\ 0 \\ 1 \end{pmatrix} (01000001) = \begin{pmatrix} 0 & 0 & 0 & 0 & 0 & 0 & 0 & 0 \\ 0 & 1 & 0 & 0 & 0 & 0 & 0 & 1 \\ 0 & 0 & 0 & 0 & 0 & 0 & 0 & 0 \\ 0 & 0 & 0 & 0 & 0 & 0 & 0 & 0 \\ 0 & 0 & 0 & 0 & 0 & 0 & 0 & 0 \\ 0 & 0 & 0 & 0 & 0 & 0 & 0 & 0 \\ 0 & 0 & 0 & 0 & 0 & 0 & 0 & 0 \\ 0 & 1 & 0 & 0 & 0 & 0 & 0 & 1 \end{pmatrix}$$

This storage algorithm can be represented in algebraic terms as

$$W = m^t m$$

W is the information storage matrix, m is the memory vector, and m^t is the transpose of vector m.

This model does not consider an important point. The elements W_{ii} should be set to 0. Hopfield (1982, 1984) and McEliece, et. al. (1985), have shown that if W_{ii} is not 0, then the hardware implementation of the model can result in chaotic oscillations. The correct algebraic relation is

$$W = m^t m - I_n$$

where I_n is the n x n identity matrix. So the storage algorithm consists of the outer product of the memory vector with itself, except that 0's are placed on the diagonal. Next is an example using the ASCII code for the letter "Z":

$$\begin{pmatrix} 0 \\ 1 \\ 0 \\ 1 \\ 1 \\ 0 \\ 1 \\ 0 \end{pmatrix} (01000001) = \begin{pmatrix} 0 & 0 & 0 & 0 & 0 & 0 & 0 & 0 \\ 0 & 0 & 0 & 1 & 1 & 0 & 1 & 0 \\ 0 & 0 & 0 & 0 & 0 & 0 & 0 & 0 \\ 0 & 1 & 0 & 0 & 0 & 0 & 1 & 0 \\ 0 & 1 & 0 & 1 & 1 & 0 & 1 & 0 \\ 0 & 0 & 0 & 0 & 0 & 0 & 0 & 0 \\ 0 & 1 & 0 & 1 & 1 & 0 & 0 & 0 \\ 0 & 0 & 0 & 0 & 0 & 0 & 0 & 0 \end{pmatrix}$$

Hopfield (1982, 1984) has shown that the storage matrix must be symmetric ($W_{ij} = W_{ji}$), that $W_{ii} = 0$, and that the matrix must be dilute (i.e., there must be less 1's than 0's in the matrix).

To show you how this storage matrix can produce the correct memory state, an eight-dimensional binary vector with bit errors—when multiplied by this storage matrix to give the inner product—will generate the correct memory state. The number of bit errors cannot be too great, but a partial memory will certainly work to give the complete memory state. (The correct memory state is known as an *eigenvector*.) For example, in Boolean, we have:

$$\begin{pmatrix} 0 & 0 & 0 & 0 & 0 & 0 & 0 & 0 \\ 0 & 0 & 0 & 1 & 1 & 0 & 1 & 0 \\ 0 & 0 & 0 & 0 & 0 & 0 & 0 & 0 \\ 0 & 1 & 0 & 0 & 1 & 0 & 1 & 0 \\ 0 & 1 & 0 & 1 & 0 & 0 & 1 & 0 \\ 0 & 0 & 0 & 0 & 0 & 0 & 0 & 0 \\ 0 & 1 & 0 & 1 & 1 & 0 & 0 & 0 \\ 0 & 0 & 0 & 0 & 0 & 0 & 0 & 0 \end{pmatrix} \begin{pmatrix} 0 \\ 1 \\ 0 \\ 0 \\ 1 \\ 0 \\ 0 \\ 1 \end{pmatrix} = \begin{pmatrix} 0 \\ 1 \\ 0 \\ 1 \\ 1 \\ 0 \\ 1 \\ 0 \end{pmatrix}$$

The number of bit errors is called the Hamming distance. Given the two vectors

$$v = (0\ 1\ 0\ 0\ 1\ 0\ 0\ 1)$$
$$u = (0\ 1\ 0\ 1\ 1\ 0\ 1\ 0)$$

the Hamming distance in this example is 3. Only vectors of equal dimensionality can be compared.

Hopfield (1982, 1984) has shown that if the matrix is symmetric and dilute with $W_{ii} = 0$, and if we define the dimension of the matrix as n, then m memories can be stored (where $m = 0.15n$). Table 2-1 is a list of the number of memories for a given matrix size. Notice in the table the actual number of memories has been rounded down to the nearest whole number; it doesn't make sense to store a fraction of a memory state.

Table 2-1 Number of memories for a given matrix.

Memories	Neurons (matrix size)
1	8
2	16
3	24
4	32
5	40

All of this can be expressed more formally to assist in writing code for a digital computer simulation.

The Hebb rule is used to determine the values of the W matrix. This is a vector outer product rule.

$$W_{ij} = 1 \text{ if } \sum_{s} v_i^s u_i^s > 0$$

$$W_{ij} = 0 \text{ otherwise.}$$

This states that the element W_{ij} is found by summing the outer product of the input vector element j and output vector element i. It is a simple outer product of these vectors. The W_{ij} element is then found by summing the W matrices. In other words the outer product of the input vector us and the output vector vs results in a matrix W^s. The elements of the final W matrix are found by summing the W^s matrices.

$$W = \sum_{s} W^s$$

The sum is over all memory states, with each memory producing one matrix. The total memory matrix is the sum of these matrices. Under the Hopfield conditions, stable states will exist and the network will not oscillate chaotically. If you are given the W matrix, the matrix vector product (the inner product) of this W with u_s—the input vector state—will result in the output vector v_s.

$$v^s = Wu^s$$

The elements of this vector vs are given by the following:

$$v_i^s = \sum_{j=1}^{N} W_{ij} u_j^s$$

What this says is that output vector element v_i is given by the sum of the products of elements $W_{ij}u_j$ summed over all j.

For example:

$$W = \begin{pmatrix} W_{11} & W_{12} & W_{13} \\ W_{21} & W_{22} & W_{23} \\ W_{31} & W_{32} & W_{33} \end{pmatrix}$$

$$u^s = (\ u^s_1,\ \ u^s_2,\ \ u^s_3\ )$$

then

$$v^s = \begin{pmatrix} W_{11} & W_{12} & W_{13} \\ W_{21} & W_{22} & W_{23} \\ W_{31} & W_{32} & W_{33} \end{pmatrix} \begin{pmatrix} u^s_1 \\ u^s_2 \\ u^s_3 \end{pmatrix} = \begin{pmatrix} v^s_1 \\ v^s_2 \\ v^s_3 \end{pmatrix}$$

$$v^s_1 = W_{11}u^s_1 + W_{12}u^s_2 + W_{13}u^s_3$$
$$v^s_2 = W_{21}u^s_1 + W_{22}u^s_2 + W_{23}u^s_3$$
$$v^s_3 = W_{31}u^s_1 + W_{32}u^s_2 + W_{33}u^s_3$$

This output vector should include a term for the information input, bias, and noise threshold. If these terms are added together to give one term I_i, then we get

$$v_i = \sum_j W_{ij}u_j + I_i$$

The actual vector is given by

$$v_i = 1 \text{ if } \sum_j W_{ij}u_j + I_i < I_t$$

and 0 otherwise, where I_t is threshold (see Fig. 2-2) and W_{ij} is the conductance of the connection between threshold logic units i and j. In other words, W_{ij} is the synaptic strength.

The energy corresponding to the stable states as given by Hopfield (1982, 1984), is

$$E = \frac{1}{2}\sum_{i=1}^{N}\sum_{j=1}^{N} W_{ij}\, v_i v_j$$

where $W_{ij} = W_x$ and $W_{ii} = 0$.

Goles and Vichniac (1986) write this equation in a form that clearly shows how to calculate the energy function.

$$E = -\frac{1}{2}\sum_{i=1}^{N} v_i^{t+1} \sum_{j=1}^{N} W_{ij} v_j^t$$

Content-addressable memories and energy calculations

When the inner product between a vector and a matrix is found, a vector is generated. If the starting vector doesn't differ too much from the stored memory state in the matrix, then the resulting vector is the correct memory state. This has obvious applications as a content-addressable memory.

Some examples might make this more clear. Figure 2-3a shows a partial memory state for the complete memory state of Fig. 2-3b. If the partial memory state is digitized and operated on by an appropriate memory matrix, then the correct memory is generated or recalled.

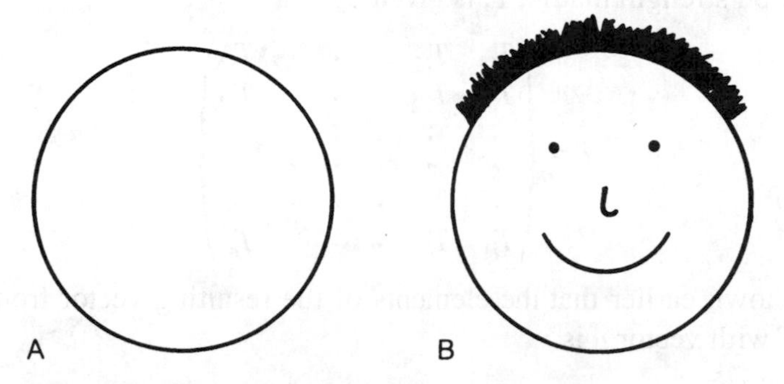

2-3 Partial memory state; complete memory state.

Take another example. Figure 2-4a shows a partial spectral pattern. In this partial memory state, no fine structure is observed. But when this spectrum is digitized and operated on by the appropriate storage matrix, then the spectrum of Fig. 2-4b is recalled or generated.

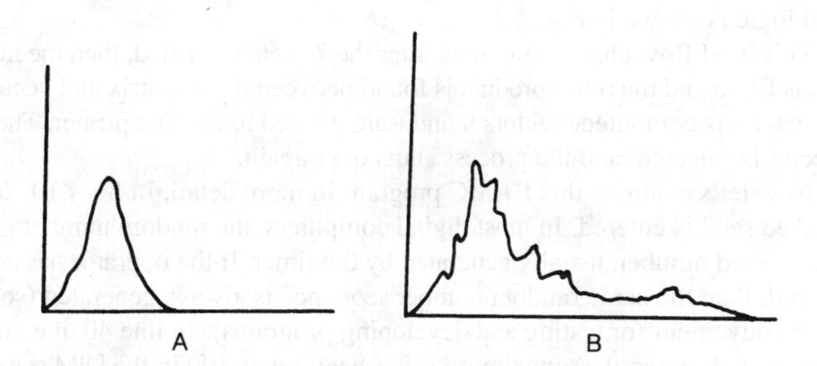

2-4 Partial spectral pattern; complete spectral pattern.

This memory recall often happens to people, also. We see a person in the distance with lime-green socks, but other details are not clear. Then we recall our friend had lime-green socks. This produces, in our mind, the entire picture of our friend.

Further examples of content-addressable memory are easy to conceive. Looking in a tool box, we might see only five percent of a wrench handle because the rest of the wrench is hidden by other tools. However, this is still enough of a pattern for us to recognize it as the wrench we need.

The first step in an algorithm for content-addressing is to find the inner product of a vector and a matrix.

$$v^s = Tu^s$$

The connection strength matrix, T, is given by

$$T = \begin{pmatrix} T_{11} & T_{12} & \cdot & \cdot & \cdot & T_{1N} \\ T_{21} & T_{22} & \cdot & \cdot & \cdot & T_{2N} \\ \cdot & \cdot & \cdot & \cdot & \cdot & \cdot \\ \cdot & \cdot & \cdot & \cdot & \cdot & \cdot \\ \cdot & \cdot & \cdot & \cdot & \cdot & \cdot \\ T_{i1} & T_{i2} & \cdot & \cdot & \cdot & T_{\epsilon} \end{pmatrix}$$

It was shown earlier that the elements of the resulting vector from the inner product of T with vector u is

$$v_i^s = \sum_{j=1}^{N} T_{ij} u_j^s$$

It was further shown that the diagonal elements of the T matrix must be 0, and the matrix should be symmetric and dilute.

The program NEURON4P implements these ideas and calculates the inner product of a vector with a matrix. This simple program is not too useful by itself; but it will be used to build other programs. A simplified flow diagram of the program logic is shown in Fig. 2-5.

This brief flow chart shows that, after the T matrix is filled, then the input vector u is filled and the inner product is found between the T matrix and vector u. After vector v is computed, vectors u and v are printed to the line printer. Then a new u vector is selected, and the process starts over again.

Now let's examine this BASIC program in more detail. In lines 10–20, a randomized seed is entered. In most digital computers, the random number generator needs a seed number, usually generated by the timer. If the operator has control of the seed, then the same random number sequence is always generated (something that is convenient for testing and developing programs). In line 40, the number of neurons, N, is entered. A maximum N has been set as 100 in the DIM statement of line 90. This can be changed as desired by the user.

Let's continue with the theory. Recall that Fig. 2-2 defines the threshold value of I_t. The output of the i-th neuron is logic 1 if the sum of the products of the elements $T_{ij}u_j$ and I_i is greater than the threshold. Otherwise, the output is logic 0.

The threshold is input in line 50 as variable IO. For small networks, it is sufficient to choose 0 or 1 for the threshold (which is equivalent to shifting the curve of Fig. 2-2 to the left or right). The information input to the neuron is entered in line 60. You can think of this information to the neuron as a bias from another signal

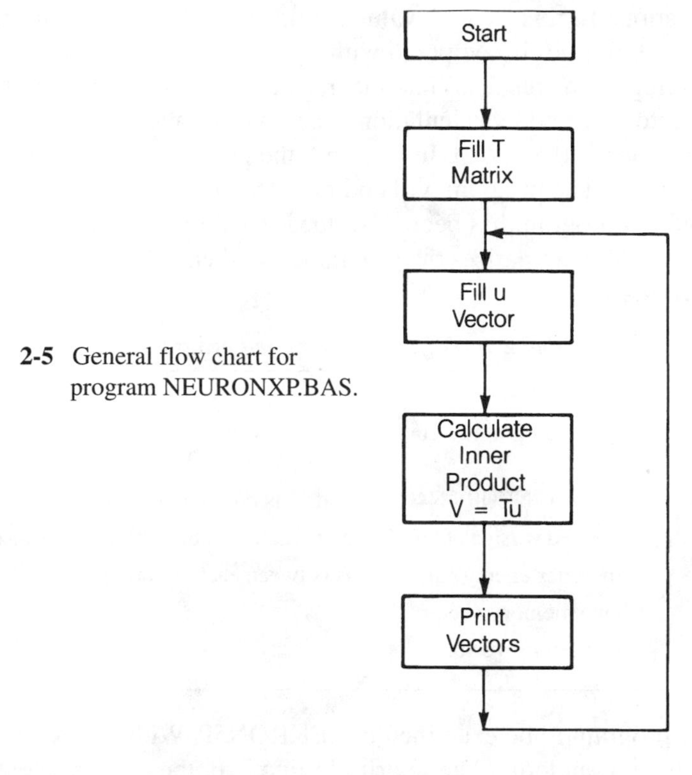

2-5 General flow chart for program NEURONXP.BAS.

source and/or noise. It is simply called INFO in the program. By experimenting with the INFO and IO variables, the user can get very different results.

Line 70 and 80 ask if the user would like to enter the input vector and synapse matrix from the keyboard; otherwise, the program will select a random binary vector for input, as well as a random, dilute, symmetric matrix with $T_{ii} = 0$. Line 110 is a decision point for operator-entered T matrix or machine-generated matrix. Lines 120–170 allow the user to enter the matrix from the keyboard. Lines 190–350 are for a machine-generated matrix. A random variable R, between 0 and 1, is selected in line 210. Line 220 dilutes the selection. If $R < 0.75$, then $R = 0$, else $R = +1$. The T_{ij} element being addressed is set equal to this R value in line 230. At the completion of line 260, the matrix is now filled and dilute. Line 270 begins a routine to diagonalize the matrix, and in line 290, T_{ii} is set to 0. Line 300 symmetrizes the matrix with $T_{ij} = T_{ji}$. Finally, in line 310, the matrix is printed out to the line printer.

In line 360, the process to fill the input vector, u, is started. If the operator chooses to enter the vector from the keyboard, then lines 380–410 are executed; otherwise, the machine selects a random binary vector in lines 430–460. This routine calls a subroutine, lines 670–690, to select a random number and decide if the vector element should be a 0 or 1.

In line 470, the calculation of the inner product of the input vector and the T matrix is started. In line 500, the variable named SIGMA is assigned to the result of this calculation. To this SIGMA value is added the INFO value in line 520. Finally in line 530, SIGMA is compared with the threshold value IO and set equal to 0 or 1, depending on the results. This final result becomes output vector element v_i. After the v vector is filled by calculations, the u vector and v vector are sent to the line printer in lines 580 and 620. In line 660, the process continues by selecting a new input vector u. The program will end only by a BREAK.

Now that this program has been described in detail, we can move on to other programs. Table 2-2 summarizes the programs and their differences; I'll give more details afterwards.

Table 2-2 Differences between programs.

BASIC program	Comments
NEURON4P	Basic program to find inner product of a vector and a matrix. $T_{ij}=T_{ji}$ and $T_{ii}=0$
NEURON5P	Same as NEURON4P, but matrix is random and a little more dilute. $T_{ij}\#T_{ji}$
NEURON6P	Iterated version of NEURON5P. Eight iterations; then prints resulting vector.
NEURON8P	Includes energy calculations between each iteration.
HEBB2P	One memory vector.
HEBB3P	m memory vectors.

The next program to be examined is NEURON5P. With this program, we can see the effects of a random dilute matrix. In line 210, the random element is chosen, and in line 220 the matrix is diluted. In this program, the matrix will be more dilute than in NEURON4P. Here, if $R < 0.85$, then $R = 0$, else R is set equal to +1; this results in less than 15% of the elements being set to 1. In NEURON4P, the dilution was 25%. The rest of this program is similar to NEURON4P.

A run of this program will produce some stable states, but it also produces many superfluous states. Stable states can be thought of as low points on an energy hyperplane in an n-dimensional hypercube. This will be explained more clearly later in this chapter.

It is interesting to note that a run of program NEURON5P results in some spurious states. There are two problems: the matrix is nonsymmetric, and an iterative dynamics has not been implemented in this program. By iterative dynamics, I mean that the results from one vector-matrix product should be sent back into the network and operated on again by the same matrix. Only by iterative operation can these massively parallel networks compute stable states.

As pointed out earlier in this chapter, the neurons or threshold logic processors are connected to each other through a synapse or conductance matrix. The output of a processor may be connected to the input of several other processors. Each threshold logic device sums the inputs it receives. So the signal travels around this feedback loop, but not to itself, many times in a second before the stable state is reached.

Iterative dynamics has been introduced in the next program.

In NEURON6P, the new variable ITERATE has been introduced. This variable is a counter in a loop that starts in line 480. The output vector form the inner product of the input vector and the matrix is set equal to a new input vector in line 640. The inner product of this vector and the matrix is then found, and the process is repeated eight times. Later we will see that eight times is a few too many; only three or four repetitions are needed. This program prints the matrix, and then prints the initial input vector followed by the eighth iterated resulting vector as the output. A new initial vector is then selected, and the process started over again.

A run of this program is shown in Fig. 2-6.

Notice we still appear to have more than one stable state. From Table 2-1, we know that, for an eight-neuron circuit, we can have only one stable state (this can be explained from an energy consideration). By the iterative dynamics, the algorithm computes the minimal states in an n-dimensional hyperspace. There will be more than one minimum on this hypersurface. In fact, all the corners of the hypercube are stable states. These do not all have the same degree of stability and may only be metastable states. These minima are attractors that may be strange attractors.

It might be convenient for you to think of the energy surface as a sheet of rubber being pulled in many directions, from many points on its surface. This results in a surface with hills, valleys, and wells. If a small marble is dropped on this surface, it will be attracted to the nearest lowest point. This point might not be the lowest point in the entire hypersurface; it is just a local minima or attractor. If the marble is kicked around hard enough, it will jump out of this local minima and settle to the next. Repeating this process will result in the marble settling in the deepest basin of attraction, which is equivalent to the most stable memory state of the network.

The computed result shows several minima. Because of Johnson noise and other component noise in real neural network hardware circuits, we will see that only one stable state results for an eight-neuron circuit. This noise is the analogous effect of kicking the marble around till it settles in the deepest basin of attraction.

In the next program, we calculate the energy after each iteration. As shown earlier, we can use the following equation:

$$E(t) = -\frac{1}{2} \sum_{i=1}^{N} v_i^{t+1} \sum_{j=1}^{N} T_{ij} v_j^t$$

What this algorithm says is that the inner product of vector v at time t with connection strength matrix T is multiplied with the vector v at time $t + 1$. Let time t count the number of sweeps through the network, and $t + 1$ is the next sweep. This second multiplication is an outer product of two vectors and results in a scalar. This scalar value is proportional to the energy.

From this dynamical equation, it is clear that the energy reaches a minimum when

$$v^{t+1} = v^t$$

```
0 0 0 0 0 0 0 1
0 0 1 0 0 0 0 0
0 1 0 0 0 0 0 0
0 0 0 0 0 1 0 0
0 0 0 0 0 0 0 0
0 0 0 1 0 0 0 0
0 0 0 0 0 0 0 1
1 0 0 0 0 0 1 0

0 0 0 1 0 1 1 0
1 0 0 1 0 1 1 0

0 0 0 0 1 1 0 1
0 0 0 0 0 1 0 1

0 0 1 0 0 1 0 1
0 0 1 0 0 1 0 0

1 0 1 0 1 1 0 1
1 0 1 0 0 1 1 1

0 1 1 0 0 0 1 1
1 1 1 0 0 0 1 1

1 1 1 0 0 1 1 0
1 1 1 0 0 1 1 0

0 0 1 1 0 1 0 1
0 0 1 1 0 1 0 1

0 1 0 1 1 1 1 1
1 1 0 1 0 1 1 1

0 1 0 1 0 0 0 1
0 1 0 1 0 0 0 1

0 1 0 0 0 1 1 1
1 1 0 0 0 1 1 1

1 0 1 1 0 1 1 0
1 0 1 1 0 1 1 0

0 1 1 1 1 1 1 0
1 1 1 1 0 1 1 0

1 1 0 0 0 0 0 1
1 1 0 0 0 0 1 1

1 0 1 0 0 0 0 0
1 0 1 0 0 0 1 0

1 1 0 1 0 0 1 1
1 1 0 1 0 0 1 1

1 0 1 0 0 0 1 1
1 0 1 0 0 0 1 1

1 1 1 1 1 1 1 0
1 1 1 1 0 1 1 0

0 1 0 0 0 0 0 1
0 1 0 0 0 0 0 1

1 0 1 0 1 0 0 0
1 0 1 0 0 0 1 0

0 0 1 1 0 0 1 1
1 0 1 1 0 0 1 1

1 1 0 1 0 0 1 1
1 1 0 1 0 0 1 1

1 0 1 0 0 1 0 1
1 0 1 0 0 1 1 1
```

2-6 Example run of program NEURON6P.BAS. Seed 72873, threshold 1, information 1.

This is clearly seen in a computer simulation. After a few iterations, usually four or less, the energy settles to a stable point. Now let us examine the program, NEURON8P.

The program NEURON8P introduces the new variable ENERGY. The energy calculation takes place within the ITERATE loop. The energy is set equal to 0 in line 660, and calculations take place in a loop starting at 670 and ending at 690. The final energy is printed out in line 700. In line 750, the energy is set equal to 0 again and the next iteration begins. Figure 2-7 shows a run of this program. After the matrix is printed, the random binary vector is printed. Then the resulting inner product vector and energy is printed. The program then begins the next iteration by feeding the resulting vector back into the matrix calculation.

```
0  0  0  0  1  0  0  0  0  0  1  1  0  0  0  0
0  0  0  0  0  0  0  0  0  0  0  1  0  0  0  0
0  0  0  0  0  0  1  0  0  0  0  0  1  0  0  1
0  0  0  0  0  0  0  0  1  0  1  0  0  0  0  0
1  0  0  0  0  1  0  0  0  1  0  0  0  0  1  1
0  0  0  0  1  0  0  0  0  0  0  0  0  0  0  0
0  0  1  0  0  0  0  0  0  0  0  1  0  0  0  0
0  0  0  0  0  0  0  0  1  0  1  1  0  1  0  0
0  0  0  1  0  0  0  1  0  0  0  1  0  1  0  1
0  0  0  0  1  0  0  0  0  0  1  1  0  0  0  0
1  0  0  1  0  0  0  1  0  1  0  0  0  0  0  1
1  1  0  0  0  0  1  1  1  1  0  0  0  0  0  0
0  0  1  0  0  0  0  0  0  0  0  0  0  0  0  1
0  0  0  0  0  0  0  1  1  0  0  0  0  0  0  0
0  0  0  0  1  0  0  0  0  0  0  0  0  0  0  0
0  0  1  0  1  0  0  0  1  0  1  0  1  0  0  0


1  0  1  0  1  0  1  1  1  1  0  1  1  0  0  0
1  0  1  0  1  0  1  1  1  1  1  1  0  1  0  1
    ENERGY -4
1  0  1  0  1  0  1  1  1  1  1  1  0  1  0  1
1  0  1  1  1  0  1  1  1  1  1  1  1  1  0  1
    ENERGY -5.5
1  0  1  1  1  0  1  1  1  1  1  1  1  1  0  1
1  0  1  1  1  0  1  1  1  1  1  1  1  1  0  1
    ENERGY -6.5
1  0  1  1  1  0  1  1  1  1  1  1  1  1  0  1
1  0  1  1  1  0  1  1  1  1  1  1  1  1  0  1
    ENERGY -6.5
1  0  1  1  1  0  1  1  1  1  1  1  1  1  0  1
1  0  1  1  1  0  1  1  1  1  1  1  1  1  0  1
    ENERGY -6.5
1  0  1  1  1  0  1  1  1  1  1  1  1  1  0  1
1  0  1  1  1  0  1  1  1  1  1  1  1  1  0  1
    ENERGY -6.5
1  0  1  1  1  0  1  1  1  1  1  1  1  1  0  1
1  0  1  1  1  0  1  1  1  1  1  1  1  1  0  1
    ENERGY -6.5
1  0  1  1  1  0  1  1  1  1  1  1  1  1  0  1
1  0  1  1  1  0  1  1  1  1  1  1  1  1  0  1
    ENERGY -6.5
```

2-7 Example run of program NEURON8P.BAS. Seed 0, threshold 1, information 0.

2-7 Continued.

```
1  0  1  1  1  0  1  0  1  1  0  1  0  1  1  1
1  0  1  0  1  0  1  1  1  1  1  1  1  0  0  1
    ENERGY -4
1  0  1  0  1  0  1  1  1  1  1  1  1  0  0  1
1  0  1  1  1  0  1  1  1  1  1  1  1  1  0  1
    ENERGY -5.5
1  0  1  1  1  0  1  1  1  1  1  1  1  1  0  1
1  0  1  1  1  0  1  1  1  1  1  1  1  1  0  1
    ENERGY -6.5
1  0  1  1  1  0  1  1  1  1  1  1  1  1  0  1
1  0  1  1  1  0  1  1  1  1  1  1  1  1  0  1
    ENERGY -6.5
1  0  1  1  1  0  1  1  1  1  1  1  1  1  0  1
1  0  1  1  1  0  1  1  1  1  1  1  1  1  0  1
    ENERGY -6.5
1  0  1  1  1  0  1  1  1  1  1  1  1  1  0  1
1  0  1  1  1  0  1  1  1  1  1  1  1  1  0  1
    ENERGY -6.5
1  0  1  1  1  0  1  1  1  1  1  1  1  1  0  1
1  0  1  1  1  0  1  1  1  1  1  1  1  1  0  1
    ENERGY -6.5
1  0  1  1  1  0  1  1  1  1  1  1  1  1  0  1
1  0  1  1  1  0  1  1  1  1  1  1  1  1  0  1
    ENERGY -6.5


1  1  0  1  1  1  0  0  1  1  0  0  0  1  0  0
0  0  0  0  1  0  0  1  1  0  1  1  0  0  0  1
    ENERGY -1
0  0  0  0  1  0  0  1  1  0  1  1  0  0  0  1
1  0  0  1  0  0  0  1  1  1  1  1  0  1  0  1
    ENERGY -2.5
1  0  0  1  0  0  0  1  1  1  1  1  0  1  0  1
1  0  0  1  1  0  0  1  1  1  1  1  0  1  0  1
    ENERGY -4.5
1  0  0  1  1  0  0  1  1  1  1  1  0  1  0  1
1  0  0  1  1  0  0  1  1  1  1  1  0  1  0  1
    ENERGY -5
1  0  0  1  1  0  0  1  1  1  1  1  0  1  0  1
1  0  0  1  1  0  0  1  1  1  1  1  0  1  0  1
    ENERGY -5
1  0  0  1  1  0  0  1  1  1  1  1  0  1  0  1
1  0  0  1  1  0  0  1  1  1  1  1  0  1  0  1
    ENERGY -5
1  0  0  1  1  0  0  1  1  1  1  1  0  1  0  1
1  0  0  1  1  0  0  1  1  1  1  1  0  1  0  1
    ENERGY -5
1  0  0  1  1  0  0  1  1  1  1  1  0  1  0  1
1  0  0  1  1  0  0  1  1  1  1  1  0  1  0  1
    ENERGY -5
```

Looking at Fig. 2-7 we see that, in the case of the first random binary vector, the energy has a value of –4 after the first iteration and –5.5 after the second iteration, finally settling to a stable state at –6.5 energy units. In the next two programs, we will examine associative learning using the Hebb learning rule.

Associative learning: The Hebb learning rule

The program HEBB2P is a basic building unit for the Hebb learning rule. The original Hebb learning rule (Hebb, 1949) was not quantitative enough to build a good model. In its original version, the rule stated that if neuron A and neuron B are simultaneously excited, then the synaptic connection strength between them is increased.

An excellent example of associative learning in humans is when we hold a red apple in front of a baby and repeatedly say "red." Synaptic connection strengths will be increased when the appropriate neurons from the optic center are simultaneously activated with those from the auditory center for the sound of the word "red." Another example is Pavlov's experiments in which, after repeated trials, a dog learned to associate the sound of a bell with food. Using this learning rule, we could train a simple network such as that shown in Fig. 2-8. An input vector would be presented to both the auditory and optic neurons. The appropriate synaptic connections would then be strengthened.

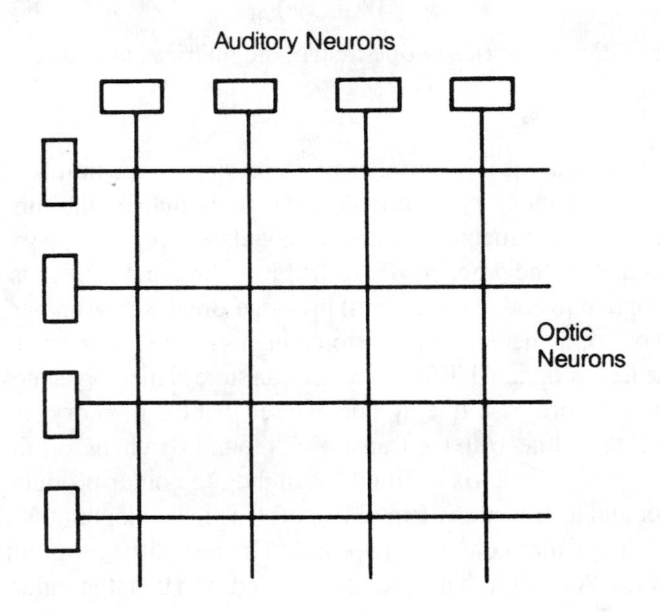

2-8 Simple network of optic and auditory neurons.

In digital simulations of this model, the outer product of two vectors is found to produce a synaptic connection strength matrix. This is given symbolically as

$$W = v^t u$$

For a one-dimensional vector of length 4, we would get the matrix W:

$$u = (3 \quad 1 \quad 2 \quad 4)$$
$$v = (0 \quad 1 \quad 1 \quad 6)$$

$$W = \begin{pmatrix} 0 \\ 1 \\ 1 \\ 6 \end{pmatrix} (3 \quad 1 \quad 2 \quad 4) = \begin{pmatrix} 0 & 0 & 0 & 0 \\ 3 & 1 & 2 & 4 \\ 3 & 1 & 2 & 4 \\ 18 & 6 & 12 & 24 \end{pmatrix}$$

Notice I used the transpose of the vector v. Symbolically, to find an element of the matrix, we write

$$W = v_i^t u_j$$

The connection strengths are the elements of the matrix. These represent the stored memory state or states. If the storage matrix is small, then only one memory state can be stored. For larger storage matrices, more than one memory can be stored. In that case, each memory state will generate one matrix.

$$W^s = (v^t)^s u^s$$

To store all the memories in one matrix the matrices are added over all states.

$$W = \sum_s W^s$$

Table 2-1 summarizes the number of memories versus number of neurons. In order to recover the memory state from the storage matrix, the inner product of a partial memory and the memory matrix is calculated.

Now let's discuss the program HEBB2P. First, observe that there is no END. In order for the program to end, the user must press the Break key. Another obvious observation is that only one memory state is stored in this matrix. This program will be used to develop the next program, HEBB3P, which can store m memory states for N neurons.

Looking at the program line by line, we see that the memory state is entered in lines 130 to 170. In line 160, the transpose is found by changing the label. Beginning in line 180, the T matrix is filled by finding the outer product of the desired storage vector and its transpose. Line 210 puts 0's on the diagonal of the matrix, and line 220 prints the T matrix at the line printer. The rest of the program is exactly like earlier programs. A random binary vector is filled, and then the inner product of this vector with the T matrix is calculated. This calculation is iterated eight times, and the random vector and final product vector is printed at the line printer.

Program HEBB3P includes a routine to allow the user to choose the number of memory states to store. This program asks the user how many neurons are to be simulated. It then prompts the user for the number of memory states to be stored and reminds the user that the number of memory states stored is given by

```
m = INT(0.15N)
```

where N is the number of neurons.

This equation for the number of memories is empirically derived. You can experiment with these programs and deduce more or less the same equation for the number of memory states. After the user enters the first memory vector, the first T matrix is found and printed at the line printer. The next memory vector is entered, and the new T matrix is printed. This new T matrix includes the sum of the T matrix for state one and state two. This continues through m memory states. Then a random binary vector is chosen or entered from the keyboard, and the inner product of this vector and the summed T matrix is found and printed at the line printer along with the random vector.

Figure 2-9 is a run of this program with 16 neurons and 2 memory states.

The two states entered were

$$(1\ 0\ 0\ 0\ 1\ 0\ 0\ 0\ 1\ 0\ 0\ 0\ 1\ 0\ 0\ 0)$$
$$(0\ 1\ 1\ 0\ 0\ 0\ 1\ 1\ 0\ 0\ 0\ 0\ 0\ 1\ 1\ 0)$$

Notice that these states did not come out from random vectors except twice for the second vector and once for the first vector. In the other cases, the Hamming distance is too great to result in a correct memory state. The end result is stable spurious state. We see that this same stable spurious state arises many times, indicating that it is probably a deeper energy minima than the two stored states. These spurious states can be caused by overlapping vectors in Hamming space (i.e, too many interconnections among the neurons).

Some interesting programming experiments would be to include energy calculations between each iteration and see if there are in fact deeper stable states than the stored memory states. Another experiment would be a study of Hamming distance to see how far off one can be in Hamming space and still "pull in" to one of the stored states.

Boltzman machines and statistical learning

Boltzman machines are neural-network-like architectures with a Boltzman statistical algorithm for updating the processing elements. Hinton and Sejnowski (1986) were early developers of the mathematical technique for Boltzman machines. Wasserman (1989) has presented a very readable account of the methods. The method is derived form Metropolis et al. (1953).

The Boltzman distribution is shown in Fig. 2-10 and is given by the relation

$$P(x) = e^{-\frac{x}{kT}}$$

where $P(x)$ is the probability of change in magnitude of x, k is a constant, and T is an artificial temperature.

This probability distribution is used to select the state of the neurons. In the case shown in Fig. 2-2, we wanted to approach the ideal step function for threshold logic. The Boltzman machine output function is related to the probability, which is a function of pseudotemperature (as shown in Fig. 2-11).

```
0  0  0  0  1  0  0  0  1  0  0  0  1  0  0  0
0  0  0  0  0  0  0  0  0  0  0  0  0  0  0  0
0  0  0  0  0  0  0  0  0  0  0  0  0  0  0  0
0  0  0  0  0  0  0  0  0  0  0  0  0  0  0  0
1  0  0  0  0  0  0  0  1  0  0  0  1  0  0  0
0  0  0  0  0  0  0  0  0  0  0  0  0  0  0  0
0  0  0  0  0  0  0  0  0  0  0  0  0  0  0  0
0  0  0  0  0  0  0  0  0  0  0  0  0  0  0  0
1  0  0  0  1  0  0  0  0  0  0  0  1  0  0  0
0  0  0  0  0  0  0  0  0  0  0  0  0  0  0  0
0  0  0  0  0  0  0  0  0  0  0  0  0  0  0  0
0  0  0  0  0  0  0  0  0  0  0  0  0  0  0  0
1  0  0  0  1  0  0  0  1  0  0  0  0  0  0  0
0  0  0  0  0  0  0  0  0  0  0  0  0  0  0  0
0  0  0  0  0  0  0  0  0  0  0  0  0  0  0  0
0  0  0  0  0  0  0  0  0  0  0  0  0  0  0  0
```

```
0  0  0  0  1  0  0  0  1  0  0  0  1  0  0  0
0  0  1  0  0  0  1  1  0  0  0  0  0  1  1  0
0  1  0  0  0  0  1  1  0  0  0  0  0  1  1  0
0  0  0  0  0  0  0  0  0  0  0  0  0  0  0  0
1  0  0  0  0  0  0  0  1  0  0  0  1  0  0  0
0  0  0  0  0  0  0  0  0  0  0  0  0  0  0  0
0  1  1  0  0  0  0  1  0  0  0  0  0  1  1  0
0  1  1  0  0  0  1  0  0  0  0  0  0  1  1  0
1  0  0  0  1  0  0  0  0  0  0  0  1  0  0  0
0  0  0  0  0  0  0  0  0  0  0  0  0  0  0  0
0  0  0  0  0  0  0  0  0  0  0  0  0  0  0  0
0  0  0  0  0  0  0  0  0  0  0  0  0  0  0  0
1  0  0  0  1  0  0  0  1  0  0  0  0  0  0  0
0  1  1  0  0  0  1  1  0  0  0  0  0  0  1  0
0  1  1  0  0  0  1  1  0  0  0  0  0  1  0  0
0  0  0  0  0  0  0  0  0  0  0  0  0  0  0  0
```

```
1  1  0  0  0  1  0  1  1  1  1  1  1  0  1  0
1  1  1  0  1  0  1  1  1  0  0  0  1  1  1  0
```

```
0  1  0  0  1  1  0  1  1  0  0  0  0  1  1  0
1  1  1  0  1  0  1  1  1  0  0  0  1  1  1  0
```

```
1  0  0  0  1  0  1  1  0  0  0  0  1  1  0  0
1  1  1  0  1  0  1  1  1  0  0  0  1  1  1  0
```

2-9 Example run of HEBB3P.BAS. Seed 72873.

```
0 0 0 1 1 1 1 1 0 1 1 1 0 0 1 0
1 1 1 0 1 0 1 1 1 0 0 0 1 1 1 0

0 0 0 1 0 1 1 0 0 0 0 0 1 1 0 1
1 1 1 0 1 0 1 1 1 0 0 0 1 1 1 0

0 0 1 0 0 1 0 0 1 0 1 0 1 1 0 1
1 1 1 0 1 0 1 1 1 0 0 0 1 1 1 0

0 1 1 0 0 0 1 1 1 1 1 0 0 1 1 0
1 1 1 0 1 0 1 1 1 0 0 0 1 1 1 0

0 0 1 1 0 1 0 1 0 1 0 1 1 1 1 1
1 1 1 0 1 0 1 1 1 0 0 0 1 1 1 0

0 1 0 1 0 0 0 1 0 1 0 0 0 1 1 1
0 1 1 0 0 0 1 1 0 0 0 0 0 1 1 0

1 0 1 1 0 1 1 0 0 1 1 1 1 1 1 0
1 1 1 0 1 0 1 1 1 0 0 0 1 1 1 0

1 1 0 0 0 0 0 1 1 0 1 0 0 0 0 0
1 1 1 0 1 0 1 1 1 0 0 0 1 1 1 0

1 1 0 1 0 0 1 1 1 0 1 0 0 0 1 1
1 1 1 0 1 0 1 1 1 0 0 0 1 1 1 0

1 1 1 1 1 1 1 0 0 1 0 0 0 0 0 1
1 1 1 0 1 0 1 1 1 0 0 0 1 1 1 0

1 0 1 0 1 0 0 0 0 0 1 1 0 0 1 1
1 1 1 0 1 0 1 1 1 0 0 0 1 1 1 0

1 1 0 1 0 0 1 1 1 0 1 0 0 1 0 1
1 1 1 0 1 0 1 1 1 0 0 0 1 1 1 0

1 1 1 1 0 1 0 0 1 1 1 0 0 0 1 0
1 1 1 0 1 0 1 1 1 0 0 0 1 1 1 0
```

2-9 Continued.

```
1  1  1  0  1  1  0  1  0  1  0  0  1  0  1  0
1  1  1  0  1  0  1  1  1  0  0  0  1  1  1  0

1  1  1  0  1  0  1  0  0  0  0  0  0  1  1  0
1  1  1  0  1  0  1  1  1  0  0  0  1  1  1  0

1  1  1  1  0  1  0  0  1  0  1  0  0  1  0  0
1  1  1  0  1  0  1  1  1  0  0  0  1  1  1  0

1  1  1  1  0  0  1  1  1  1  1  0  0  0  1  1
1  1  1  0  1  0  1  1  1  0  0  0  1  1  1  0

1  0  0  1  1  1  0  0  1  1  0  1  0  0  0  0
1  0  0  0  1  0  0  0  1  0  0  0  1  0  0  0

1  0  1  0  0  0  0  1  1  1  1  0  1  0  1  0
1  1  1  0  1  0  1  1  1  0  0  0  1  1  1  0

1  0  1  1  0  0  1  0  1  1  0  1  0  1  1  0
1  1  1  0  1  0  1  1  1  0  0  0  1  1  1  0

1  1  1  1  1  0  1  1  1  1  1  1  1  0  0  0
1  1  1  0  1  0  1  1  1  0  0  0  1  1  1  0

1  0  0  0  1  1  1  1  1  1  1  0  1  1  1  0
1  1  1  0  1  0  1  1  1  0  0  0  1  1  1  0

1  0  1  0  1  1  1  1  1  0  0  1  1  1  1  0
1  1  1  0  1  0  1  1  1  0  0  0  1  1  1  0

1  0  0  0  0  1  0  1  0  0  0  0  0  1  0  1
1  1  1  0  1  0  1  1  1  0  0  0  1  1  1  0

0  1  1  0  0  0  1  1  0  0  0  0  1  0  1  0
1  1  1  0  1  0  1  1  1  0  0  0  1  1  1  0

1  1  1  0  0  0  0  1  0  1  0  1  1  0  0  1
1  1  1  0  1  0  1  1  1  0  0  0  1  1  1  0
```

1 1 0 1 0 0 0 1 0 0 0 0 0 0 1 1
1 1 1 0 1 0 1 1 1 0 0 0 1 1 1 0

1 1 1 0 1 1 1 0 1 0 1 0 1 0 1 1
1 1 1 0 1 0 1 1 1 0 0 0 1 1 1 0

0 1 1 1 0 1 0 0 0 1 0 0 0 1 0 0
0 1 1 0 0 0 1 1 0 0 0 0 0 1 1 0

1 0 1 0 0 1 0 0 0 0 1 0 0 1 1 0
1 1 1 0 1 0 1 1 1 0 0 0 1 1 1 0

0 1 0 1 1 0 0 0 0 0 0 1 0 1 0 0
1 1 1 0 1 0 1 1 1 0 0 0 1 1 1 0

1 0 0 0 0 1 1 0 0 0 0 1 1 1 0 0
1 1 1 0 1 0 1 1 1 0 0 0 1 1 1 0

1 0 0 1 0 0 0 1 0 1 0 0 0 0 1 1
1 1 1 0 1 0 1 1 1 0 0 0 1 1 1 0

1 1 1 1 1 0 0 1 0 0 0 0 0 0 0 1
1 1 1 0 1 0 1 1 1 0 0 0 1 1 1 0

1 1 1 1 1 0 1 0 1 1 0 1 0 1 0 1
1 1 1 0 1 0 1 1 1 0 0 0 1 1 1 0

0 1 1 0 0 0 0 1 1 1 1 1 0 0 0 1
1 1 1 0 1 0 1 1 1 0 0 0 1 1 1 0

0 1 1 0 0 1 0 1 1 0 0 0 1 1 0 1
1 1 1 0 1 0 1 1 1 0 0 0 1 1 1 0

0 0 1 0 0 0 0 0 1 0 1 0 1 0 0 0
1 1 1 0 1 0 1 1 1 0 0 0 1 1 1 0

0 0 0 1 1 1 1 1 0 1 1 1 1 0 0 0

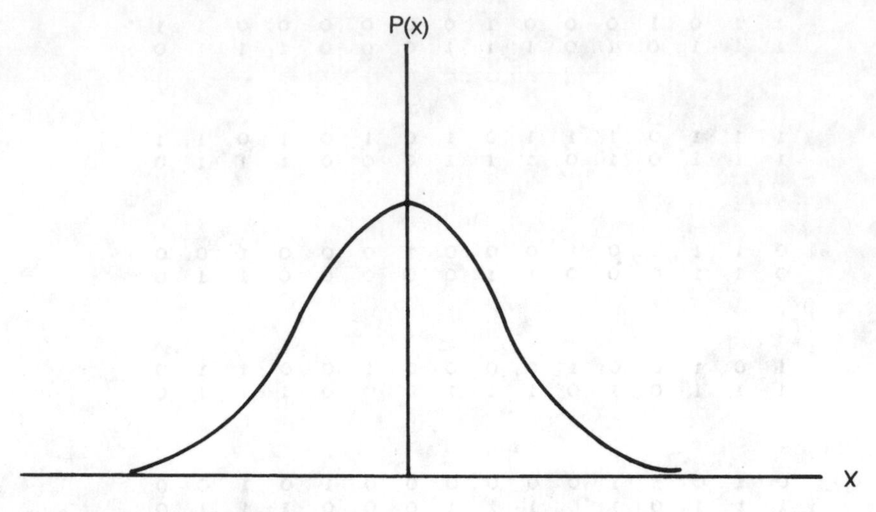

2-10 Probability distribution.

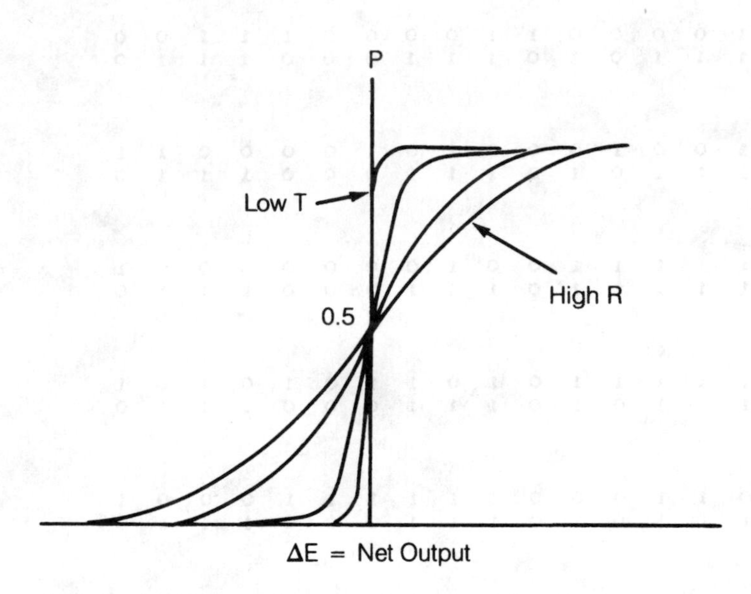

2-11 Probability transfer curves for high and low *T*.

The Boltzman updating starts at a high pseudotemperature and is slowly cooled to a much lower temperature. This results in preventing the output state of the system from being confined to a local minimum in the energy space. The higher temperature acts as a noise to provide enough energy to escape from local minima. Table 2-3 is actual values for the relation

$$P(\Delta E) = \frac{1}{1 + e^{-\frac{\Delta E}{T}}}$$

Table 2-3
Values for the logistic equation.

T	E	P
1.0	1.0	0.731
1.0	2.0	0.880
1.0	3.0	0.952
1.0	–1.0	0.269
1.0	–2.0	0.119
0.1	1.0	0.999
0.1	2.0	0.999
0.1	3.0	1.000
0.1	–1.0	4.5×10^{-5}
0.1	–2.0	2.0×10^{-9}

This relation is sometimes called the logistic equation and has a sigmodial-shaped curve.

The actual plot of this data is given in Fig. 2-12. Notice that, at lower temperatures, the curve has a steeper slope where it crosses the P axis at $P = 0.5$. In actual practice, a random number r is selected in the range $0 <= r <= 1$ and state selection is based on a comparison of the result of the logistic equation and the random number r. The C code in Fig. 2-13 shows how this decision is implemented in software.

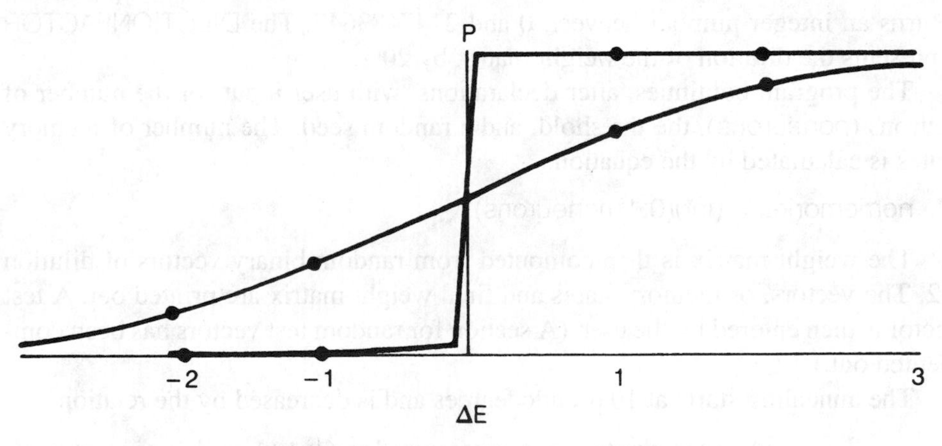

2-12 Plot of data from Table 2-3.

The Boltzman machine is capable of demonstrating learning behavior. By clamping the input units to a pattern and allowing the network to anneal from a high temperature to a low temperature, the network can effectively learn to generate a specific output pattern for a given input pattern. For details on how to use a Boltzman machine for learning, I recommend Wasserman (1989). To demonstrate

```
if((r < p) && (p > 0.5))
        return (1);
else
        return (0);
```

2-13 Random updating algorithm in C.

r is a random number

$p = 1/(1 + \exp(\Delta E/T))$

the basic ideas of Boltzman annealing, I have written a program that is a hybrid of a Hopfield network and a Boltzman machine.

The program tboltz9.c is a C code program that simulates a Boltzman updating of a Hopfield network on a T414 transputer. Simulations of this type are called Monte Carlo simulations because they are probabilistic. Furthermore, these simulations can take very long. I therefore use a Transputer board for these long compute intensive simulations. For a discussion of the Transputer, see my earlier book on parallel processing (Rietman, 1990). The code could be modified for another type of CPU; you don't need to use a Transputer.

Let's look at each block of the code, to help you understand the Boltzman method of updating a neural network. After some sharp includes, RNDMAX is defined as 2147483647.0. This value reflects the fact that the 32-bit transputer chip returns an integer number between 0 and 2147483647. The DILUTIONFACTOR represents the dilution of the weight matrix by 20%.

The program continues, after declarations, with user input for the number of neurons (noneurons), the threshold, and a random seed. The number of memory states is calculated by the equation

```
nomemories = (int)(0.1*noneurons).
```

The weight matrix is then computed from random binary vectors of dilution 0.2. The vectors, or memory states and final weight matrix are printed out. A test vector is then entered by the user. (A section for random test vectors has been commented out.)

The annealing starts at 10 pseudodegrees and is decreased by the relation

$$\text{temperature} = \text{temperature}/\log_e(1 + t)$$

where t is an increasing variable. (Any annealing schedule is possible.)

The neuron to update is selected at random, which results in asynchronous parallel updating, and the vector matrix product is found for a net input to that neuron. The output state, activation[i], of neuron i, is then selected based on the C code in Fig. 2-13. Then, after energy calculations, the main loop ends.

Several functions in the program are well annotated. The function rnd() returns a 0 or 1 to use on the binary vector generation. The function random(n) returns a random integer between 0 and the total number of neurons selected by the user, which allows asynchronous updating. The function expt(net,tem) returns an exponent of a quotient. The last function loge(temp) returns the natural logarithm of the temperature for use in updating the cooling cycle.

The program is very slow, as are all Boltzman-Monti Carlo simulations. I have not included a test run because it's similar to the Hopfield network. For more details on the Boltzman machine, I recommend the book by Aarts and Korst (1989).

Learning by back-propagation of errors

The major learning algorithm for neural networks is the back-propagation of errors discussed by Rumelhart et al. (1986), which is a generalization of the delta rule developed by Widrow and Hoff (1960, 1962) and by LeCun (1986). The method seeks to minimize the error in the output of the network, as compared to a target or a desired response. For a network having multiple outputs, the rms error is given by

$$E = \left(\sum_j (o_j - t_j)^2 \right)^{\frac{1}{2}}$$

where t_j and o_j are the target and actual output values for the j-th component of the vectors. The goal of the back-propagation learning procedure is to minimize this error. If the network is time invariant, then its output will depend only on its inputs–ii–and the current value of the connection weight matrices, w_{ij}. For a given input vector, therefore, the error is determined by the values of the weighting coefficients that connect the network. The approach used in the adaptive procedure is to modify these connections by an amount proportional to the gradient of the error in weight space:

$$\Delta W_{ij} \propto - \frac{\partial E}{\partial W_{ij}}$$

This procedure generally results in reductions in the average error as the weight matrices in the network evolve. The changes in the weights after each trial are proportional to the error itself, which naturally leads to a system that settles out to a stable weight configuration as the errors become small.

The adaptive technique just described has been applied to layered feed-forward networks of the type shown in Fig. 2-14, rather than the Hopfield feedback networks.

Figure 2-15 is a generic layer within a layered, feed-forward network.

In this representation, the output of the layer is a vector of signals: o_j. Its input vector–$o_{\overline{i}}$–may itself be an output from a preceding layer, and its output vector–$o_{\overline{j}}$–may, in turn, provide input to a subsequent layer. The neurons have an activation function f_j and are coupled to the input vector by a weight matrix W_{ij}. The net input to each neuron is given by

$$net_j = \sum_i o_i W_{ij}$$

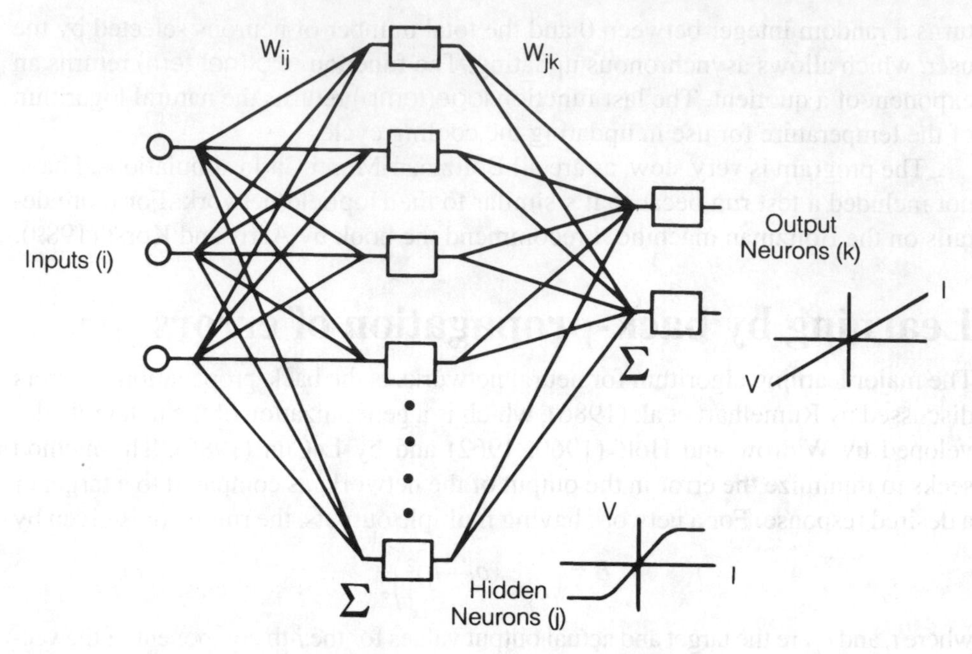

2-14 Layered network architecture.

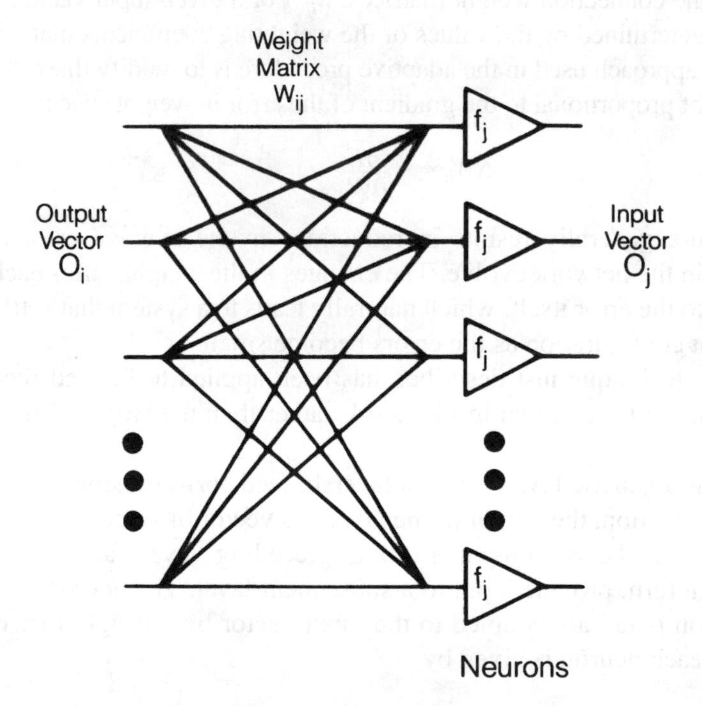

2-15 Diagram of generic layer in feed forward network.

and the output vector is given by

$$o_j = f_j(net_j)$$

By generalized delta rule, the weight changes are

$$\Delta W_{ij} = \eta o_i \delta_j$$

In this relationship, η is the learning rate. If the layer in question is an output layer, then δ_j is given by

$$\delta_j = (t_j - o_j) f'_j(net_j)$$

where t_j is the target, or desired output vector and f'_j denotes differentiation of the neuron's activation function with respect to the input signal. If the layer is hidden inside the network, it is not immediately apparent what its target response should be. In this case, δ_j is computed iteratively using

$$\delta_j = f'_j(net_j) \sum_k \delta_k W_{jk}$$

where δ_k and W_{jk} refer to the layer immediately after the one in question.

Now let's look at a simple neural network program to demonstrate learning by back-propagation. The C program back8.c simulates the back-propagation learning in a neural network. The network has three input nodes, ten hidden nodes, and two output nodes. The objective of this simple simulation is for the neural network to learn Newton's laws of gravity and predict where a projectile will be in ten time units. The input to the network is the angle and velocity of the projectile, which are chosen randomly with boundary conditions to keep the projectile confined to the given space. A third input is a constant that acts as a threshold to adjust the neurons. The two outputs are the x,y coordinates of the predicted position of the projectile.

The program begins with some sharp defines after the sharp includes. The first sharp define is PI 3.14159. The second, third, and fourth are the numbers of neurons in the input, hidden, and output layers of the network. The next six sharp defines are concerned with graphics, while the following two are the maximum time or number of iterations and the variable FREQUENCY, which is the frequency of updating the graphic display. Three additional sharp defines are used in a random number generation routine; the random number generator is taken from the book *Numerical Recipes in C*, by Press et al. (1988).

The program continues by declaring several arrays. The array itohweight [IUNITS][HUNITS] is the weight matrix from the input units to hidden units. The array htooweight[HUNITS][OUNITS] is the matrix from hidden units to output units. After these declarations, an input vector, a net-input-to-hidden vector, an output vector, an error vector, and three more vectors are declared. These last three are examined later. After main() starts, there are many more declarations, including the counting variables i, j, iterate, and counter. Then d1, d2, and d—which are distances used later in the program—are declared. The variables eta1 and eta2 are the learning rates for the input units and the hidden units. The variable a is a pseudo-acceleration for the projectile. The variables t1 and t2 are used in the input vector. The variables vel and theta are the initial random velocity and angle. The next series of declarations are for graphics.

The program then continues with a test of the rnd function and various setups. The first block of setups are for entering small random numbers in all the matrix positions, of the weight matrices. These small numbers are used to break symmetry and prevent the network from diverging. The network must start with random weights. The next large section of the program sets up MicroSoft C graphics and could be modified for other computers or languages.

The program learning loop begins with the counter iterate. The first block of lines allows the user to interrupt the program by hitting any key, which clears the screen and returns to DOS. Random numbers chosen for velocity and angle are used to compute the target vector. The input vector is then set up from the same random numbers, and the vector is fed forward into the hidden layer of neurons. Each hidden neuron has a hyperbolic tangent output. The output from the hidden layer is then fed forward into the output layer. The output layer neurons have a linear response. The error vector is also computed from the difference of the target and the output vectors.

The weight correction for the hidden-to-output weight matrix elements are computed using the algorithm outlined above. The formula for the correction is

$$\Delta W_{ij} = \eta o_j \delta_j$$
$$\delta_j = (t_j - o_j) f_j'(net_j)$$

Also, when you use this algorithm, the weight correction for the input-to-hidden matrix elements are computed from the relation

$$\Delta W_{ij} = \eta o_j \delta_j$$
$$\delta_j = f_j'(net_j) \sum_k \delta_k W_{jk}$$

An error vector or distance vector is then computed from the magnitude of the difference of the output vector and the target vectro. This distance vector is used in the graphic update, which is the next section of the program. Several counters are set back to 0, and the learning loop continues with another random projectile.

Figure 2-16 is a rough sketch of the computer screen during a run of this program. The upper graph represents the learning curve; the curve is a function of the magnitude of the distance vector between the target and output points and the iterations or learning cycles and, in addition, is a plot of the percent of correct guesses by the neural network. The graph on the bottom represents the projectile and the neural network prediction, while the bargraph on the right represents the analog output of the ten hidden neurons. This emphasizes the fact that neural networks are parallel analog processing systems and not token processors.

As a second application/example program for back-propagation neural net-

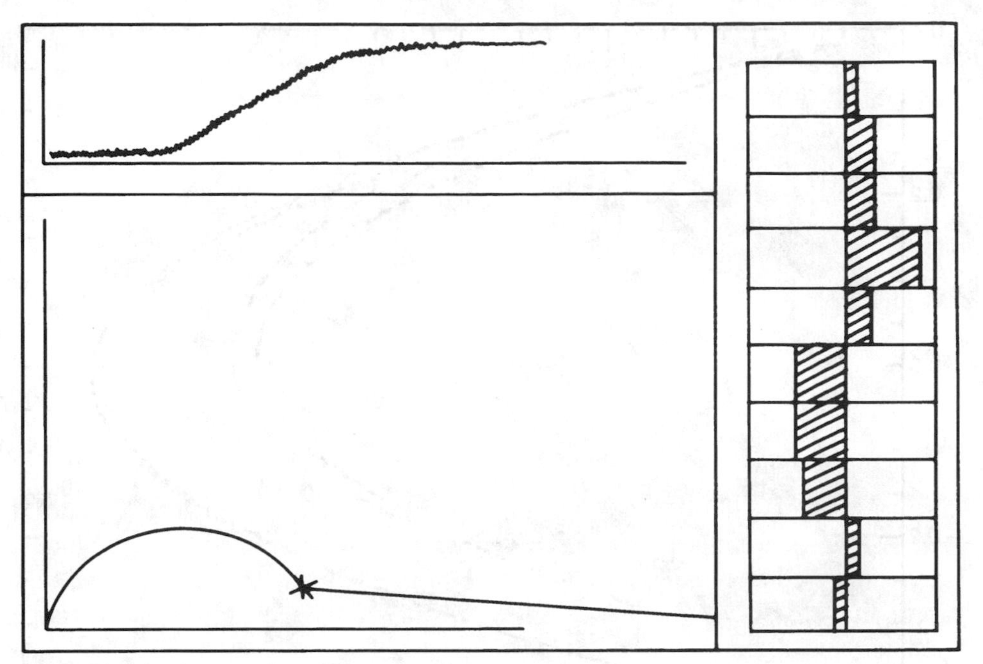

2-16 Sketch of screen dump from the program back8.c.

works, I will discuss the strange attractor known as the Henon attractor discussed in Chapter 1.

The system is shown in Fig. 2-17; it is an area-contracting mapping of a strange attractor. Iterating to infinity will not cause the points to diverge; rather, they will always wander on the chaotic attractor.

Figure 2-18 shows the x and y plots as functions of time. (These chaotic time plots were made with the program henont.c.) You can clearly see that attempting to predict the value for the next iteration is almost impossible. Yet this task can be done reasonably well with a neural network.

Incidentally, neural networks can be used to discriminate real statistical noise from deterministic chaos. For detailed discussion of this aspect of neural networks with respect to real analog processing hardware, I recommend Frye, et al. (1989).

The learning curve shown in Fig. 2-19 is an example of a neural network learning the Henon attractor prediction problem. This figure was prepared with the program henon1.c.

Let's examine the program. Up to main() should be obvious, because of its resemblance to back8.c. After declarations and setups similar to the previous program, an endless learning loop begins. The input vector is selected from random numbers, and the target vector is computed. The input vector is feed-forwarded to the hidden layer and then through to the output layer. The error vector and weight changes are computed, and finally a distance vector is computed. In this case, the

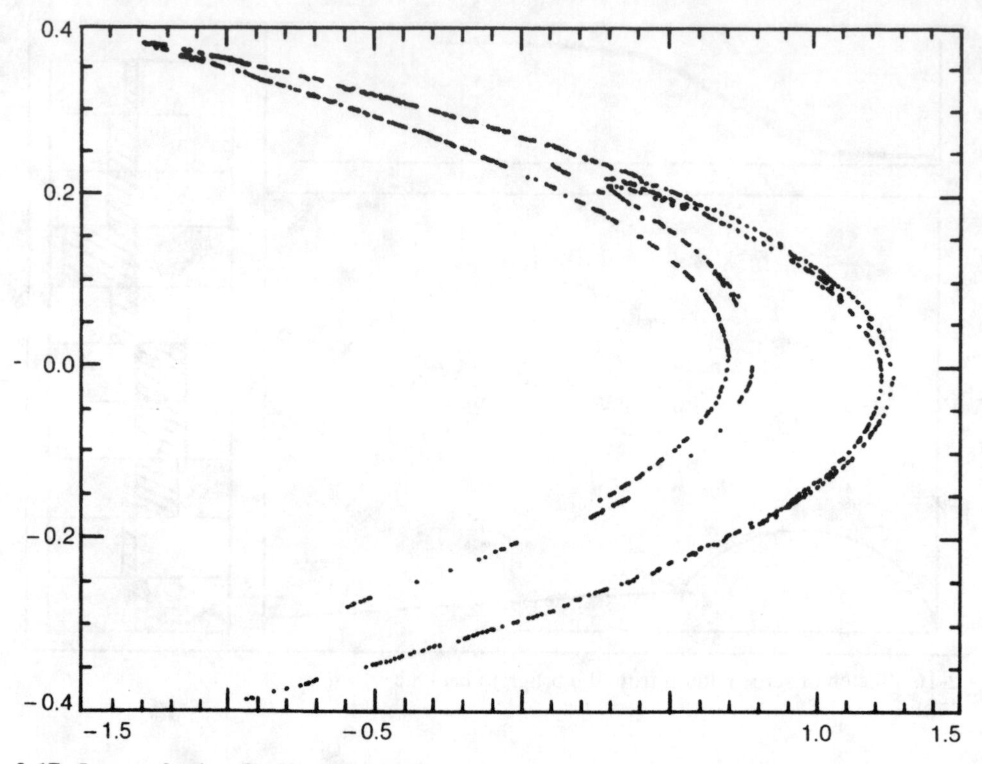

2-17 Locus of points for Henon attractor.

distance vector is simply the magnitude of the difference between the output vector and the target vector. As the network improves, this magnitude decreases (as shown in Fig. 2-19.)

Competitive learning networks

We have seen in the backpropagation of errors that a teacher is needed for the learning. In that case, the output of the network was compared with the desired output. The computer was the teacher to correct the errors by adjusting the synaptic weights. With competitive learning, no teachers are needed.

Much of the original research on these models were developed by Kohonen (1984) and is discussed in Dayhoff (1990) and Wasserman (1989). A Kohonen network is a winner-take-all circuit or an *n*-flop. But rather than hardwire the *n*-flop, which is easily done, we will apply a learning algorithm.

Before the input vector is applied to the input nodes, we do some preprocessing to normalize the input vector elements. Divide the element by the magnitude of the vector; the net result of this is to create a binary vector from the original analog vector. In applying the normalized input vector to the network, the dot product is found between the vector and the weight matrix. The neuron with the largest

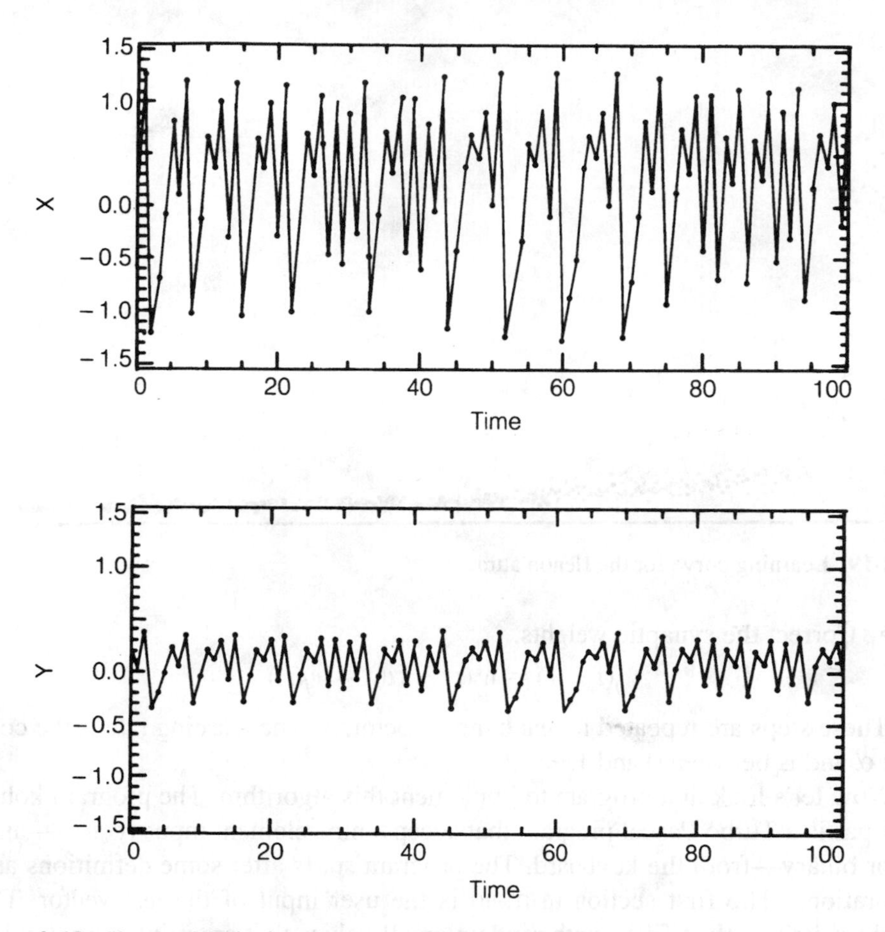

2-18 Simple chaotic time sequences for the Henon attractor.

value of the dot product is the winner, and its weights are adjusted. The weights are adjusted to change by an amount proportional to the difference between its value and the input vector.

These steps are summarized algebraically as follows:

- Preprocess the input vector.

$$x_i' = \frac{x_i}{(x_1^2 + x_2^2 + \ldots + x_i^2)^{\frac{1}{2}}}$$

- Find the Euclidean distance.

$$D_j = \sqrt{\sum_j (x_i - w_{ij})^2}$$

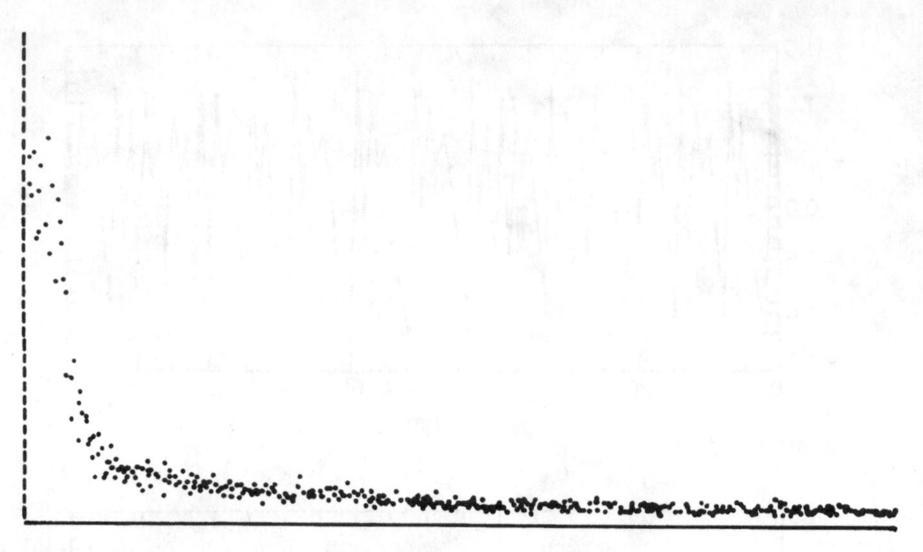

2-19 Learning curve for the Henon attractor.

- Correct the synaptic weights.

$$w_j(t + 1) = w_j(t) + \alpha(X - w_j(t))$$

These steps are repeated for each input vector, X. The learning rate is the constant α and is between 0 and 1.

Now let's look at a program to implement this algorithm. The program kohonen1.pas is a Turbo Pascal program that accepts nine element input vectors—analog or binary—from the keyboard. The program starts after some definitions and declarations. The first section in `main` is the user input of the test vector. The weight matrix is then filled with random small values that are printed on the display. The input vector is then normalized.

The next section is commented out and would be the section of the program to insert input vector code for other desired operations. The distance is then computed between the input vector and the weight vector for each neuron. After selecting the winner neuron, the weight matrix is printed on the display, and the output vector is computed and displayed.

Figure 2-20 shows the input and output vectors from using this program. Notice that the winner neuron has selected the highest input element. Interestingly, the output elements are large analog numbers.

Comments on neural networks

At this point, there is a great deal of research on neural network algorithms. However, the clear winner is the back-propagation algorithm. This algorithm is being applied to many areas of science and technology. Its chief advantage is that it is very robust and—importantly—that it works. (Nothing succeeds like success.)

```
input vector
(1, 2, 3, 4, 5, 4, 3, 2, 1)

output vector
(2.7323, 5.4375, 8.1464, 10.8431, 13,5524,
 10.8500, 8.1469, 5.4345, 2.7443)
```

2-20 I/O vectors for Kohonen program.

Networks can learn in the presence of noise; and, in hardware implementations, it has been shown that the system can learn with missing neurons and connections. Most current research on backpropagation is concerned with speed improvements, but they have only a minor effect.

Another area of current research activity is statistical learning; for a good introduction to this, read Weisbach (1991). The method consists of extending the ideas of energy and pseudotemperature, as previously described, and apply advanced concepts of statistical thermodynamics. These ideas could provide more insight into complex systems and networks. It is hard at this time to predict if they will extend learning algorithms. What is interesting to note about this approach is that the capacity limit of Hopfield networks can be exactly calculated by the theory. The results are the same as those we found empirically.

back8.c

```
/* back propagation — neural network simulation
                120888                                    */

#include "\c\quick\include\stdio.h"
#include "\c\quick\include\math.h"
#include "\c\quick\include\float.h"
#include "\c\quick\include\stdlib.h"
#include "\c\quick\include\graph.h"


#define PI 3.14159
#define IUNITS  3
#define HUNITS  10
#define OUNITS  2
#define XMAXSCREEN 320
#define YMAXSCREEN 200
#define YDOWN 60 /* upper limit of lower box */
#define XOVER 240 /* left limit of RH box */
#define WBORD 6 /* width of border */
#define ERMAX 100 /* maximum error on plot */
#define TRMAX (long int)(50000) /* maximum time */
#define FREQUENCY 100

#define M 714025 /* see Press, et al. 1988 */
#define IA 1366
#define IC 150889

static  double  itohweight[IUNITS][HUNITS];
static  double  htooweight[HUNITS][OUNITS];
static  double  target[OUNITS];
static  double  invector[IUNITS];
static  double  netih[HUNITS];
static  double  oout[OUNITS];
```

```
static  double  error[OUNITS];
static  double  sigma[HUNITS];
static  double  fprime[HUNITS];
static  double  hout[HUNITS];


struct videoconfig vc;
char error_message[] = "this video mode is not supported";

void main()
{
        /* declarations */
        int      i,j,k;
        long int      iterate = 0;
        int counter = 0;
        double  d1,d2,d;
        double  eta1 = 0.33;    /* 1/IUNITS  */
        double  eta2 = 0.1;     /* 1/HUNITS */
        double  a = 0.02;       double  t1,t2,vel,theta;
        double  temp,delta;
        double  rnd();
        int exmin,exmax,eymin,eymax; /* error box mins and maxs */
        int nxmin,nxmax,nymin,nymax; /* neuron box */
        int sxmin,sxmax,symin,symax; /* space box */
        int u;
        double miss, percent = 0.0;
        double oldxt,oldyt,xt,yt,xo,yo; /* used in space plot */
        long dummy;
        /* rnd test stuff */
        for(i=1;i<=5;i++)
        {
                printf("%lf  ",rnd(dummy));
        }
/*        exit(0);
*/
        /* set ups */
        for(i=0;i<IUNITS;i++)
        {
                for(j=0;j<HUNITS;j++)
                {
                        itohweight[i][j] = 0.1*rnd(dummy);
                }
        }

        for(j=0;j<HUNITS;j++)
        {
                for(k=0;k<OUNITS;k++)
                {
                        htooweight[j][k] = 0.1*rnd(dummy);
                }
        }

                /* setup graphics screen */
        if(_setvideomode(_MRES4COLOR) == 0)
        {
                printf("%s\n",error_message);
                exit(0);
        }

        _setbkcolor(_BLACK);
        _selectpalette(3);
        _setviewport(0,0,320,200);
        _setcolor(1);
        _rectangle(_GFILLINTERIOR,0,0,XMAXSCREEN,YMAXSCREEN);

        _setcolor(3);
                        /* error rect */
```

```
_rectangle(_GFILLINTERIOR,WBORD,WBORD,XOVER-WBORD,YDOWN-WBORD);
                /* neurons rect */
_rectangle(_GFILLINTERIOR,XOVER,WBORD,XMAXSCREEN-WBORD,
                                           YMAXSCREEN-WBORD);
                /* space rect */
_rectangle(_GFILLINTERIOR,WBORD,YDOWN,XOVER-WBORD,
                                           YMAXSCREEN-WBORD);
        /* set up error box axes */
exmin = 2*WBORD;
exmax = XOVER-2*WBORD;
eymin = YDOWN-2*WBORD;
eymax = 2*WBORD;

_setcolor(0);
_moveto(exmin,eymin);
_lineto(exmin,eymax);
_moveto(exmin,eymin);
_lineto(exmax,eymin);

for(u=1;u<11;u++)
{
        _moveto(exmin+u*(exmax-exmin)/10,eymin);
        _lineto(exmin+u*(exmax-exmin)/10,eymin+(eymax-eymin)/15);
}
for(u=1;u<6;u++)
{
        _moveto(exmin,eymin+u*(eymax-eymin)/5);
        _lineto(exmin+(exmax-exmin)/75,eymin+u*(eymax-eymin)/5);
}
        /* setup axes for neuron box */
nxmin = XOVER + WBORD;
nxmax = XMAXSCREEN - 2*WBORD;
nymin = YMAXSCREEN - 2*WBORD;
nymax = 4*WBORD;

                /* blue boxes */
_setcolor(1);
for(u=0;u<10;u++)
{
        _rectangle(_GBORDER,nxmin,nymin+u*(nymax-nymin)/10,
                            nxmax,nymin+(u+1)*(nymax-nymin)/10);
}

                /* line */
                /* repeat this code after update */
_setcolor(0);
_moveto((nxmax+nxmin)/2,nymin);
_lineto((nxmax+nxmin)/2,nymax);

                /* + and - signs */
_moveto(nxmin+0.2*(nxmax-nxmin),nymax-WBORD);
_lineto(nxmin+0.3*(nxmax-nxmin),nymax-WBORD);
_moveto(nxmin+0.7*(nxmax-nxmin),nymax-WBORD);
_lineto(nxmin+0.8*(nxmax-nxmin),nymax-WBORD);
_moveto(nxmin+0.75*(nxmax-nxmin),nymax-WBORD+0.05*(nxmax-nxmin));
_lineto(nxmin+0.75*(nxmax-nxmin),nymax-WBORD-0.05*(nxmax-nxmin));

                /* space axes */
sxmin = 2*WBORD;
sxmax = 0.75*(XOVER - 4*WBORD) + 2*WBORD;
symin = YMAXSCREEN - 2*WBORD;
symax = YDOWN + WBORD;

_setcolor(0);
_moveto(sxmin,symax);
_lineto(sxmin,symin);
_lineto(sxmax,symin);
for(u=1;u<11;u++)
{
```

```
        _moveto(sxmin+u*(sxmax-sxmin)/10,symin);
        _lineto(sxmin+u*(sxmax-sxmin)/10,symin+(symax-symin)/40);
}
for(u=1;u<13;u++)
{
        _moveto(sxmin,symin+u*(symax-symin)/12);
        _lineto(sxmin+(sxmax-sxmin)/60,symin+u*(symax-symin)/12);
}



/* learning loop */
for(iterate=0;iterate<TRMAX;iterate++)
{
        if(kbhit())
        {
                _setvideomode(_DEFAULTMODE);
                exit(0);
        }

        /* select random numbers */
        t1 = rnd(dummy);
        t2 = rnd(dummy);
        vel = 0.116 + (0.2 - 0.116)*t1;
        theta = (60 + 30*t2)*PI/180;

        /* compute the target vector using the model */
        target[0] = 10*vel*cos(theta);
        target[1] = 10*vel*sin(theta)-0.5*a*100;


        /* select input vector */
        invector[0] = t1;
        invector[1] = t2;
        invector[2] = 1.0;


        /* feed forward  --  input to hidden */
        for(j=0;j<HUNITS;j++)
        {
                netih[j] = 0;
                for(i=0;i<IUNITS;i++)
                {
                        netih[j] = netih[j] + itohweight[i][j]*invector[i];
                }
                hout[j] = tanh(netih[j]);

         }


        /* feed forward  --  hidden to output */
        for(k=0;k<OUNITS;k++)
        {
                oout[k] = 0;
                for(j=0;j<HUNITS;j++)
                {
                        oout[k] = oout[k] + htooweight[j][k]*hout[j];
                }
                error[k] = target[k] - oout[k];
        }


        /* compute the error correction for htooweight matrix elements */
        for(k=0;k<OUNITS;k++)
        {
                for(j=0;j<HUNITS;j++)
                {
                        delta = error[k]*hout[j]*eta2;
```

```
                        htooweight[j][k] = htooweight[j][k] + delta;
                }
        }

        /* compute the error correction for itohweitht matrix elements */
        for(j=0;j<HUNITS;j++)    /* first find sigma(e(k)u(kj)) */
        {
                sigma[j] = 0;
                for(k=0;k<OUNITS;k++)
                {
                        sigma[j] = sigma[j] + error[k]*htooweight[j][k];
                }
                temp = (double)(1.0/cosh(netih[j]));
                fprime[j] = temp*temp;
        }

        for(i=0;i<IUNITS;i++)
        {
                for(j=0;j<HUNITS;j++)
                {

                delta = eta1*fprime[j]*sigma[j]*invector[i];
                itohweight[i][j] = itohweight[i][j] + delta;
        }       }
d1 = pow((oout[0]-target[0]),2.0);
d2 = pow((oout[1]-target[1]),2.0);
miss = sqrt(d1 + d2);
d = d + (sqrt(d1 + d2))/FREQUENCY;
if( d>0.5)
{
        d = 0.5;
}
if(miss <= 0.01)
{
        percent = percent + 100/FREQUENCY;
}
counter = counter + 1;
if(counter == FREQUENCY)
{
        /* plot stuff here */
                /* error plot */
        _setcolor(0);
        _setpixel(exmin+iterate*(exmax-exmin)/TRMAX,
                        eymin+percent*(eymax-eymin)/ERMAX);
                /* neurons */
        _setcolor(3);
        _rectangle(_GFILLINTERIOR,nxmin,nymin,nxmax,nymax);
        _setcolor(1);
        for(u=0;u<10;u++)
        {
                _rectangle(_GBORDER,nxmin,nymin+u*(nymax-nymin)/10,
                                nxmax,nymin+(u+1)*(nymax-nymin)/10);
        }
        _setcolor(2);
        for(u=0;u<10;u++)
        {
                _rectangle(_GFILLINTERIOR,(nxmax+nxmin)/2,
                        nymin+u*(nymax-nymin)/10,
                        (nxmax+nxmin)/2 + (nxmax-nxmin)*hout[u]/2,
                        nymin+(u+1)*(nymax-nymin)/10);
        }
        _setcolor(0);
        _moveto((nxmax+nxmin)/2,nymin);
        _lineto((nxmax+nxmin)/2,nymax);
                /* space data plot */
        _setcliprgn(sxmin+3,symin-3,XOVER-WBORD,YDOWN+WBORD);
        _setcolor(3);
        _rectangle(_GFILLINTERIOR,sxmin+3,symin-3,XOVER-WBORD,
                                        YDOWN+WBORD);
```

```
				oldxt=0;
				oldyt=0;
				_moveto(sxmin,symin);
				for(u=0;u<51;u++)
				{
					_setcolor(2);
					xo = 1.3 - u*(1.3 - oout[0])/50.0;
					yo = u*oout[1]/50.0;
					xt = vel*u*cos(theta)/5.0;
					yt = (vel*u*sin(theta)/5.0) - 0.5*a*u*u/25.0;
					_setpixel(sxmin+xo*(sxmax-sxmin),
						symin+yo*(symax-symin)/1.2);
					_setcolor(0);
					_moveto(sxmin+oldxt*(sxmax-sxmin),
						symin+oldyt*(symax-symin)/1.2);
					_lineto(sxmin+xt*(sxmax-sxmin),
						symin+yt*(symax-symin)/1.2);
					oldxt = xt;
					oldyt = yt;
				}
				if(miss <= 0.01)
				{
					_setcolor(2);
					_moveto(sxmin+xt*(sxmax-sxmin),
							symin+yt*(symax-symin)/1.2+3);
					_lineto(sxmin+xt*(sxmax-sxmin),
							symin+yt*(symax-symin)/1.2-3);
					_moveto(sxmin+xt*(sxmax-sxmin)-3,
							symin+yt*(symax-symin)/1.2);
					_lineto(sxmin+xt*(sxmax-sxmin)+3,
							symin+yt*(symax-symin)/1.2);
					_moveto(sxmin+xt*(sxmax-sxmin)-2,
							symin+yt*(symax-symin)/1.2-2);
					_lineto(sxmin+xt*(sxmax-sxmin)+2,
							symin+yt*(symax-symin)/1.2+2);
					_moveto(sxmin+xt*(sxmax-sxmin)+2,
							symin+yt*(symax-symin)/1.2-2);
					_lineto(sxmin+xt*(sxmax-sxmin)-2,
							symin+yt*(symax-symin)/1.2+2);
				}

				_setcliprgn(0,0,XMAXSCREEN,YMAXSCREEN);



				percent = 0;
				d = 0;
				counter = 0;

			}

		}
		while(!kbhit());
		_setvideomode(_DEFAULTMODE);

}

double rnd(idum)
long *idum;
{
		static long iy,ir[98];
		static int iff=0;
		int j;

		if(*idum < 0 || iff == 0)
		{
				iff=1;
				if((*idum=(IC-(*idum)) % M) < 0) *idum = -(*idum);
```

```
		for(j=1;j<=97;j++)
		{
			*idum = (IA*(*idum)+IC) % M;
			ir[j] = (*idum);
		}
		*idum=(IA*(*idum)+IC) % M;
		iy=(*idum);
	}
	j=1 + 97.0*iy/M;
	iy=ir[j];
	*idum=(IA*(*idum)+IC) % M;
	ir[j] = (*idum);
	return (double) iy/M;

}
```

hebb2p.bas

```
 10 CLS
 20 INPUT "INPUT RANDOM SEED ";SEED
 30 RANDOMIZE SEED
 40 INPUT "ENTER THE NUMBER OF NEURONS (100 MAXIMUM) ";N
 50 INPUT "INPUT THE THRESHOLD VALUE (0 TO 2 ARE REASONABLE VALUES) ";IO
 60 INPUT "ENTER THE VALUE OF THE INFORMATION (0 TO 1 IS A GOOD VALUE ) ";INFO
 70 INPUT "DO YOU WANT TO ENTER THE INPUT VECTOR YOURSELF (1/YES 0/NO)? ";VECTOR
 80 PRINT "BINARY MATRIX WITH Tii=0 AND Tij=Tji."
 90 DIM T(100,100),V(100),U(100)
100 REM FILL T(I,J) MATRIX
110 PRINT:PRINT:PRINT
120 PRINT "INPUT THE MEMORY VECTOR FOR THE HEBB MATRIX"
130 FOR I=1 TO N
140 PRINT "V(";I;")"
150 INPUT V(I)
160 U(I)=V(I)
170 NEXT I
180 FOR I=1 TO N
190 FOR J=1 TO N
200 T(I,J)=V(I)*U(J)
210 IF I=J THEN T(I,J)=0
220 LPRINT T(I,J);
230 NEXT J
240 LPRINT
250 NEXT I
260 LPRINT:LPRINT:LPRINT
270 REM FILL INPUT VECTOR U
280 IF VECTOR=0 THEN 340
290 FOR I=1 TO N
300 PRINT "INPUT U(";I;")"
310 INPUT U(I)
320 NEXT I
330 GOTO 380 : 'BEGIN CALCULATIONS OF OUTPUT VECTOR
340 FOR I=1 TO N
350 GOSUB 630
360 U(I)=R
370 NEXT I
380 REM BEGIN CALCULATION
390 FOR ITERATE=1 TO 8: REM THIS ALLOWS THE OUTPUT VECTOR TO BE FEED BACK
400 FOR I=1 TO N
410 FOR J=1 TO N
420 SIGMA=T(I,J)*U(J)+SIGMA
430 NEXT J
440 SIGMA=SIGMA+INFO
450 IF SIGMA > IO THEN SIGMA=1 ELSE SIGMA=0
460 V(I)=SIGMA
470 SIGMA=0
480 NEXT I
490 IF ITERATE=1 THEN 500 ELSE 540
500 FOR I=1 TO N
```

```
510 LPRINT U(I);
520 NEXT I
530 LPRINT
540 FOR I=1 TO N
550 U(I)=V(I): REM FOR FEEDBACK
560 NEXT I
570 NEXT ITERATE
580 FOR I=1 TO N
590 LPRINT V(I);
600 NEXT I
610 LPRINT:LPRINT:LPRINT:LPRINT
620 GOTO 270
630 R=RND(1)
640 IF R<.5 THEN R=0 ELSE R=+1
650 RETURN
```

hebb3p.bas

```
 10 CLS
 20 INPUT "INPUT RANDOM SEED ";SEED
 30 RANDOMIZE SEED
 40 INPUT "ENTER THE NUMBER OF NEURONS (100 MAXIMUM) ";N
 50 INPUT "DO YOU WANT TO ENTER THE INPUT VECTOR YOURSELF (1/YES 0/NO)? ";VECTOR
 60 DIM T(100,100),V(100),U(100)
 70 REM FILL T(I,J) MATRIX
 80 PRINT:PRINT:PRINT
 90 INPUT "INPUT THE NUMBER OF MEMORY VECTORS (M=INT(.15*N) ";M
100 FOR MEMS=1 TO M
110 PRINT "INPUT THE MEMORY VECTOR ";MEMS;"FOR THE HEBB MATRIX."
120 FOR I=1 TO N
130 PRINT "V(";I;")"
140 INPUT V(I)
150 U(I)=V(I)
160 NEXT I
170 FOR I=1 TO N
180 FOR J=1 TO N
190 T(I,J)=T(I,J)+V(I)*U(J)
200 IF I=J THEN T(I,J)=0
210 IF T(I,J)>1 THEN T(I,J)=1
220 LPRINT T(I,J);
230 NEXT J
240 LPRINT
250 NEXT I
255 LPRINT:LPRINT:LPRINT:LPRINT:LPRINT
260 NEXT MEMS
270 LPRINT:LPRINT:LPRINT
280 REM FILL INPUT VECTOR U
290 IF VECTOR=0 THEN 350
300 FOR I=1 TO N
310 PRINT "INPUT U(";I;")"
320 INPUT U(I)
330 NEXT I
340 GOTO 390 : 'BEGIN CALCULATIONS OF OUTPUT VECTOR
350 FOR I=1 TO N
360 GOSUB 640
370 U(I)=R
380 NEXT I
390 REM BEGIN CALCULATION
400 FOR ITERATE=1 TO 8: REM THIS ALLOWS THE OUTPUT VECTOR TO BE FEED BACK
410 FOR I=1 TO N
420 FOR J=1 TO N
430 SIGMA=T(I,J)*U(J)+SIGMA
440 NEXT J
450 SIGMA=SIGMA
460 IF SIGMA > 0  THEN SIGMA=1 ELSE SIGMA=0
470 V(I)=SIGMA
480 SIGMA=0
490 NEXT I
```

```
500 IF ITERATE=1 THEN 510 ELSE 550
510 FOR I=1 TO N
520 LPRINT U(I);
530 NEXT I
540 LPRINT
550 FOR I=1 TO N
560 U(I)=V(I): REM FOR FEEDBACK
570 NEXT I
580 NEXT ITERATE
590 FOR I=1 TO N
600 LPRINT V(I);
610 NEXT I
620 LPRINT:LPRINT:LPRINT:LPRINT
630 GOTO 280
640 R=RND(1)
650 IF R<.5 THEN R=0 ELSE R=+1
660 RETURN
```

henon1.c

```
/* back propagation — neural network simulation */

#include "\c\quick\include\stdio.h"
#include "\c\quick\include\math.h"
#include "\c\quick\include\float.h"
#include "\c\quick\include\stdlib.h"
#include "\c\quick\include\graph.h"
#include "\c\quick\include\conio.h"


#define IUNITS  3
#define HUNITS  10
#define OUNITS  2
#define XMAXSCREEN (double)(280)
#define YMAXSCREEN (double)(160)
#define YMAXTEXT (double)(30)
#define XMIN (double)(0)
#define XMAX (double)(10000)
#define YMIN (double)(0.0)
#define YMAX (double)(1.0)

static  double  itohweight[IUNITS][HUNITS];
static  double  htooweight[HUNITS][OUNITS];
static  double  target[OUNITS];
static  double  invector[IUNITS];
static  double  netih[HUNITS];
static  double  oout[OUNITS];
static  double  error[OUNITS];
static  double  sigma[HUNITS];
static  double  fprime[HUNITS];
static  double  hout[HUNITS];


struct videoconfig vc;
char error_message[] = "this video mode is not supported";


void main()
{
        /* declarations */
        int     i,j,k;
        long int     iterate = 0;
        int counter = 0;
        double  d1,d2,d = 0;
        double  eta = 0.1;
        double  a = 3.95;
        double  deltat = 1.0;
        double  t1;
```

```
double  temp,delta;
double  rnd();
short int  xcoord,ycoord;
double xin=0.5,yin=0.5;

/* set ups */
for(i=0;i<IUNITS;i++)
{
        for(j=0;j<HUNITS;j++)
        {
                itohweight[i][j] = 0.1*rnd();
        }
}

for(j=0;j<HUNITS;j++)
{
        for(k=0;k<OUNITS;k++)
        {
                htooweight[j][k] = 0.1*rnd();
        }
}

if(_setvideomode(_MRESNOCOLOR) == 0)
{
        printf("%s\n",error_message);
        exit(0);
}

_setviewport(20,20,300,180);
_moveto(0,0);
_lineto(0,YMAXSCREEN);
_moveto(0,YMAXSCREEN);
_lineto(XMAXSCREEN,YMAXSCREEN);

target[0]=0.5;
target[1]=0.5;

/* endless learning loop */
for(;;)
{

        if(kbhit())
        {
                _setvideomode(_DEFAULTMODE);
                exit(0);
        }

        /* select input numbers */
        xin = target[0];
        yin = target[1];

        /* compute the target vector using the model */
        target[0] = yin + 1.0 - 1.4*xin*xin;
        target[1] = 0.3*xin;


        /* select input vector */
        invector[0] = xin;
        invector[1] = yin;
        invector[2] = 1.0;

        /* feed forward  --  input to hidden */
        for(j=0;j<HUNITS;j++)
        {
              netih[j] = 0;
              for(i=0;i<IUNITS;i++)
              {
                      netih[j] = netih[j] + itohweight[i][j]*invector[i];
              }
```

```
                hout[j] = tanh(netih[j]);

        }


        /* feed forward  --  hidden to output */
        for(k=0;k<OUNITS;k++)
        {
                oout[k] = 0;
                for(j=0;j<HUNITS;j++)
                {
                        oout[k] = oout[k] + htooweight[j][k]*hout[j];
                }
                error[k] = target[k] - oout[k];
        }


        /* compute the error correction for htooweight matrix elements */
        for(k=0;k<OUNITS;k++)
        {
                for(j=0;j<HUNITS;j++)
                {
                        delta = error[k]*hout[j]*eta;
                        htooweight[j][k] = htooweight[j][k] + delta;
                }
        }

        /* compute the error correction for itohweitht matrix elements */
        for(j=0;j<HUNITS;j++)   /* first find sigma(e(k)u(kj)) */
        {
                sigma[j] = 0;
                for(k=0;k<OUNITS;k++)
                {
                        sigma[j] = sigma[j] + error[k]*htooweight[j][k];
                }
                temp = (double)(1.0/cosh(netih[j]));
                fprime[j] = temp*temp;
        }

        for(i=0;i<IUNITS;i++)
        {
                for(j=0;j<HUNITS;j++)
                {

                        delta = eta*fprime[j]*sigma[j]*invector[i];
                        itohweight[i][j] = itohweight[i][j] + delta;
                }
        }

        d1 = pow((oout[0] - target[0]), 2.0);
        d2 = pow((oout[1] - target[1]), 2.0);
        d = d + (sqrt(d1 + d2))/25.0;

        iterate = iterate + 1;
        counter = counter + 1;
        if(counter == 24)
        {
                xcoord = (((double)(iterate) - XMIN)/(XMAX- XMIN))*XMAXSCREEN;
                ycoord = YMAXSCREEN - ((d-YMIN)/(YMAX- YMIN))*YMAXSCREEN;
                _setpixel(xcoord,ycoord);


                d = 0;
                counter = 0;

        }

    }
}
```

```
double rnd()
{
        double result;
        result = (double)(rand())/(double)(32767.0);

        return (result);
}
```

henont.c

```
#include <float.h>
#include <stdio.h>
#include <graph.h>
#include <math.h>

struct videoconfig vc;
char error_message[] = "this video mode is not suported";
float xdata[1000],ydata[1000],tdata[1000];


void main()
{

        /* declarations */
        int i,t,count,n;
        float x0,y0;
        double xmax,ymax,xmin,ymin;
        int xmaxscreen=280,ymaxscreen=160;
        int ymaxtext=30;
        float xcoord,ycoord;
        float text[80];

        /* user input */
        printf("input number of data points\n");
        scanf("%d",&n);

        /* set mode of screen */
        if (_setvideomode(_MRESNOCOLOR) == 0)
        {
                printf("%s\n",error_message);
                exit(0);
        }
        _getvideoconfig(&vc);
        _setcolor(1);
        _clearscreen(_GCLEARSCREEN);

        /* computation here */
        xdata[0] = 0.05;
        ydata[0] = 0.05;
        count = 0;
        for(t=0;t<n;t++)
        {
                xdata[t+1] = ydata[t] + 1 - 1.4*xdata[t]*xdata[t];
                ydata[t+1] = 0.3*xdata[t];
                xdata[count] = xdata[t+1];
                ydata[count] = ydata[t+1];
                tdata[count] = count;
                count++;
        }


        /* find min and max of x and y
           (tdata is set as xdata)      */
        xmax= -1e+20;
        xmin= -xmax;
        ymax= -1e20;
```

```
        ymin= -ymax;
        for(i=0;i<n;++i)
        {
                if(ymin < ydata[i])
                        ymin=ydata[i];
                if(ymax < ydata[i])
                        ymax=ydata[i];
                if(xmax < tdata[i])
                        xmax=tdata[i];
                if(xmin < tdata[i])
                        xmin=tdata[i];
        }
        /* printf("xmin, xmax, ymin, ymax:%lf %lf %lf %lf\n",xmin,xmax,ymin,ymax); */


        /* draw axes */
        _setviewport(20,20,300,180);
        _moveto(0,0);
        _lineto(0,ymaxscreen);
        _moveto(0,ymaxscreen);
        _lineto(xmaxscreen,ymaxscreen);
        for(i=0;i<=ymaxscreen; i=i+ymaxscreen/10)
        {
                _moveto(0,i);
                _lineto(5,i);
        }
        /* tic marks */
        for(i=0;i<=xmaxscreen; i=i+xmaxscreen/10)
        {
                _moveto(i,ymaxscreen-5);
                _lineto(i,ymaxscreen);
        }

        /* plot data */
        _moveto(xmin,ymin);
        for(i=0;i<n;++i)
        {
                xcoord=((tdata[i]-xmin)/(xmax-xmin))*xmaxscreen;
                ycoord=ymaxscreen-((ydata[i]-ymin)/(ymax-ymin))*ymaxscreen;
                _lineto(xcoord,ycoord);
        }


       /* clear screen & return control hit enter */
       while(!kbhit());
       _clearscreen(_GCLEARSCREEN);
       _setvideomode(_DEFAULTMODE);


}
```

kohonen1.pas

```
{$N+}
{$R+}
(* simple kohonen neural network *)

program kohonen1;

(* definitions *)

const
        iunits = 9;  (* it is a good idea to make iunits = ounits *)
        ounits = 9;  (* number of neurons *)
        nvectors = 5; (* number of vectors in the training set *)
        alpha = 0.7;

type
```

```
        weight_matrix = array[1..iunits,1..ounits] of real; (* weight matrix *)
        invector_matrix = array[1..iunits] of real; (* input vector *)
        ninvector_matrix = array[1..iunits] of real; (* normalized input vector *)
        dvector_matrix = array[1..iunits] of real; (* difference vector *)
        ovector_matrix = array[1..ounits] of real; (* output vector *)

(* declarations *)

var
        i,j,k,v: integer;
        weight : weight_matrix;
        invector : invector_matrix;
        ninvector : ninvector_matrix;
        dvector : dvector_matrix;
        ovector : ovector_matrix;
        counter : integer;
        temp,sigmadif: real;

(* statements *)

begin (* main *)

        (* input a test vector - for test purposes only *)
        for i := 1 to iunits do
        begin (* i loop *)
                write(output, 'input vector element ',i:4, '   ');
                read(input, invector[i]);
        end; (* i loop *)
        writeln;
        writeln;


        (* fill matrix with random numbers *)
        for i := 1 to iunits do
        begin (* i loop *)
                for j := 1 to ounits do
                begin (* j loop *)
                        weight[i,j] := random;
                        writeln(output, weight[i,j]:12:4); (* for test purposes only *)
                end; (* j loop *)
        end; (* i loop *)

        for v := 1 to nvectors do
        begin (* v loop to apply vectors and learn *)

                (* normalize each input vector *)
                for k := 1 to nvectors do
                begin (* k loop for number of vectors *)
                        for i := 1 to iunits do
                        begin (* i loop *)
                                temp := 0;
                                for j := 1 to iunits do
                                begin (* j loop *)
                                        temp := temp + sqr(invector[j]);
                                end; (* j loop *)
                                ninvector[i] := invector[i]/sqrt(temp);
                        end; (* i loop *)
                end; (* k loop *)

                (* begin training the kohonen network *)
                (* apply the normalized input vector *)

                {       ninvector[i]            }


                (* calculate the distance in weight space *)
```

```
        for i := 1 to ounits do (* for each neuron *)
        begin (* i loop *)
                sigmadif := 0;
                for j := 1 to iunits do (* for each connection *)
                begin (* j loop *)
                        sigmadif := sigmadif + sqr((ninvector[j] - weight[i][j]));
                end; (* j loop *)
                dvector[i] := sigmadif;
        end; (* i loop *)

        (* select winner -- find the largest element in the dvector *)
        temp := 0;
        counter := 0;
        for i := 1 to ounits do
        begin (* i loop *)
                if (temp < dvector[i]) then
                begin (* if test *)
                        temp := dvector[i];
                        counter := counter + 1; (* represents the number of the winner *)
                end; (* if test *)
        end; (* i loop *)

        (* adjust the weights *)
        for i := 1 to ounits do (* for each neuron *)
        begin (* i loop *)
                for j := 1 to iunits do (* for each connection *)
                begin (* j loop *)
                        weight[i,j] := weight[i,j] + alpha*(ninvector[j] - weight[i,j]);
                end; (* j loop *)
        end; (* i loop *)

    end; (* end v loop -- go back to apply a new/same vector for learning *)

    (* print out the weight matrix *)
    for i := 1 to ounits do (* for each neuron *)
    begin (* i loop *)
        for j := 1 to iunits do (* for each connection *)
                begin (* j loop *)
                        write(output, weight[i,j]:6:3,'   ');
                end; (* j loop *)
        writeln;
    end; (* i loop *)


    (* test the network -- how good is it? *)
    for i := 1 to ounits do (* for each neuron *)
    begin (* i loop *)
        ovector[i] := 0;
        for j := 1 to iunits do (* for each connection *)
        begin (* j loop *)
                ovector[i] := ovector[i] + invector[j]*weight[j,i];
        end; (* j loop *)
        writeln(output, ovector[i]:12:4);
    end; (* i loop *)

end. (* main *)
```

neuron4p.pas

```
10 CLS
20 INPUT "INPUT RANDOM SEED ";SEED
30 RANDOMIZE SEED
40 INPUT "ENTER THE NUMBER OF NEURONS (100 MAXIMUM) ";N
50 INPUT "INPUT THE THRESHOLD VALUE (0 TO 2 ARE REASONABLE VALUES) ";IO
60 INPUT "ENTER THE VALUE OF THE INFORMATION (0 TO 1 IS A GOOD VALUE)";INFO
70 INPUT "DO YOU WANT TO ENTER THE INPUT VECTOR YOURSELF (1/YES 0/NO)? ";VECTOR
80 INPUT "DO YOU WANT TO INPUT THE T MATRIX (1/Y 0/NO) ";MATRIX
90 DIM T(100,100),V(100),U(100)
```

```
100 REM FILL T(I,J) MATRIX
110 IF MATRIX=0 THEN 190
120 FOR I=1 TO N
130 FOR J=1 TO N
140 PRINT "T(";I;",";J;") "
150 INPUT T(I,J)
160 NEXT J
170 NEXT I
180 GOTO 360 : 'FILL INPUT VECTOR
190 FOR I=1 TO N
200 FOR J=I TO N
210 R=RND(1)
220 IF R<.75 THEN R=0 ELSE R=+1: REM DILUTE MATRIX
230 T(I,J)=R
240 NEXT J
250 LPRINT
260 NEXT I
270 FOR I=1 TO N
280 FOR J=1 TO N
290 IF I=J THEN T(I,J)=0
300 T(J,I)=T(I,J)
310 LPRINT T(I,J);
320 NEXT J
330 LPRINT
340 NEXT I
350 LPRINT:LPRINT:LPRINT
360 REM FILL INPUT VECTOR U
370 IF VECTOR=0 THEN 430
380 FOR I=1 TO N
390 PRINT "INPUT U(";I;")"
400 INPUT U(I)
410 NEXT I
420 GOTO 470 : 'BEGIN CALCULATIONS OF OUTPUT VECTOR
430 FOR I=1 TO N
440 GOSUB 670
450 U(I)=R
460 NEXT I
470 REM BEGIN CALCULATION
480 FOR I=1 TO N
490 FOR J=1 TO N
500 SIGMA=T(I,J)*U(J)+SIGMA
510 NEXT J
520 SIGMA=SIGMA+INFO
530 IF SIGMA > IO THEN SIGMA=1 ELSE SIGMA=0
540 V(I)=SIGMA
550 SIGMA=0
560 NEXT I
570 FOR I=1 TO N
580 LPRINT U(I);
590 NEXT I
600 LPRINT:LPRINT
610 FOR I=1 TO N
620 LPRINT V(I);
630 NEXT I
640 LPRINT:LPRINT
650 LPRINT:LPRINT
660 GOTO 360
670 R=RND(1)
680 IF R<.5 THEN R=0 ELSE R=+1
690 RETURN
```

neuron5p.bas

```
10 CLS
20 INPUT "INPUT RANDOM SEED ";SEED
30 RANDOMIZE SEED
```

```
 40 INPUT "ENTER THE NUMBER OF NEURONS (100 MAXIMUM) ";N
 50 INPUT "INPUT THE THRESHOLD VALUE (0 TO 2 ARE REASONABLE VALUES) ";IO
 60 INPUT "ENTER THE VALUE OF THE INFORMATION (0 TO 1 IS A GOOD VALUE ) ";INFO
 70 INPUT "DO YOU WANT TO ENTER THE INPUT VECTOR YOURSELF (1/YES 0/NO)? ";VECTOR
 80 INPUT "DO YOU WANT TO INPUT THE T MATRIX (1/Y 0/NO) ";MATRIX
 90 DIM T(100,100),V(100),U(100)
100 REM FILL T(I,J) MATRIX
110 IF MATRIX=0 THEN 190
120 FOR I=1 TO N
130 FOR J=1 TO N
140 PRINT "T(";I;",";J;") "
150 INPUT T(I,J)
160 NEXT J
170 NEXT I
180 GOTO 360 : 'FILL INPUT VECTOR
190 FOR I=1 TO N
200 FOR J=1 TO N
210 R=RND(1)
220 IF R<.85 THEN R=0 ELSE R=+1: REM DILUTE MATRIX
230 T(I,J)=R
240 NEXT J
250 LPRINT
260 NEXT I
270 FOR I=1 TO N
280 FOR J=1 TO N
290 IF I=J THEN T(I,J)=0
300 REM   T(J,I)=T(I,J)
310 LPRINT T(I,J);
320 NEXT J
330 LPRINT
340 NEXT I
350 LPRINT:LPRINT:LPRINT
360 REM FILL INPUT VECTOR U
370 IF VECTOR=0 THEN 430
380 FOR I=1 TO N
390 PRINT "INPUT U(";I;")"
400 INPUT U(I)
410 NEXT I
420 GOTO 470 : 'BEGIN CALCULATIONS OF OUTPUT VECTOR
430 FOR I=1 TO N
440 GOSUB 670
450 U(I)=R
460 NEXT I
470 REM BEGIN CALCULATION
480 FOR I=1 TO N
490 FOR J=1 TO N
500 SIGMA=T(I,J)*U(J)+SIGMA
510 NEXT J
520 SIGMA=SIGMA+INFO
530 IF SIGMA > IO THEN SIGMA=1 ELSE SIGMA=0
540 V(I)=SIGMA
550 SIGMA=0
560 NEXT I
570 FOR I=1 TO N
580 LPRINT U(I);
590 NEXT I
600 LPRINT:LPRINT
610 FOR I=1 TO N
620 LPRINT V(I);
630 NEXT I
640 LPRINT:LPRINT
650 LPRINT:LPRINT
660 GOTO 360
670 R=RND(1)
680 IF R<.5 THEN R=0 ELSE R=+1
690 RETURN
```

neuron6p.bas

```
 10 CLS
 20 INPUT "INPUT RANDOM SEED ";SEED
 30 RANDOMIZE SEED
 40 INPUT "ENTER THE NUMBER OF NEURONS (100 MAXIMUM) ";N
 50 INPUT "INPUT THE THRESHOLD VALUE (0 TO 2 ARE REASONABLE VALUES) ";IO
 60 INPUT "ENTER THE VALUE OF THE INFORMATION (0 TO 1 IS A GOOD VALUE ) ";INFO
 70 INPUT "DO YOU WANT TO ENTER THE INPUT VECTOR YOURSELF (1/YES 0/NO)? ";VECTOR
 80 INPUT "DO YOU WANT TO INPUT THE T MATRIX (1/Y 0/NO) ";MATRIX
 90 DIM T(100,100),V(100),U(100)
100 REM FILL T(I,J) MATRIX
110 IF MATRIX=0 THEN 190
120 FOR I=1 TO N
130 FOR J=1 TO N
140 PRINT "T(";I;",";J;") "
150 INPUT T(I,J)
160 NEXT J
170 NEXT I
180 GOTO 360 : 'FILL INPUT VECTOR
190 FOR I=1 TO N
200 FOR J=1 TO N
210 R=RND(1)
220 IF R<.8 THEN R=0 ELSE R=+1: REM DILUTE MATRIX
230 T(I,J)=R
240 NEXT J
250 LPRINT
260 NEXT I
270 FOR I=1 TO N
280 FOR J=1 TO N
290 IF I=J THEN T(I,J)=0
300           T(J,I)=T(I,J)
310 LPRINT T(I,J);
320 NEXT J
330 LPRINT
340 NEXT I
350 LPRINT:LPRINT:LPRINT
360 REM FILL INPUT VECTOR U
370 IF VECTOR=0 THEN 430
380 FOR I=1 TO N
390 PRINT "INPUT U(";I;")"
400 INPUT U(I)
410 NEXT I
420 GOTO 470 : 'BEGIN CALCULATIONS OF OUTPUT VECTOR
430 FOR I=1 TO N
440 GOSUB 720
450 U(I)=R
460 NEXT I
470 REM BEGIN CALCULATION
480 FOR ITERATE=1 TO 8: REM THIS ALLOWS THE OUTPUT VECTOR TO BE FEED BACK
490 FOR I=1 TO N
500 FOR J=1 TO N
510 SIGMA=T(I,J)*U(J)+SIGMA
520 NEXT J
530 SIGMA=SIGMA+INFO
540 IF SIGMA > IO THEN SIGMA=1 ELSE SIGMA=0
550 V(I)=SIGMA
560 SIGMA=0
570 NEXT I
580 IF ITERATE=1 THEN 590 ELSE 630
590 FOR I=1 TO N
600 LPRINT U(I);
610 NEXT I
620 LPRINT
630 FOR I=1 TO N
640 U(I)=V(I): REM FOR FEEDBACK
650 NEXT I
660 NEXT ITERATE
670 FOR I=1 TO N
```

```
680 LPRINT V(I);
690 NEXT I
700 LPRINT:LPRINT:LPRINT:LPRINT
710 GOTO 360
720 R=RND(1)
730 IF R<.5 THEN R=0 ELSE R=+1
740 RETURN
```

neuron8p.bas

```
 10 CLS
 20 INPUT "INPUT RANDOM SEED ";SEED
 30 RANDOMIZE SEED
 40 INPUT "ENTER THE NUMBER OF NEURONS (100 MAXIMUM) ";N
 50 INPUT "INPUT THRESHOLD VALUE (0 TO 2 ARE REASONABLE VALUES) ";IO
 60 INPUT "ENTER THE VALUE OF THE INFORMATION (0 TO 1 IS A GOOD VALUE ) ";INFO
 70 INPUT "DO YOU WANT TO ENTER THE INPUT VECTOR YOURSELF (1/YES 0/NO)? ";VECTOR
 80 INPUT "DO YOU WANT TO INPUT THE T MATRIX (1/Y 0/NO) ";MATRIX
 90 DIM T(100,100),V(100),U(100)
100 REM FILL T(I,J) MATRIX
110 IF MATRIX=0 THEN 190
120 FOR I=1 TO N
130 FOR J=1 TO N
140 PRINT "T(";I;",";J;") "
150 INPUT T(I,J)
160 NEXT J
170 NEXT I
180 GOTO 360 : 'FILL INPUT VECTOR
190 FOR I=1 TO N
200 FOR J=1 TO N
210 R=RND(1)
220 IF R<.8 THEN R=0 ELSE R=+1: REM DILUTE MATRIX
230 T(I,J)=R
240 NEXT J
250 LPRINT
260 NEXT I
270 FOR I=1 TO N
280 FOR J=1 TO N
290 IF I=J THEN T(I,J)=0
300           T(J,I)=T(I,J)
310 LPRINT T(I,J);
320 NEXT J
330 LPRINT
340 NEXT I
350 LPRINT:LPRINT:LPRINT
360 REM FILL INPUT VECTOR U
370 IF VECTOR=0 THEN 430
380 FOR I=1 TO N
390 PRINT "INPUT U(";I;")"
400 INPUT U(I)
410 NEXT I
420 GOTO 470 : 'BEGIN CALCULATIONS OF OUTPUT VECTOR
430 FOR I=1 TO N
440 GOSUB 790
450 U(I)=R
460 NEXT I
470 REM BEGIN CALCULATION
480 FOR ITERATE=1 TO 8:REM THIS ALLOWS THE OUTPUT VECTOR TO BE FEED BACK
490 FOR I=1 TO N
500 FOR J=1 TO N
510 SIGMA=T(I,J)*U(J)+SIGMA
520 NEXT J
530 SIGMA=SIGMA+INFO
540 IF SIGMA > IO THEN SIGMA=1 ELSE SIGMA=0
550 V(I)=SIGMA
560 SIGMA=0
570 NEXT I
580 FOR I=1 TO N
```

```
590 LPRINT U(I);
600 NEXT I
610 LPRINT
620 FOR I=1 TO N
630 LPRINT V(I);
640 NEXT I
650 LPRINT
660 ENERGY=0
670 FOR I=1 TO N : REM ENERGY CALCULATION
680 ENERGY=ENERGY+(U(I)*V(I))
690 NEXT I
700 ENERGY=-.5*ENERGY
710 LPRINT "      ENERGY ";ENERGY:PRINT
720 FOR I=1 TO N
730 U(I)=V(I): REM FOR FEEDBACK
740 NEXT I
750 ENERGY=0
760 NEXT ITERATE
770 LPRINT:LPRINT:LPRINT:LPRINT
780 GOTO 360
790 R=RND(1)
800 IF R<.5 THEN R=0 ELSE R=+1
810 RETURN
```

tboltz9.c

```
/* program to simulate a boltzman update of
        a hopfield neural network
        Transputer   T414 version  11/88    */

#include  "c:\lsc\include\stdio.h"
#include  "c:\lsc\include\math.h"
#include  "c:\lsc\include\float.h"

#define RNDMAX 2147483647.0
#define DILUTIONFACTOR 0.2

static float weight[100][100];
static long int ivector[100],activation[100],oldvalue[100];
static long int tvector[100],ttvector[100];

void main()
{
        /* declarations */
        long int i,j,k,memories,updateno,ielement = 0;
        int noneurons, nomemories;
        float hightemp,lowtemp;
        float temperature;
        double logistic;
        float netinput;
        float randomnumber;
        double t,energy;
        float threshold;
        int seed;

        int rnd();
        int random();
        double expt(); /* compute exp base e */
        double loge(); /* compute log base e */

        /* user input */
        printf("input the number of neurons \n");
        scanf("%d",&noneurons);
        printf("input and threshold\n");
        scanf("%f",&threshold);
        printf("input seed \n");
        scanf("%d",&seed);
```

```
        /* number of memory states */
        nomemories= (int)(0.1*noneurons);

        /* calculate the weight matrix form the
                product of the memory vectors with their transpose */

        srand(seed);

        for(memories=0;memories<=nomemories-1;++memories)
        {
                for(k=0;k<=noneurons-1;++k)
                {
                        tvector[k]=(int)(rnd());  /* the memory state */
                        printf("%d ",tvector[k]);
                        ttvector[k]=tvector[k];   /* the transpose state */
                }
                printf("\n\n");
                for(i=0;i<=noneurons-1;++i)     /* sum the matrices */
                        for(j=0;j<=noneurons-1;++j)
                        {
                          weight[i][j]=weight[i][j]+tvector[i]*ttvector[j];
                                if(i==j)
                                        weight[i][j]=0;
                                else
                                weight[i][j]=weight[i][j];
                         }
        }

        /* print out the final weight matrix */
        for(i=0;i<=noneurons-1;++i)
        {
                for(j=0;j<=noneurons-1;++j)
                {
                        printf("%d ",(int)(weight[i][j]));
                }
                /* printf("\n"); */
        }
        printf("\n\n");

        /* clip the matrix here -- if needed */

        /* input a vector to use as test vector */
        for(k=0;k<=noneurons-1;++k)
        {
                printf("enter the input vector element %d\n",k);
                scanf("%ld",&ielement);
                ivector[k] = ielement;
                activation[k] = ivector[k];
        }



        /* find a random vector to use as input to test the network
                ivector                                             */
/*        for(k=0;k<=noneurons-1;++k)
        {
                ivector[k]=(int)(rnd());
                activation[k]=ivector[k];
                printf("%d ",ivector[k]);
        }
        printf("\n\n");

*/

        /* begin calculation loops form high temperature to low temp.
                ovector                                             */
        temperature=10.0;
        for(t = 2; t <= 50; t = t + 2 )
        {
```

```
            /* save old value for use later in energy calculation */
            for(j=0; j<=noneurons-1;++j)
            {
                    oldvalue[j] = activation[j];
            }
            temperature = temperature/loge(1+t);
            for(updateno = 0; updateno <= t*100.0*noneurons; updateno++)
            {
                    /* code here for basic calculation */
                    i=(long int)(random(noneurons));
                    /* netinput for single neuron */
                    netinput=0;
                    for(j=0;j<=noneurons-1;++j)
                    {
                            netinput += activation[j]*weight[i][j];
                    }
                    netinput = netinput - threshold;
                    /*printf("%f ",netinput);*/
                    /* end basic code here */

                    /* update algorithm here */
                    logistic=(double)(1.0/(1.0+expt(netinput,temperature)));
                    randomnumber=(double)(rand()/(double)(RNDMAX));
                    if (randomnumber > logistic)
                            {
                            activation[i]=1;
                            }
                    else
                            {
                            activation[i]=0;
                            }
                    if(temperature < 0.05)
                    {
                            if(netinput == 0.0)
                                    activation[i] = 0;
                            if(netinput == 1.0)
                                    activation[i] = 1;
                    }


            }
            /* energy calculation */
            energy=0;
            for(j=0;j<=noneurons-1;j++)
            {
                    energy = energy + (oldvalue[j]*activation[j]);                  }
            energy = -energy;
            printf("\n");
            for(i=0;i<=noneurons-1;i++)
            {
            printf("%d ",(int)(activation[i]));
            }
            printf("\n%lf  %lf\n\n ",temperature,energy);
            }

}

/* return a random 0 or 1 for use in the vector generation */
int rnd()
{
        int result;
        result=rand();
        if(result<DILUTIONFACTOR*(double)(RNDMAX))
```

```
                result=1;
        else
                result=0;
        return(result);
}

/* return a random integer between 0 and number of neurons
        for use in random update of the network */
int random(n)
int n;
{
        int result;
        float tmp;

        tmp = rand()/RNDMAX;
        result = (int)(n * tmp);

        return(result);
}

/* claculate an exponent */
double expt(net,tem)
float net;
float tem;
{
        double answer;
        float x;

        x=-1*net/tem;

        answer = 1+x+x*x/2;
        answer = answer + x*x*x/6;
        answer = answer + x*x*x*x/24;
        answer = answer + x*x*x*x*x/120;
        answer = answer + x*x*x*x*x*x/720;
        answer = answer + x*x*x*x*x*x*x/5040;
        answer = answer + x*x*x*x*x*x*x*x/40320;
        answer = answer + x*x*x*x*x*x*x*x*x/362880;
        answer = answer + x*x*x*x*x*x*x*x*x*x/3628800;

        return (answer);
}

/* calculate log base e */
double loge(temp)
double temp;
{
        int ct, lt;
        double t,y,sqrt();

        ct = 1;
        lt = 1;

        if(temp < 1)
        {
                temp = 1/temp;
                lt = -1;
        }

        while(temp > 2)
        {
                temp = sqrt(temp);
                ct *= 2;
        }

        t = (temp-1)/(temp+1);
        y = 0.868591718 * t;
```

```
        y = y + 0.289335524 * (t*t*t);
        y = y + 0.177522071 * (t*t*t*t*t);
        y = y + 0.094376476 * (t*t*t*t*t*t*t);
        y = y + 0.179337714 * (t*t*t*t*t*t*t*t*t);

        return(y/0.43429466*ct*lt);
}
```

3
CHAPTER

Robot simulation and modeling

When building hardware autonomous mobile robots, you must concentrate much effort on the subsystems, such as power supplies, rechargers, motor drive and control circuits, motor speed control, and mechanical subsystems (just to mention a few). If the real goal of the project is to study the behavior of the mobile robot, you can easily be overwhelmed by the subsystem development. Thus, many workers have simulated the mobile robot in a simulated environment. Using this approach, you can easily concentrate on the primary goal of mobile robot behavior in environments of varying complexity.

In this chapter, we'll review some work that has been done on simulation of autonomous mobile agents, hereinafter called robots. We will examine robots with simple memory structures, as well as robots with neural network brains. We will also examine robots with feed-forward control in which the sensors are connected directly to the actuators. All of these will include computer programs written in C. Following this, we'll review modeling more advanced autonomous mobile agents and attempt to draw some conclusions on what we have learned and the relevance to artificial life.

Heiserman robots

Heiserman (1976, 1979) has written two books about building autonomous mobile robots. The later robot had a microprocessor on board, and the robot was able to build a "map" of its environment while encountering objects in its world. It started with a random motion vector (speed and direction); when an object in its path was detected by bumper switches, the robot stored the motion vector in a memory matrix. After the encounter with the object, the robot would again select a random motion vector. When a good selection was made (i.e., removing the robot from the vicinity of the obstacle), the selected motion vector was also stored in a memory matrix and indexed to the motion code that caused the robot to encounter the object. Using this approach, you can enable a robot to build a map of the environment. When the robot encounters an obstacle, it will automatically select a good motion vector to remove it from the vicinity of the object.

While the robot is building this environmental map, it is on a learning curve. In other words, it makes some bad motion code selections that do not take it away from the obstacle. Eventually it will get better at selecting motion vectors. In order to study the learning behavior of these robots, you would have to examine a statistically meaningful population of them; this would require either a huge amount of time with only one robot or a huge amount of money with a population of robots all learning in parallel. Furthermore it would require much time for a human observer to record the learning curves in real time. The obvious and most convenient solution is to use a simulation and let the computer record and analyze the statistical results.

Heiserman (1981) has discussed the simulation of what he calls alpha-, beta-, and gamma-class robots. Each of the three classes of robots starts its learning by acting like a bumper-car (i.e., like a very simple life-form). One of its learning constraints is that each organism must keep moving at all times. Each robot (the terms life-form, organism, and robot will be used interchangeably) starts its life with no knowledge of its environment and must learn to avoid obstacles.

When the alpha-class robot encounters a wall or obstacle, it simply selects a new motion vector at random from its allowed vectors. This could result in the robot again encountering the wall. These repeated encounters with the wall will do nothing to help its learning curve. The learning curve is a plot of the number of contacts versus the number of good selected vectors that remove the robot from the obstacle. This robot is a simple reflexive organism.

The beta-class robot starts out like the alpha robot but, as good motion codes are selected, these are placed in memory and can be retrieved when the robot again encounters an obstacle. Thus, the beta-class robot has the ability to learn.

The alpha and beta robots are an old idea that has been in existence since the 1940's, when people like Wiener (1948) and Shannon (ca. 1954) built relay-based learning machines. They didn't have the advantage of the digital equipment we have for simulations.

The next logical step in evolution of robot intelligence is the gamma-class robot developed by Heiserman (1981). The alpha robot lives in the present only and the beta robot has a memory of the past, but the gamma robot has the ability to predict the future or anticipate encounters with obstacles.

The fourth class of robot, which we call the delta robot, has a neural network to assist it in learning the behavior that will keep it away from obstacles. Each of these classes of robots will be explored in subsequent sections. In the next section, I'll describe the motion codes.

Table 3-1 Motion codes.

(–2,–2)	(–1,–2)	(0,–2)	(1,–2)	(2,–2)
(–2,–1)	(–1,–1)	(0,–1)	(1,–1)	(2,–1)
(–2,0)	(–1,0)	(0,0)	(1,0)	(2,0)
(–2,1)	(–1,1)	(0,1)	(1,1)	(2,1)
(–2,2)	(–1,2)	(0,2)	(1,2)	(2,2)

Motion codes and scoring

It's important that you understand the motion codes so that you can also understand both how these robots are programmed and their reflexive motion. Each robot is, in essence, a finite-state machine. The motion codes in the Motion Code Table are the allowed states. The robot is always in position (0,0) at any given moment.

Each of these states are vectors. The (x,y) coordinates of the desired state is given as (x,y) pairs in the table. If the robot is in state (0,0) and heading toward (–2,2), then this indicates not only the target coordinate but also the speed. For example, the amplitude of the vector from (0,0) to (–2,2) is greater than to (–1,1). There are thus 24 motion vectors that the robot can select when an obstacle is encountered.

A programmed constraint imposed on the robot is that it must keep moving in straight lines. Once a vector has been selected that takes the robot away from the obstacle, then the robot must keep moving in this direction/speed until another obstacle is encountered.

These encounters are not actual collisions with the obstacle. Each robot is given a finite sense of sight. When an obstacle is seen by the robot, this is considered to be a contact. Each robot can see up to two pixels in the direction it is traveling, with the exception to this being the delta-class robot, which can see two pixels in all directions. This visual information is used by the robot to decide if an encounter is imminent.

The robot shown in Fig. 3-1 is in the upper-left corner of the figure. The black border is four pixels thick and is the world or play-pen for the robot. The robot is in actually only one pixel in size but is shown here as a block of 2×2. Because the robot is in the corner, only about ¼ of its motion codes will get it out of this position. When it is up against a wall, about ½ of the motion codes will allow it to escape from its position. Traveling along a wall is allowed.

The probability of selecting a good motion code when up against the wall is 0.500 to 0.583, depending on the angle and speed of contact. The probability of selecting a good motion code when in the corner is between 0.167 and 0.333, again depending on the angle and speed of contact. The net result is that the reflexive alpha robot will score about 0.46.

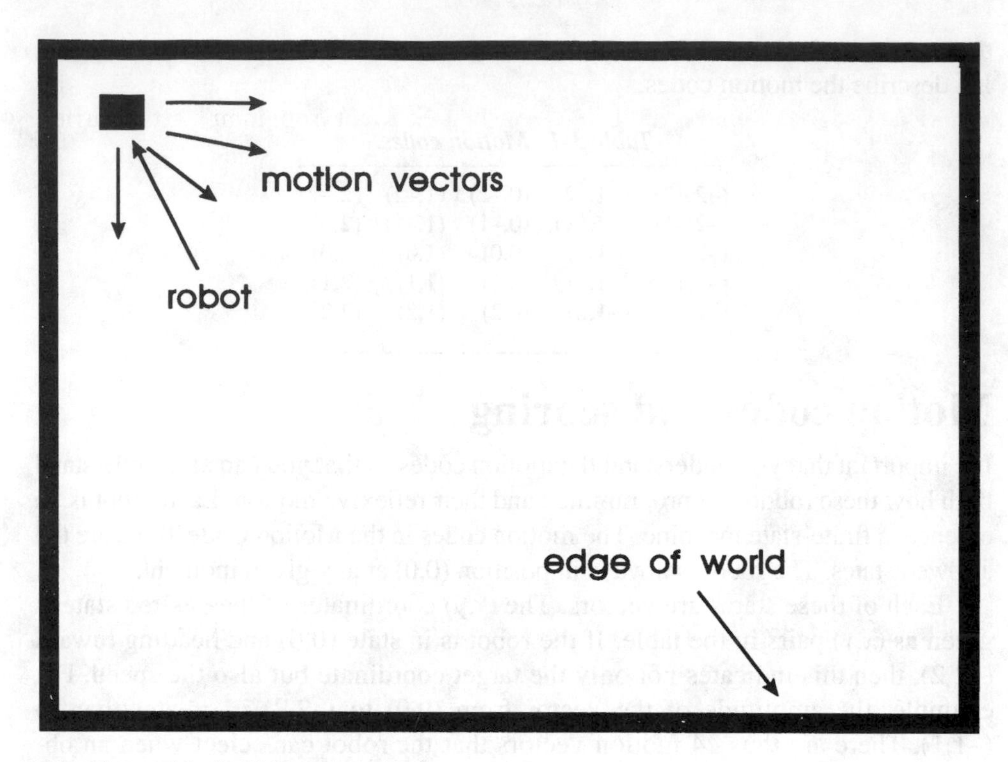

3-1 The world of a Heiserman robot.

In actual practice, though, this is not the case. If the robot selects a bad motion code while in the corner, this is counted against it. In fact, this is just a random number generator and a bad selection was made. Despite these errors in scoring, some classes of robots do show a learning ability, as will soon be demonstrated.

In summary, the alpha robot counts good moves as vectors that take it away from the wall. If the selected random vector does not take it away, then it counts the move against itself. The end result is that the statistics make the robot appear not as good as it should. Graphs presented here will demonstrate that these statistical errors can be ignored.

Alpha robots

The program alpha1.c was designed to emulate a simple reflexive creature that makes random decisions based on speed and direction. In this program, the function newcoord() is used to select new coordinates—random integers in the set {–2, –1, 1, 2}. The function point() is used to check one or two pixels ahead of the robot and basically returns the color of that pixel. This allows the robot to have sight and determine if it is about to encounter a barrier to its straight line motion. Such a situation is known as a *contact*.

The alpha robot then selects a new random vector for its motion. After declarations, the screen mode is setup and the world/cage is drawn. Notice that the boundary must be at least 3 pixels thick to keep the robot from jumping the barrier. I have made the wall 4 pixels thick.

The robot position starts at (160,100), which is the center of the screen. A new move, one or two pixels forward, is then computed by

```
newx = x + vx
newy = y + vy
```

The coordinate (newx,newy) would be the new pixel position of the robot, with the motion vector (vx,vy). Before moving, the robot must use the point function to check if the proposed position is clear. If point returns a value > 0, then this is not a good position and a new vector must be selected. This new vector must also be checked if it will remove the robot from the boundary.

Checking is repeated until a good vector is selected. At that point, the robot moves forward with the new vector. The function sgn() at the end of the program is old code from the development of the program and may be deleted at will. As before, this alpha1 robot counts good moves as vectors that finally take it away from a wall. If it selects (at random) a vector that will not take it away from the boundary, then the bad choice is counted against it. As mentioned earlier, the chance of getting out of a corner are between 0.167 and 0.333, depending on the motion vector selected. (Not all motion vectors are equal.) For the flat wall, encounter the chances are 0.5 to 0.583.

In the program alpha1s.c, the only change has been to provide a means of scoring a population of robots. The data is stored in a data file. Initially, while the program is running through the population of robots, the data is stored in an array. For a contact life of up to 100 contacts, every 10 contacts the score is stored in the array score[][]. The score[][0] elements are running totals of the good moves, and score[][1] is a running total of the contacts. These values, the counting index, and the score (a ratio of the two numbers) is stored in the data file.

The first learning curve in Fig. 3-2a shows the score along the ordinate and the number of contacts along the abscissa. The score was computed, as described earlier, as the ratio of good moves to contacts. The good moves are defined as those that actually take the robot away from a contact situation. A contact is defined as sensing that an obstacle is one or two pixels away from the robot. A key feature of this graph is that the learning curve starts at about 0.4 and soon drops—because of counting statistical errors—to about 0.34. A second feature of this curve is that it's flat. The robot doesn't learn to improve its score. The curve was prepared from a population of 1000 robots, each of which lived for a 100-contact lifetime.

The second plot in Fig. 3-2b is contacts versus bad moves (as explained earlier). Any changes in the slope of the curve indicates learning. Large positive slope changes indicate large learning increase. This curve shows no change in the slope. The second plot doesn't tell us much about the learning ability, or lack thereof, of

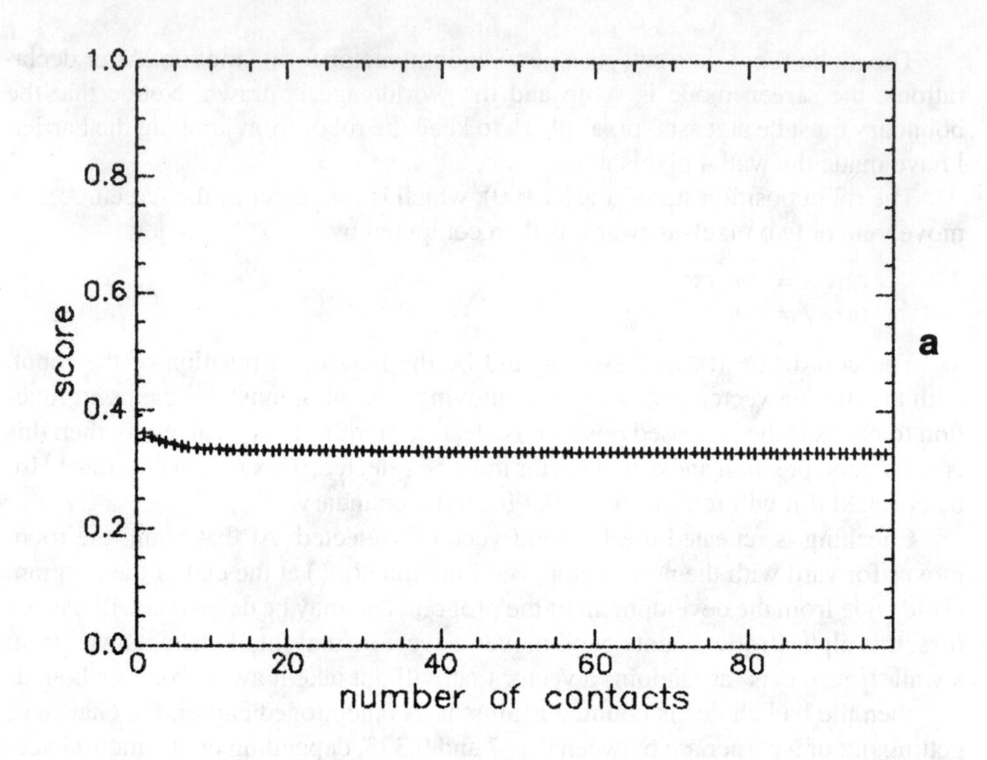

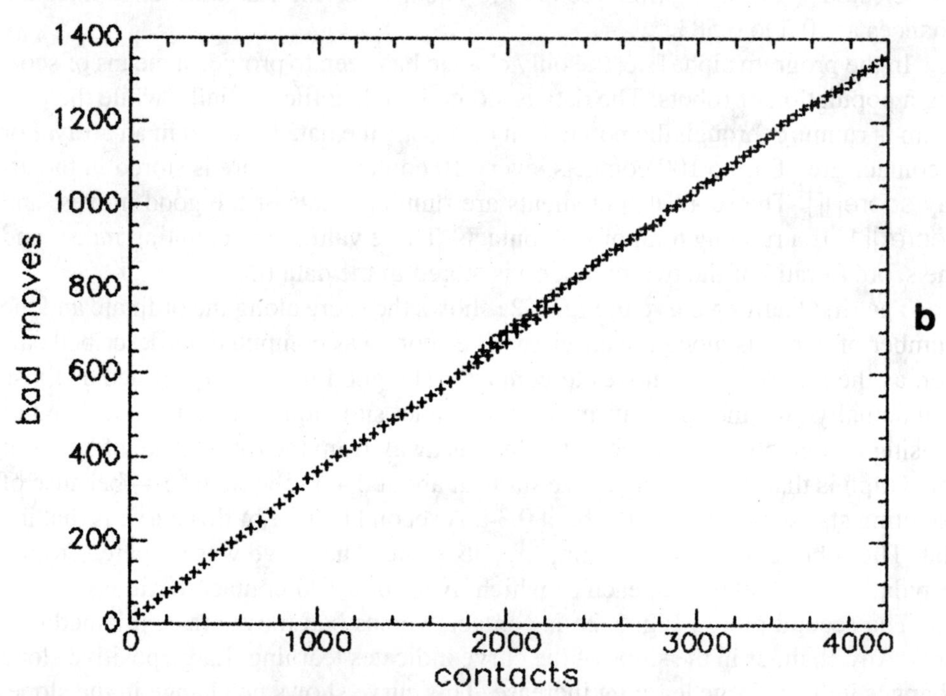

3-2a, b Learning curves for the alpha class robot.

the alpha robots. However this second type plot will show great changes in the learning for delta robots.

Beta robots

The alpha robots are simple reflexive organisms, but the beta robots have the ability to make decisions based on past experiences. In this section, we'll see how this is implemented, and we'll examine the learning curves and results for the beta robots.

Referring to the table of motion codes, if a robot is traveling with vector (–2,0), for example, and the robot senses an obstacle, then the coordinates (–2,0) become the address in a memory matrix. The robot will then look in this address for a vector to get it away from the obstacle. If it is the first time that the robot met an obstacle while traveling with this vector, then the robot must select a motion code or new vector at random, just like an alpha robot, because there's nothing in the memory yet. When a good vector is found, it's stored at the address (–2,0). By using this technique, the robot can build up a memory matrix of the entire space (or at least the space visited).

The program beta1.c contains many of the features of alpha1.c, and the major difference can be explained as follows. If the robot is moving along with a vector of (*Vx*, *Vy*) and it encounters a barrier, then it looks in a 3-D memory matrix with the dimensions M(5)(5)(2). At the time of contact, the robots check for vectors at the address of the contact. The memory matrix is built up as follows:

$M(V_x)(V_y)(X)$
$M(V_x)(V_y)(Y)$

If the matrix elements are 0, the robot must resort to an alpha-like behavior and select a new vector. This resorting to a lower behavior is what Heiserman calls *evolutionary adaptive machine intelligence*. When it finally has a vector that will get it out of the contact situation, then that vector is stored at the address of the old vector (*Vx*,*Vy*). In the future, when the robot is traveling with that motion vector and it encounters a contact, it will have a good selection in the memory matrix and not have to resort to alpha behavior.

Beta1s.c is a scoring program for a population of robots. The program itself is a hybrid of alpha1s.c and beta1.c.

In order to account for the counting errors (as described earlier), the programs alpha3.c and beta3.c were developed. If the robot makes a contact, then it must select a new vector at random. The initial contact is called a bad move and counted as before. Bad moves and contacts are stored in a data file for every ten contacts. A plot of the bad moves versus contacts is a straight line representing the learning accomplished.

Figure 3-3a shows the learning curve for a population of 1000 robots each with a lifetime of 100 contacts. The abscissa is the number of contacts, and the ordinate is the score computed from the ratio of good moves to contacts. These robots also

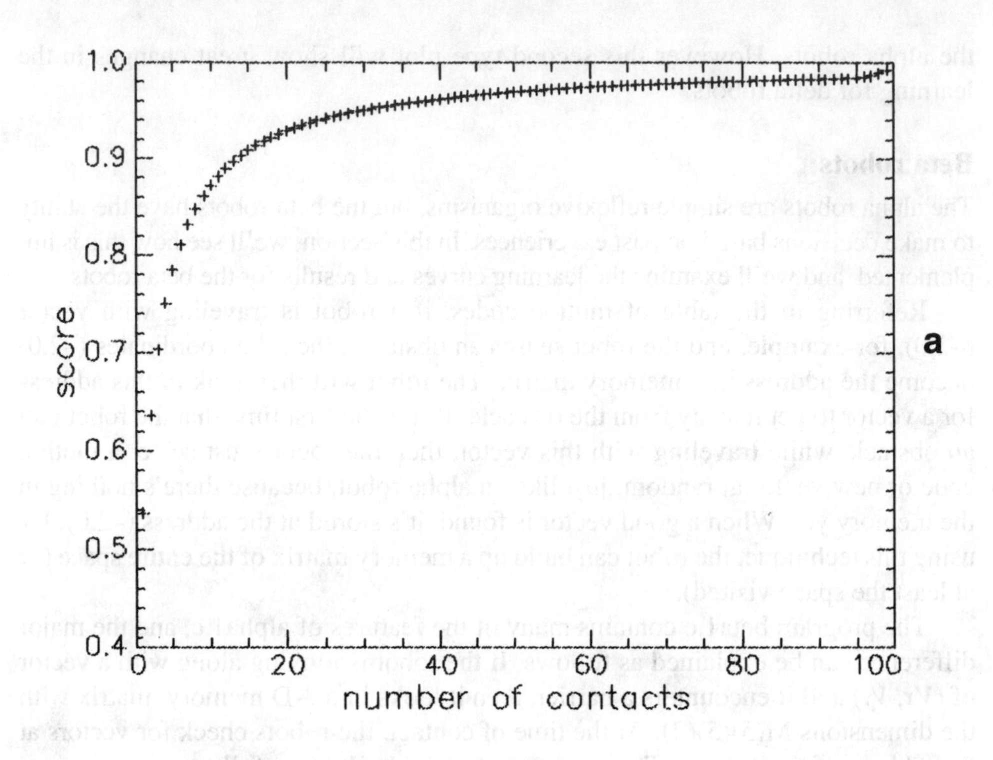

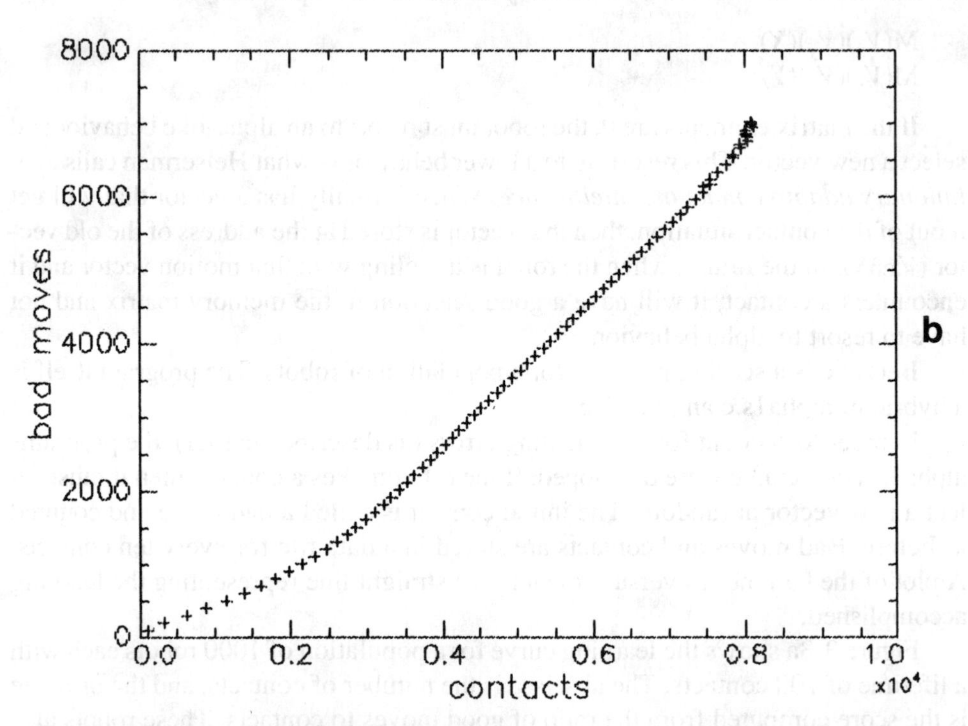

3-3a, b Learning curves for the beta class robot.

start at about 0.4, as can be seen in the curve. This is what you would expect because the memory matrix is empty at first, thus forcing the robot to resort to reflexive behavior. However, the robots quickly learn and achieve a score of greater than 0.98. The robots actually can be seen to lock into a limit cycle.

For example, if a robot first traveling with motion code (–2,0) and contacts a wall and then randomly selects (2,0) as a new motion code, it will store this because it is a good selection. Then traveling with motion code (2,0), it will encounter a wall; and if it randomly selects motion code (–2,0) it will store this. The robot is now locked into a simple back-and-forth cycle or limit cycle. A robot in such a limit cycle is not using its full potential (i.e., most of the memory matrix is empty). If the robot is disturbed, by perhaps selecting a new random motion code, it can be forced to cover a larger area of learning space, thereby making a robot that can more easily adapt to changes in its environment. These ideas of disturbing the beta robots are discussed by Heiserman and will not be covered here. You should note that the limit cycle is a complex behavior for such a simple organism.

The plot in Fig. 3-3b shows large changes in the learning rate. These robots are very fast learners.

Gamma robots

By extending the beta robot memory matrix, M(5,5,2), by one more element, M(5,5,3), it is possible for this to act as a confidence counter. When each of the responses involved in the limit cycle is repeated with a good move, the confidence counter can be incremented. The net effect is that, as the good responses increase, the robot is said to gain confidence.

However, there is little point in increasing the confidence beyond about 4. Moreover, if a robot is given a disturbance and a learned habit pattern is found to not work, then the confidence counter can be decreased. The gamma robot will use this confidence counter to discover the relevant components of its memory matrix (habit pattern) and attempt to make generalizations. It then places a confidence level of 1 in its memory for the generalized direction. Looking at an example might make this more clear.

If a robot is traveling with a (–2,–2) vector when it encounters a barrier and it previously has selected (2,–2) as an escape vector (i.e., confidence level of 4 for this vector), then it will make a generalization that (1,–1) is also a good escape vector and it will also theorize that (1,–2) and (2,–1) may also be good vectors. The robot will only be able to test its theories if it is disturbed or placed in a new environment. For details of the algorithm on how the gamma robots are able to generalize, the interested reader should consult Heiserman (1981).

Delta robots

I have extended the work of Heiserman to include robots with neural networks. These robots are called *delta robots*, and they contain a neural network of the feed-forward type described in Chapter 7.

The task involved in writing a program for the learning is far less than the rule driven approach. Furthermore, the neural network robot can generalize for new obstacles, whereas the alpha, beta, and gamma class robots must resort to reflexive behavior. The input to the neural network is a 25-element vector representing the point function of each of the surrounding pixels, up to two pixels in radii from the robot. The point function is a computer graphics function that returns a Boolean value for a pixel being checked. If the pixel is "on," then a TRUE is returned; otherwise, a FALSE is returned. This 25-element vector is represented schematically in Fig. 3-4. It is the entire 360-degree visual field of the robot and includes the pixel value of the robot itself.

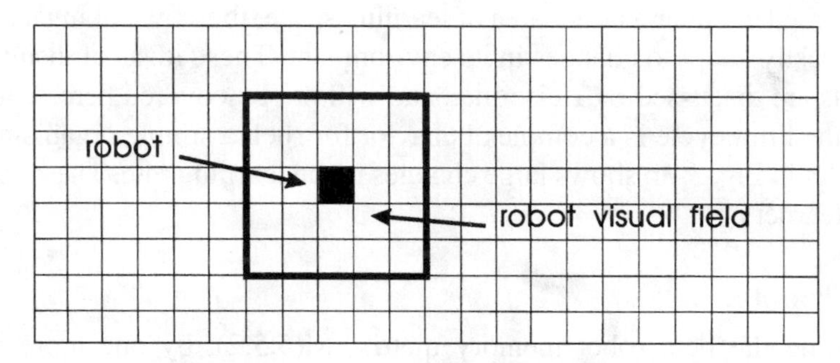

3-4 Schematic diagram of the visual field input for the delta class robot.

The architecture of the neural network consists of the 25 inputs, one hidden layer with ten nonlinear neurons, and an output layer with two linear neurons. The output from the neurons, found by vector matrix multiplication of the input vector and the weight matrix, is a two-element threshold vector. The elements of the output vector are clipped to stay in the range (–2 to +2) and represent the change in the direction vector. If the outputs are $(x,y) = (0,0)$, then there is no change in the direction and the robot continues at the same speed and direction. This updating occurs for each step the robot takes.

When the network output vector is (0,0), then the entire visual field of the robot is clear and it can proceed. These robots are not constrained to travel in straight lines, but they must keep moving. The error vector for updating the synaptic matrix is computed from the difference in the new vector and the output vector. The new vector is the vector selected by the random function. This vector difference represents an error to be fed back to adjust all the matrix elements. Learning only occurs at contact situations, because in these situations, new vectors must be selected by the reflexive random procedure used in the alpha, beta, and gamma class robots.

The delta robot program is delta3.c. After the output of the neural network is computed, the results are postprocessed to represent motion vectors:

```
Vx = (int)oout[0]
Vy = (int)oout[1]
```

These integers are clipped to stay in the range [–2, 2]. A new vector is then selected by

```
newx = x + Vx
newy = y + Vy
```

If (Vx,Vy) = (0,0), no change occurs in the vector and the robot continues in the forward direction. Given these conditions, the robot's visual field indicated that the path ahead was clear. The new position (newx, newy) is checked with the point function, and new vectors are selected and checked, if necessary. From this, we can compute the error vector for training the network as follows:

```
error[0] = (double)(Vx – (int)(oout[0]))
error[1] = (double)(Vy – (int)(oout[1]))
```

If (Vx, Vy) didn't change in the selected new vector, then the error is 0 and no learning takes place; you can only have learning if you have errors. When errors do occur, the feedback algorithm of Chapter 2 is applied to adjust the synaptic strengths.

Figure 3-5 shows a learning curve for a delta class robot. This plot is the correct statistical counting method and shows bad moves versus contacts. It was nec-

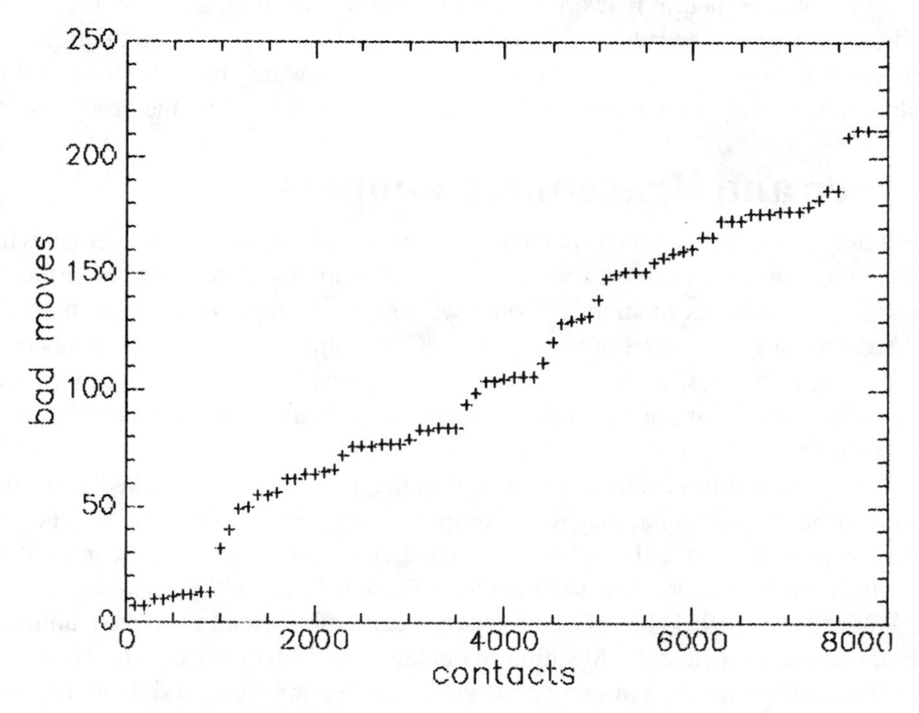

3-5 Learning curves for the delta class robot.

essary to train each robot for a contact life of 100000, and a population of 1000 robots was trained. The graph represents the average robot of this population. The program took about eight days to run on a IBM 286/287 clone.

Notice in the graph that there are jumps in the learning curve. These jumps were a result of random translations and rotations of the robot every 1000 contacts. This was done to allow the robot to explore other regions of space and not get stuck in a local minimum. (Heiserman (1981) tested the beta robots by giving them a random disturbance. His results were alluded to earlier in this chapter.) It can clearly be seen that the robot was able to quickly recover, and the learning curve would flatten out.

The robot, after tens of thousands of contacts for its learning, can be seen to be in a limit cycle similar to the beta robots. The limit cycle learned was a simple back-and-forth motion between two parallel obstacles; this is a valid limit cycle. If however, we constrain the robot to not make 180 degree turns when it encounters an obstacle, then it will—after tens of thousands of contacts—be found in a tight limit cycle of only a few pixel radius. This is an example of a "creative" solution evolved by the robot itself.

We have shown that a robot neural network is capable of generalizing and evolving "creative" solutions on its own. The main advantage in the neural network approach is that it can easily be expanded to many types of robots with many sensors. Furthermore, the algorithm could be easily burned into ROM and placed on board a mobile robot. The neural network robots program themselves, and humans do not need to know about the rules for the rule-driven approach. A neural network robot, on a space mission, could adapt to an environment in which the rule-driven robot could not because the programmers hadn't been able to foresee all the problems.

Walter and Braitenberg animats

Much of the program overhead in the Heiserman robot programs is concerned with memory operation, learning curve statistics, and graphics. Can we "prewire" simple organisms without memories? Could we construct simple robots with ballistic or feed-forward control? For example, how realistic would it be to connect a robot's eyes to its legs, so that when the robot sees a particular object, it moves forward or backward, giving simple reflexive behavior? (By realistic, I mean in comparison to real biological organisms, however simple.)

Ewert and Arbib (1989) edited a conference proceeding concerned with just this question. In particular, they were comparing frogs and robots. The frog neural wiring is such that, if a fly crosses the visual field, the tongue will reach out to catch it. If a large shadow crossed the visual field, the frog will jump away.

Walter (1950, 1951) has assembled two species of artificial animals or animats (animal-robot) that he calls Mechina Speculatrix and Mechina Docilis. The animats are nothing more than two photosensors, two amplifiers, and drive motors. The amplifiers act as neurons performing simple summing operations. The robot is designed to move in the direction of light, but if the light is too intense, it will move away toward shadow. Two such animats with lights attached will tend to "dance" around each other.

More recently, Braitenberg (1984) has written a small book of gedanken experiments describing a variety of these simple feed-forward control robots. He shows that using a few sensors, simple amplifiers, and actuators (motors), it is possible to simulate complex behavior such as love, fear, and aggression.

Let's examine some of these Braitenberg machines. The vehicles (a) and (b) in Fig. 3-6 have excitatory inputs, and vehicles (c) and (d) have inhibitory inputs. If a light source is directly in front of the (a) vehicle, it will speed forward and attack the light source. If the source is off to one side, the vehicle will approach and then turn away because one drive wheel will receive more current than the other. This could be called curiosity with fear. Vehicles (c) and (d) will race toward the light source, but when they get close, they will stop and tend to watch the light source in an almost love-like behavior.

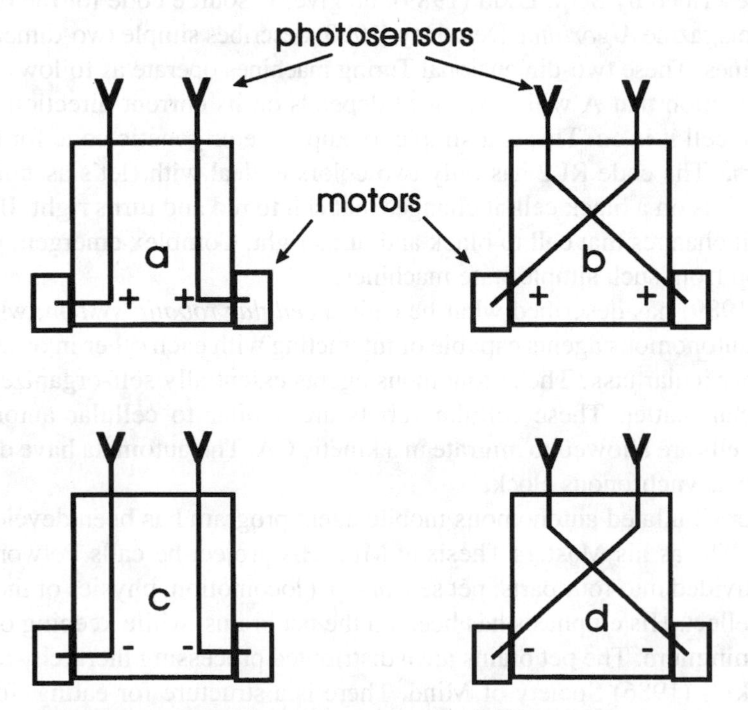

3-6 Simple Braitenberg animats (after Braitenberg, 1984).

Dewdney (The Magic Machine, 1990) talks about these Braitenberg vehicles and suggests a computer simulation of them. The program brait1.c is designed to emulate one of these animats. The program is a simple derivative of the Heiserman alpha1.c program. There is no statistical analyses in the program. Because of the resemblance of these program and alpha1.c, I won't describe the algorithm any further.

Related animat simulations

Dewdney (May, 1989) describes a computer program designed to simulate the evolution of protozoa feeding on bacteria. The same work is also reproduced in the Magic Machine by Dewdney (1990).

The program starts with ten protozoa, each with its own genetic code. The genetic code acts as a motion vector look-up table. When the protozoa lands on a patch of bacteria, the energy reserve of the protozoa is increased by 40 points. If the protozoa has been alive for 800 clock cycles and the energy reserve is 1000 or more, then it reproduces by fission. The two protozoa now at the same spot each have an energy reserve of 500 units. Further, one of the bugs has a new genetic code in which one gene (motion code) has been mutated. A very similar program has been described by Scott Ladd (1989); he gives C source code for the project.

In the magazine *Algorithm*, Dewdney (1990) describes simple two-dimensional-state machines. These two-dimensional Turing machines operate as follows:

The direction that A will move next depends on its current direction and the color of the cell it is on. This is a simple lookup table or genetic code for the motion vectors. The code RL2 has only two colors to deal with (let's assume black and red). If it is on a black cell, it changes that cell to red and turns right. If it is on a red cell, it changes that cell to black and turns right. Complex emergent patterns can develop from such simple state machines.

Beni (1989) has described what he calls a *cellular robotic system*, which are simulated autonomous agents capable of interacting with each other in order to undertake a particular task. The autonomous agents essentially self-organize to process artificial matter. These cellular robots are similar to cellular automata in which the cells are allowed to migrate in a kinetic CA. The automata have dynamic rules and an asynchronous clock.

Another simulated autonomous mobile agent program has been developed by Caderre (1989) as his Masters Thesis at MIT. His project he calls Petworld. The system is divided into four parts: pet sensors, pet locomotion, physics of the world, and pet intellect. His emphasis has been on the pet brains, while keeping other aspects to a minimum. The pet brains are a distributed processing hierarchy modeled after Minsky's (1986) Society of Mind. There is a structure for eating, foraging, nest building, exploring, and combat.

Travers (1989), as part of his Thesis work at MIT, has developed an animal construction kit. Using a menu, you can select various levels of construction. The simplest is to design the brain. Then you can turn the simulated animal loose in its simulated world and observe the behavior. The animals can undergo evolution within the simulation.

The study of cooperative behavior among autonomous agents isn't really the focus to this chapter, but I'd like to mention the work of Manderick and Moyson (1990) and Deneubourg et al. (1983). These workers simulated the cooperative behavior of food gathering of a colony of ants. When a wandering ant finds food, it leaves a chemical trail to attract other ants and lead them to the food source. The mean flow of ants toward the food source is $J_+ = aX(N - X)$, where X is the number of workers at the source and N is the number of ants able to participate. The flow of departure from the food source is given by $J_- = -bX$. The first of these equations is the logistic equation studied in Chapter 1. The main point I want to make here is that the autonomous agents are assumed to be nothing more than a parameter in a difference equation.

Let's get back to the simulation of autonomous agents. Getting and Dekin (1985) have studied at length the swimming behavior of the sea slug—Tritonia Diomedea—and have developed a model sea slug to emulate the neural circuitry involved in the swimming behavior. A simplified circuit diagram of the central pattern generator is shown in Fig. 3-7. The circuit element, DSI, is a few neurons connected to dorsal swimming muscle fibers, the VSI element is connected to ventral swimming muscle fibers, and the element C2 is an interneuron simply called cerebral cell 2. The connections in the circuit diagram marked *I* are inhibitory, while those marked *E* are excitatory.

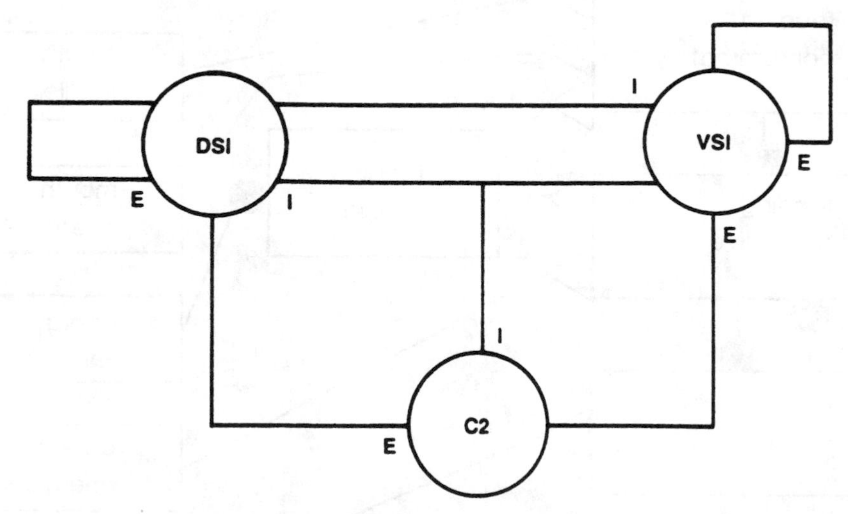

3-7 Simplified circuit diagram for central pattern generator for sea slug. I-inhibit, E-excite.

When DSI depolarizes, it begins to fire. This depolarization can be caused by the escape response initiated by contact with the tube feet of a predatory star fish, or by noxious chemical stimuli. When DSI fires a pulse burst, it inhibits VSI and excites C2. Then DSI and C2 fire together for a short period of time. During the firing, VSI is being inhibited from DSI but excited by C2. The inputs are summed, and VSI begins to fire. When the threshold is reached, VSI fires, inhibiting both DSI and C2, thus decreasing the firing of VSI and completing a cycle. Silverston (1985) and Koch and Segev (1989) should be consulted for more neural models of actual biological circuits.

The last work I want to review is the most extensive. Beer (1990), for his Ph.D. thesis, simulated an artificial insect. Expanding on Minsky's Society of Mind idea (1986), Beer designed several neural circuits, all of which interacted to give an intelligent behavior to the simulated insect.

Figure 3-8 is the overall circuit diagram. Both the following and wandering behaviors are connected to the leg lifting and walking as well as to antennas tactile circuits. The mouth tactile and chemical circuits are connected by excitatory inputs to the feeding consummatory circuits. The energy level circuit has an inhibitory connection to a decision circuit called *feeding arousal*. This circuit has two outputs: one for feeding consummatory, and one for feeding appetitive. Antenna chemical sensors also connect by excitatory connections to feeding appetitive circuits. This is a true distributed computation similar to a local area network.

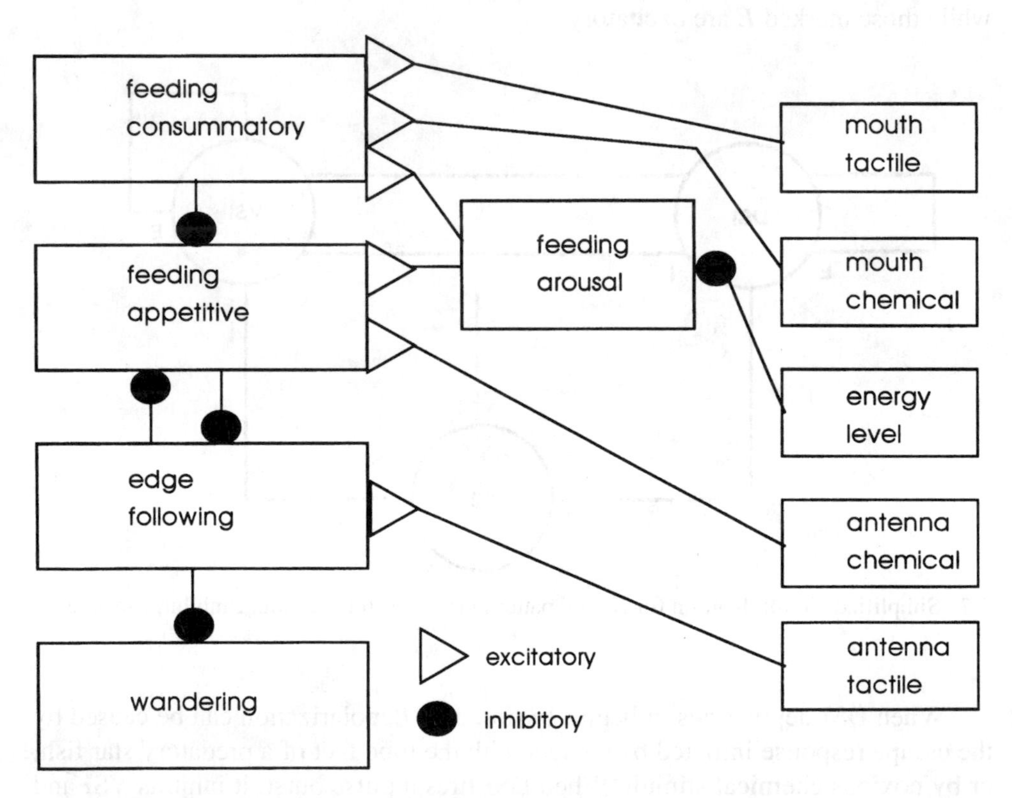

3-8 Circuit diagram for the Beer animat (after Beer, 1990).

alpha1.c

```
#include "\quickc\include\dos.h"
#include "\quickc\include\float.h"
#include "\quickc\include\stdio.h"
#include "\quickc\include\graph.h"
#include "\quickc\include\math.h"

struct videoconfig vc;
char error_message[] = "this video mode is not suported";

void main()
{

        /* declarations */
        int newcoord();
        int point();
        int sgn();

        int nx,ny;
        int x,y,newx,newy;
        int vx,vy;
        long int contact = 0, move = 0;
        long int delay;

        long int newselected = 0;

        /* set mode of screen */
        if (_setvideomode(_MRES4COLOR) == 0)
        {
                printf("%s\n",error_message);
                exit(0);
        }
        _getvideoconfig(&vc);
        _clearscreen(_GCLEARSCREEN);

        _setviewport(0,0,320,200);
        _selectpalette(3);

        /* draw cage/world */
        _setcolor(3);
        _rectangle(_GBORDER,5,5,315,195);
        _rectangle(_GBORDER,6,6,314,194);
        _rectangle(_GBORDER,7,7,313,193);
        _rectangle(_GBORDER,8,8,312,192);

        _rectangle(_GFILLINTERIOR,40,40,60,60);
        _ellipse(_GFILLINTERIOR,150,180,140,170);

        /* initialize creature position and select a starting vector */
        vx = newcoord();
        vy = newcoord();
        x = 160 + vx;
        y = 100 + vy;
        _setcolor(1);
        _setpixel(x,y);
        /* begin moving the creature */
        for(;;)
        {
                /* select new move */
                newx = x + vx;
                newy = y + vy;

                /* check if contact with object */
                if(point(newx,newy) > 0)
                {
                        select_new_vector:
                        {
                                contact++;
```

```
                    vx = newcoord();
                    vy = newcoord();
                    newx = x + vx;
                    newy = y + vy;
                }
                /* check again */
                if(point(newx,newy) > 0)
                {
                    goto select_new_vector;
                }
                else
                {
                    move++;
                }
            }


            /* move */
            _setcolor(0);
            _setpixel(x,y);
            _setcolor(1);
            _setpixel(newx,newy);
            x = newx;
            y = newy;


            /* slow the display down */
            for(delay = 0; delay < 100; delay++);

            if(kbhit())
            {
                break;
            }
        }

        /* dump results to a data file and/or graph the learning data */
        /* clear screen & return control hit enter */
        _clearscreen(_GCLEARSCREEN);
        _setvideomode(_DEFAULTMODE);

        printf("%ld\t%ld\n",move,contact);

}

int newcoord()
{
        int new;
        new = (int)((5*(rand()/32767.0))-3);
        return(new);
}

int point(x,y)
int x,y;
{
        union REGS r, *inregs, *outregs;

        inregs = &r;
        outregs = &r;

        r.x.cx = x;
        r.x.dx = y;
        r.h.ah = 13;

        int86(0x10, inregs, outregs);

        return((int) r.h.al);

}
```

```
int sgn(x)
int x;
{
        if(x == 0)
        {
                return(0);
        }
        else
        {
                return(x/abs(x));
        }
}
```

alpha1s.c

```
#include "\quickc\include\dos.h"
#include "\quickc\include\float.h"
#include "\quickc\include\stdio.h"
#include "\quickc\include\graph.h"
#include "\quickc\include\math.h"

struct videoconfig vc;
char error_message[] = "this video mode is not supported";

static double score[101][2];

void main()
{

        /* declarations */
        int newcoord();
        int point();
        int sgn();

        int j=0;
        int no,creature;
        int nx,ny;
        int x,y,newx,newy;
        int vx,vy;
        long int contact = 0, move = 0;
        long int delay;

        FILE *fd;                       /* file data */
        char filedata[80];

        /* user input */
        printf("input the number of robots/creatures for this expt. \n");
        scanf("%d",&no);

        printf("input file name \n");
        scanf("%s",filedata);

        /* set mode of screen */
        if (_setvideomode(_MRES4COLOR) == 0)
        {
                printf("%s\n",error_message);
                exit(0);
        }
        _getvideoconfig(&vc);
        _clearscreen(_GCLEARSCREEN);

        _setviewport(0,0,320,200);
        _selectpalette(3);

        /* draw cage/world */
        _setcolor(3);
```

```
_rectangle(_GBORDER,5,5,315,195);
_rectangle(_GBORDER,6,6,314,194);
_rectangle(_GBORDER,7,7,313,193);
_rectangle(_GBORDER,8,8,312,192);
x = 0;
y = 0;

for(creature = 0; creature <= no; creature++)
{
        /* initialize creature position and select a starting vector */
        _setcolor(0);
        _setpixel(x,y);
        vx = newcoord();
        vy = newcoord();
        x = 160 + vx;
        y = 100 + vy;
        _setcolor(1);
        _setpixel(x,y);
        contact = 0;
        j = 0;
        move = 0;

        /* begin moving the creature */
        while(contact < 1001)
        {
                /* select new move */
                newx = x + vx;
                newy = y + vy;

                /* check if contact with object */
                if(point(newx,newy) > 0)
                {
                        select_new_vector:
                        {
                                contact++;
                                if(contact % 10 == 0)
                                {
                                        score[j][0] = score[j][0] +
                                                (double)(move);
                                        score[j][1] = score[j][1] +
                                                (double)(contact);
                                        j++;
                                }
                                vx = newcoord();
                                vy = newcoord();
                                newx = x + vx;
                                newy = y + vy;
                        }
                        /* check again */
                        if(point(newx,newy) > 0)
                        {
                                goto select_new_vector;
                        }
                        else
                        {
                                move++;
                        }
                }


                /* move */
                _setcolor(0);
                _setpixel(x,y);
                _setcolor(1);
                _setpixel(newx,newy);
                x = newx;
                y = newy;
```

```
                        /* slow the display down */
                        /* for(delay = 0; delay < 100; delay++); */

                        if(kbhit())
                        {
                                break;
                        }
                }

        }
        /* clear screen & return control hit enter */
        _clearscreen(_GCLEARSCREEN);
        _setvideomode(_DEFAULTMODE);

        /* open file */
        if((fd=fopen(filedata,"w"))==NULL)
        {
                printf("\007ERROR! can't open file\n");
                exit();
        }

        for(j=0;j<100;j++)
        {
                fprintf(fd,"%d\t%lf\t%lf\t%lf\n",j,score[j][0],score[j][1],
                                                score[j][0]/score[j][1]);
        }
        fclose(fd);

}

int newcoord()
{
        int new;
        new = (int)((5*(rand()/32767.0))-3);
        return(new);
}

int point(x,y)
int x,y;
{       union REGS r, *inregs, *outregs;

        inregs = &r;
        outregs = &r;

        r.x.cx = x;
        r.x.dx = y;
        r.h.ah = 13;

       . int86(0x10, inregs, outregs);

        return((int) r.h.al);

}

int sgn(x)
int x;
{
        if(x == 0)
        {
                return(0);
        }
        else
        {
                return(x/abs(x));
        }
}
```

alpha3.c

```
#include "\c\quick\include\dos.h"
#include "\c\quick\include\float.h"
#include "\c\quick\include\stdio.h"
#include "\c\quick\include\graph.h"
#include "\c\quick\include\math.h"

struct videoconfig vc;
char error_message[] = "this video mode is not supported";

static double score[101][2];

void main()
{

        /* declarations */
        int newcoord();
        int point();
        int sgn();

        int j=0;
        int no,creature;
        int nx,ny;
        int x,y,newx,newy;
        int vx,vy;
        long int contact = 0, badmove = 0;
        long int delay;

        FILE *fd;                        /* file data */
        char filedata[80];

        /* user input */
        printf("input the number of robots/creatures for this expt. \n");
        scanf("%d",&no);

        printf("input file name \n");
        scanf("%s",filedata);

        /* set mode of screen */
        if (_setvideomode(_MRES4COLOR) == 0)
        {
                printf("%s\n",error_message);
                exit(0);
        }
        _getvideoconfig(&vc);
        _clearscreen(_GCLEARSCREEN);

        _setviewport(0,0,320,200);
        _selectpalette(3);

        /* draw cage/world */
        _setcolor(3);
        _rectangle(_GBORDER,5,5,315,195);
        _rectangle(_GBORDER,6,6,314,194);
        _rectangle(_GBORDER,7,7,313,193);
        _rectangle(_GBORDER,8,8,312,192);
        x = 0;
        y = 0;

        for(creature = 0; creature <= no; creature++)
        {
                /* initialize creature position and select a starting vector */
                _setcolor(0);
                _setpixel(x,y);
                vx = newcoord();
                vy = newcoord();
                x = 160 + vx;
                y = 100 + vy;
```

```
        _setcolor(1);
        _setpixel(x,y);
        contact = 0;
        j = 0;
        badmove = 0;

        /* begin moving the creature */
        while(contact < 1001)
        {
                /* select new move */
                newx = x + vx;
                newy = y + vy;

                /* check if contact with object */
                if(point(newx,newy) > 0)
                {
                        badmove++;
                        select_new_vector:
                        {
                                contact++;
                                if(contact % 10 == 0)
                                {
                                        score[j][0] = score[j][0] +
                                                (double)(badmove);
                                        score[j][1] = score[j][1] +
                                                (double)(contact);
                                        j++;
                                }
                                vx = newcoord();
                                vy = newcoord();
                                newx = x + vx;
                                newy = y + vy;
                        }
                        /* check again */
                        if(point(newx,newy) > 0)
                        {
                                goto select_new_vector;
                        }
                }

                /* move */
                _setcolor(0);
                _setpixel(x,y);
                _setcolor(1);
                _setpixel(newx,newy);
                x = newx;
                y = newy;


                /* slow the display down */
                /* for(delay = 0; delay < 100; delay++); */

                if(kbhit())
                {
                        break;
                }
        }

}
/* clear screen & return control hit enter */
_clearscreen(_GCLEARSCREEN);
_setvideomode(_DEFAULTMODE);

/* open file */
if((fd=fopen(filedata,"w"))==NULL)
{
        printf("\007ERROR! can't open file\n");
```

```
            exit();
        }

        for(j=0;j<100;j++)
        {
            fprintf(fd,"%d\t%lf\t%lf\n",j,score[j][0],score[j][1]);
        }
        fclose(fd);

}

int newcoord()
{
        int new;
        new = (int)((5*(rand()/32767.0))-3);
        return(new);
}

int point(x,y)
int x,y;
{
        union REGS r, *inregs, *outregs;

        inregs = &r;
        outregs = &r;

        r.x.cx = x;
        r.x.dx = y;
        r.h.ah = 13;

        int86(0x10, inregs, outregs);

        return((int) r.h.al);

}

int sgn(x)
int x;
{
        if(x == 0)
        {
            return(0);
        }
        else
        {
            return(x/abs(x));
        }
}
```

beta1.c

```
#include "\quickc\include\dos.h"
#include "\quickc\include\float.h"
#include "\quickc\include\stdio.h"
#include "\quickc\include\graph.h"
#include "\quickc\include\math.h"

struct videoconfig vc;
char error_message[] = "this video mode is not suported";
static int m[5][5][2];

void main()
{

        /* declarations */
        int newcoord();
        int point();
        int sgn();
```

```
int nx,ny;
int x,y,newx,newy,ovx,ovy;
int vx,vy;
long int contact = 0, move = 0;
long int delay;

long int newselected = 0;

/* set mode of screen */
if (_setvideomode(_MRES4COLOR) == 0)
{
        printf("%s\n",error_message);
        exit(0);
}
_getvideoconfig(&vc);
_clearscreen(_GCLEARSCREEN);

_setviewport(0,0,320,200);
_selectpalette(3);

/* draw cage/world */
_setcolor(3);
_rectangle(_GBORDER,5,5,315,190);
_rectangle(_GBORDER,6,6,314,189);
_rectangle(_GBORDER,7,7,313,188);
_rectangle(_GBORDER,8,8,312,187);

/* initialize creature position and select a starting vector */
vx = newcoord();
vy = newcoord();
x = 160 + vx;
y = 100 + vy;
_setcolor(1);
_setpixel(x,y);

/* begin moving the creature */
for(;;)
{
        /* select new move */
        newx = x + vx;
        newy = y + vy;

        /* check if contact with object */
        if(point(newx,newy) > 0)
        {
                contact++;
                /* select new vector at old vector address */
                ovx = vx;
                ovy = vy;
                vx = m[ovx+3][ovy+3][0] - 3;
                vy = m[ovx+3][ovy+3][1] - 3;
                newx = x + vx;
                newy = y + vy;
                /* check again */
                if(point(newx,newy) > 0)
                {
                        select_new_vector:
                        /* not a good vector so select at random */
                        contact++;
                        vx = newcoord();
                        vy = newcoord();
                        newx = x + vx;
                        newy = y + vy;
                        if(point(newx,newy) > 0)
                        {
                                /* still not good so try again */
                                goto select_new_vector;
                        }
                        else
```

```
                                        {
                                                move++;
                                                /* store new good vector at address of old */
                                                m[ovx+3][ovy+3][0] = vx+3;
                                                m[ovx+3][ovy+3][1] = vy+3;
                                        }
                                }
                                else
                                {
                                        move++;
                                        /* store new good vector at address of old */
                                        m[ovx+3][ovy+3][0] = vx+3;
                                        m[ovx+3][ovy+3][1] = vy+3;
                                }

                        }


                        /* move */
                        _setcolor(0);
                        _setpixel(x,y);
                        _setcolor(1);
                        _setpixel(newx,newy);
                        x = newx;
                        y = newy;


                        /* slow the display down */
                        for(delay = 0; delay < 1000; delay++);

                        if(kbhit())
                        {
                                break;
                        }
                }

                /* dump results to a data file and/or graph the learning data */
                /* clear screen & return control hit enter */
                _clearscreen(_GCLEARSCREEN);
                _setvideomode(_DEFAULTMODE);

                printf("%ld\t%ld\n",move,contact);

}

int newcoord()
{
        int new;
        new = (int)((5*(rand()/32767.0))-3);
        return(new);
}

int point(x,y)
int x,y;
{
        union REGS r, *inregs, *outregs;

        inregs = &r;
        outregs = &r;

        r.x.cx = x;
        r.x.dx = y;
        r.h.ah = 13;

        int86(0x10, inregs, outregs);

        return((int) r.h.al);

}
```

```
int sgn(x)
int x;
{
        if(x == 0)
        {
                return(0);
        }
        else
        {
                return(x/abs(x));
        }
}
```

beta1s.c

```
#include "\quickc\include\dos.h"
#include "\quickc\include\float.h"
#include "\quickc\include\stdio.h"
#include "\quickc\include\graph.h"
#include "\quickc\include\math.h"

struct videoconfig vc;
char error_message[] = "this video mode is not supported";

static double score[101][2];
static int m[5][5][2];

void main()
{

        /* declarations */
        int newcoord();
        int point();
        int sgn();

        int j=0;
        int no,creature;
        int nx,ny,w1,w2;
        int x,y,newx,newy,ovx,ovy;
        int vx,vy;
        long int contact = 0, move = 0;
        long int delay;

        FILE *fd;                               /* file data */
        char filedata[80];

        /* user input */
        printf("input the number of robots/creatures for this expt. \n");
        scanf("%d",&no);

        printf("input file name \n");
        scanf("%s",filedata);

        /* set mode of screen */
        if (_setvideomode(_MRES4COLOR) == 0)
        {
                printf("%s\n",error_message);
                exit(0);
        }
        _getvideoconfig(&vc);
        _clearscreen(_GCLEARSCREEN);

        _setviewport(0,0,320,200);
        _selectpalette(3);

        /* draw cage/world */
        _setcolor(3);
```

```
_rectangle(_GBORDER,5,5,315,195);
_rectangle(_GBORDER,6,6,314,194);
_rectangle(_GBORDER,7,7,313,193);
_rectangle(_GBORDER,8,8,312,192);
x = 0;
y = 0;

for(creature = 0; creature <= no; creature++)
{
        /* initialize creature position and select a starting vector */
        _setcolor(0);
        _setpixel(x,y);
        vx = newcoord();
        vy = newcoord();
        x = 160 + vx;
        y = 100 + vy;
        _setcolor(1);
        _setpixel(x,y);
        contact = 0;
        j = 0;
        move = 0;
        for(w1 = 0; w1 <=4; w1++)
        {
                for(w2 = 0; w2 <= 4; w2++)
                {
                        m[w1][w2][0] = 0;
                        m[w1][w2][1] = 0;
                }
        }

        /* begin moving the creature */
        while(contact < 1001)
        {
                /* select new move */
                newx = x + vx;
                newy = y + vy;


                /* check if contact with object */
                if(point(newx,newy) > 0)
                {
                        contact++;
                        if(contact % 10 == 0)
                        {
                                score[j][0] = score[j][0] +
                                        (double)(move);
                                score[j][1] = score[j][1] +
                                        (double)(contact);
                                j++;
                        }

                        /* select new vector at old vector address */
                        ovx = vx;
                        ovy = vy;
                        vx = m[ovx+3][ovy+3][0] - 3;
                        vy = m[ovx+3][ovy+3][1] - 3;
                        newx = x + vx;
                        newy = y + vy;
                        /* check again */
                        if(point(newx,newy) > 0)
                        {
                                select_new_vector:
                                /* not a good vector so select at random */
                                contact++;
                                vx = newcoord();
                                vy = newcoord();
                                newx = x + vx;
                                newy = y + vy;
                                if(point(newx,newy) > 0)
```

```
                                        {
                                                /* still not good so try again */
                                                goto select_new_vector;
                                        }
                                        else
                                        {
                                                move++;
                                                /* store new good vector at address of old */
                                                m[ovx+3][ovy+3][0] = vx+3;
                                                m[ovx+3][ovy+3][1] = vy+3;
                                        }
                                }
                                else
                                {
                                        move++;
                                        /* store new good vector at address of old */
                                        m[ovx+3][ovy+3][0] = vx+3;
                                        m[ovx+3][ovy+3][1] = vy+3;
                                }

                        }

                        /* move */
                        _setcolor(0);
                        _setpixel(x,y);
                        _setcolor(1);
                        _setpixel(newx,newy);
                        x = newx;
                        y = newy;


                        /* slow the display down */
                        /* for(delay = 0; delay < 100; delay++); */

                        if(kbhit())
                        {
                                break;
                        }
                }

        }
        /* clear screen & return control hit enter */
        _clearscreen(_GCLEARSCREEN);
        _setvideomode(_DEFAULTMODE);

        /* open file */
        if((fd=fopen(filedata,"w"))==NULL)
        {
                printf("\007ERROR! can't open file\n");
                exit();
        }

        for(j=0;j<100;j++)
        {
                fprintf(fd,"%d\t%lf\t%lf\t%lf\n",j,score[j][0],score[j][1],
                                                score[j][0]/score[j][1]);
        }
        fclose(fd);
}

int newcoord()
{
        int new;
        new = (int)((5*(rand()/32767.0))-3);
        return(new);
}

int point(x,y)
int x,y;
```

```
{
        union REGS r, *inregs, *outregs;

        inregs = &r;
        outregs = &r;

        r.x.cx = x;
        r.x.dx = y;
        r.h.ah = 13;

        int86(0x10, inregs, outregs);

        return((int) r.h.al);

}

int sgn(x)
int x;
{
        if(x == 0)
        {
                return(0);
        }
        else
        {
                return(x/abs(x));
        }
}
```

beta3.c

```
#include "\c\quick\include\dos.h"
#include "\c\quick\include\float.h"
#include "\c\quick\include\stdio.h"
#include "\c\quick\include\graph.h"
#include "\c\quick\include\math.h"

struct videoconfig vc;
char error_message[] = "this video mode is not supported";

static double score[101][2];
static int m[5][5][2];

void main()
{

        /* declarations */
        int newcoord();
        int point();
        int sgn();

        int j=0;
        int no,creature;
        int nx,ny,w1,w2;
        int x,y,newx,newy,ovx,ovy;
        int vx,vy;
        long int contact = 0, badmove = 0;
        long int delay;

        FILE *fd;                       /* file data */
        char filedata[80];

        /* user input */
        printf("input the number of robots/creatures for this expt. \n");
        scanf("%d",&no);

        printf("input file name \n");
        scanf("%s",filedata);
```

```
/* set mode of screen */
if (_setvideomode(_MRES4COLOR) == 0)
{
        printf("%s\n",error_message);
        exit(0);
}
_getvideoconfig(&vc);
_clearscreen(_GCLEARSCREEN);

_setviewport(0,0,320,200);
_selectpalette(3);

/* draw cage/world */
_setcolor(3);
_rectangle(_GBORDER,5,5,315,195);
_rectangle(_GBORDER,6,6,314,194);
_rectangle(_GBORDER,7,7,313,193);
_rectangle(_GBORDER,8,8,312,192);
x = 0;
y = 0;

for(creature = 0; creature <= no; creature++)
{
        /* initialize creature position and select a starting vector */
        _setcolor(0);
        _setpixel(x,y);
        vx = newcoord();
        vy = newcoord();
        x = 160 + vx;
        y = 100 + vy;
        _setcolor(1);
        _setpixel(x,y);
        contact = 0;
        j = 0;
        badmove = 0;
        for(w1 = 0; w1 <=4; w1++)
        {
                for(w2 = 0; w2 <= 4; w2++)
                {
                        m[w1][w2][0] = 0;
                        m[w1][w2][1] = 0;
                }
        }

        /* begin moving the creature */
        while(contact < 1001)
        {
                /* select new move */
                newx = x + vx;
                newy = y + vy;


                /* check if contact with object */
                if(point(newx,newy) > 0)
                {
                        badmove++;
                        contact++;
                        if(contact % 10 == 0)
                        {
                                score[j][0] = score[j][0] +
                                        (double)(badmove);
                                score[j][1] = score[j][1] +
                                        (double)(contact);
                                j++;
                        }

                        /* select new vector at old vector address */
                        ovx = vx;
                        ovy = vy;
```

```
                    vx = m[ovx+3][ovy+3][0] - 3;
                    vy = m[ovx+3][ovy+3][1] - 3;
                    newx = x + vx;
                    newy = y + vy;
                    /* check again */
                    if(point(newx,newy) > 0)
                    {
                            select_new_vector:
                            /* not a good vector so select at random */
                            contact++;
                            vx = newcoord();
                            vy = newcoord();
                            newx = x + vx;
                            newy = y + vy;
                            if(point(newx,newy) > 0)
                            {
                                    /* still not good so try again */
                                    goto select_new_vector;
                            }
                    }
                    /* store new good vector at address of old */
                    m[ovx+3][ovy+3][0] = vx+3;
                    m[ovx+3][ovy+3][1] = vy+3;
            }

            /* move */
            _setcolor(0);
            _setpixel(x,y);
            _setcolor(1);
            _setpixel(newx,newy);
            x = newx;
            y = newy;


            /* slow the display down */
            /* for(delay = 0; delay < 100; delay++); */

            if(kbhit())
            {
                    break;
            }
        }

    }
    /* clear screen & return control hit enter */
    _clearscreen(_GCLEARSCREEN);
    _setvideomode(_DEFAULTMODE);

    /* open file */
    if((fd=fopen(filedata,"w"))==NULL)
    {               printf("\007ERROR! can't open file\n");
            exit();
    }

    for(j=0;j<100;j++)
    {
            fprintf(fd,"%d\t%lf\t%lf\n",j,score[j][0],score[j][1]);
    }
    fclose(fd);
}

int newcoord()
{
    int new;
    new = (int)((5*(rand()/32767.0))-3);
    return(new);
}
```

```
int point(x,y)
int x,y;
{
        union REGS r, *inregs, *outregs;

        inregs = &r;
        outregs = &r;

        r.x.cx = x;
        r.x.dx = y;
        r.h.ah = 13;

        int86(0x10, inregs, outregs);

        return((int) r.h.al);

}

int sgn(x)
int x;
{
        if(x == 0)
        {
                return(0);
        }
        else
        {
                return(x/abs(x));
        }
}
```

brait1.c

```
/* simple program to emulate Braitenberg machines */
/*                    100591                      */

#include "\quickc\include\dos.h"
#include "\quickc\include\float.h"
#include "\quickc\include\stdio.h"
#include "\quickc\include\graph.h"
#include "\quickc\include\math.h"

#define WHITE 3
#define RED   2
#define BLUE 1

struct videoconfig vc;
char error_message[] = "this video mode is not suported";


void main()
{

        /* declarations */
        int newcoord();
        int point();

        int x,y;
        int newx, newy;
        int vx,vy;
        int tempx,tempy;


        long int delay;
```

```
/* set up graphics driver */
if (_setvideomode(_MRES4COLOR) == 0)
{
        printf("%s\n",error_message);
        exit(0);
}
_getvideoconfig(&vc);
_clearscreen(_GCLEARSCREEN);

_setviewport(0,0,320,200);
_selectpalette(3);


/* initialize world */
_setcolor(3);
_rectangle(_GBORDER,5,5,315,195);
_rectangle(_GBORDER,6,6,314,194);
_rectangle(_GBORDER,7,7,313,193);
_rectangle(_GBORDER,8,8,312,192);
_setcolor(1);
_rectangle(_GFILLINTERIOR,40,40,60,60);
_ellipse(_GFILLINTERIOR,50,150,60,160);
_setcolor(2);
_ellipse(_GFILLINTERIOR,160,160,180,180);
_rectangle(_GFILLINTERIOR,200,100,220,120);



/* initialize robot */
vx = newcoord();
vy = newcoord();
x = 160 + vx;
y = 100 + vy;
_setcolor(1);
_setpixel(x,y);


/* move/detect */
for(;;)
{
        /* move robot (select new coordinate) */
        newx = x + vx;
        newy = y + vy;

        /* if WHITE select at random */
        if(point(newx,newy) == WHITE)
        {
                select_new_vector:
                {
                        vx = newcoord();
                        vy = newcoord();
                        newx = x + vx;
                        newy = y + vy;
                }
                /* check again */
                if(point(newx,newy) > 0)
                {
                        goto select_new_vector;
                }
        } /* end of WHITE selection */

        /* if RED reverse direction */
        if(point(newx,newy) == RED)
        {
                vx = -vx;
                vy = -vy;
                newx = x + vx;
                newy = y + vy;
        }  /* end of RED selection */
```

```
                /* if BLUE turn 90 degrees */
                if(point(newx,newy) == BLUE)
                {

                        if((vx==0) || (vy == 0))
                        {
                                tempx = vx;
                                tempy = vy;
                                vx = tempy;
                                vy = -tempx;
                        }
                        else
                        {
                                vx = vx;
                                vy = -vy;
                        }

                        newx = x + vx;
                        newy = y + vy;

                        /* check and reselect at
                                random if necessary */
                        if(point(newx,newy) == BLUE)
                                goto select_new_vector;


                }  /* end of BLUE selection */



                /* take action */
                _setcolor(0);
                _setpixel(x,y);
                _setcolor(1);
                _setpixel(newx,newy);
                x = newx;
                y = newy;


                /* slow the display down */
                for(delay = 0; delay < 1000; delay++);


                if(kbhit())
                {
                        break;
                }


        } /* next move/detect */


        /* clear screen & return control */
        _clearscreen(_GCLEARSCREEN);
        _setvideomode(_DEFAULTMODE);


} /* end main */


int newcoord()
{
        int new;
        new = (int)((5*(rand()/32767.0))-3);
        return(new);
} /* end newcoord */


int point(x,y)
```

```
int x,y;
{
        union REGS r, *inregs, *outregs;

        inregs = &r;
        outregs = &r;

        r.x.cx = x;
        r.x.dx = y;
        r.h.ah = 13;

        int86(0x10, inregs, outregs);

        return((int) r.h.al);

} /* end point */
```

delta3.c

```
#include "\quickc\include\dos.h"
#include "\quickc\include\float.h"
#include "\quickc\include\stdio.h"
#include "\quickc\include\graph.h"
#include "\quickc\include\math.h"

#define IUNITS  25
#define HUNITS  10
#define OUNITS  2

static  double  itohweight[IUNITS][HUNITS];
static  double  htooweight[HUNITS][OUNITS];
static  double  target[OUNITS];
static  double  invector[IUNITS];
static  double  netih[HUNITS];
static  double  oout[OUNITS];
static  double  error[OUNITS];
static  double  sigma[HUNITS];
static  double  fprime[HUNITS];
static  double  hout[HUNITS];

struct videoconfig vc;
char error_message[] = "this video mode is not supported";

static double score[101][2];
static int m[5][5][2];

void main()
{

        /* declarations */
        int newcoord();
        int point();
        double rnd();

        int i,j,k;
        long int badmove = 0;
        int no,creature;
        int nx,ny,w1,w2;
        int x,y,newx,newy,ovx,ovy;
        int vx,vy;
        long int contact = 0;
        long int delay;
        double eta = 0.05;
        double delta, temp;
        int badflag;
        double sum;

        FILE *fd;                          /* file data */
```

```
        char filedata[80];

        /* user input */
        printf("input the number of robots/creatures for this expt. \n");
        scanf("%d",&no);
        printf("input file name \n");
        scanf("%s",filedata);

        if((fd=fopen(filedata,"w"))==NULL)
        {
                printf("\007ERROR! can't open file\n");
                exit();
        }

        /* set mode of screen */
        if (_setvideomode(_MRES4COLOR) == 0)
        {
                printf("%s\n",error_message);
                exit(0);
        }
        _getvideoconfig(&vc);
        _clearscreen(_GCLEARSCREEN);

        _setviewport(0,0,320,200);
        _selectpalette(3);

        /* draw cage/world */
        _setcolor(3);
        _rectangle(_GBORDER,5,5,315,195);
        _rectangle(_GBORDER,6,6,314,194);
        _rectangle(_GBORDER,7,7,313,193);
        _rectangle(_GBORDER,8,8,312,192);
/*
        _rectangle(_GFILLINTERIOR,75,187,79,191);
        _rectangle(_GFILLINTERIOR,150,187,154,191);
        _rectangle(_GFILLINTERIOR,225,187,229,191);
        _rectangle(_GFILLINTERIOR,75,9,79,13);
        _rectangle(_GFILLINTERIOR,150,9,154,13);
        _rectangle(_GFILLINTERIOR,225,9,229,13);
        _rectangle(_GFILLINTERIOR,9,95,13,99);
        _rectangle(_GFILLINTERIOR,307,95,311,99);

        _rectangle(_GFILLINTERIOR,100,95,200,99);
*/

        x = 0;
        y = 0;
        for(creature = 1; creature <= no; creature++)
        {
                /* initialize creature position and select a starting vector */
                /* set up arrays */
                for(i=0;i<IUNITS;i++)
                {
                        for(j=0;j<HUNITS;j++)
                        {
                                itohweight[i][j] = 0.2*rnd();
                        }
                }
                for(j=0;j<HUNITS;j++)
                {
                        for(k=0;k<OUNITS;k++)
                        {
                                htooweight[j][k] = 0.2*rnd();
                        }
                }

                _setcolor(0);
                _setpixel(x,y);
```

```
vx = newcoord();
vy = newcoord();
x = 160 + vx;
y = 150 + vy;
_setcolor(1);
_setpixel(x,y);
contact = 0;
badmove = 0;

/* begin moving the creature */
while(contact < 100000)
{

        /* point function on all 24 cells */
        /* set up input vector */
        for(i=-2; i<=2; i++)
        {
                for(j=-2; j<=2; j++)
                {
                        invector[(i+2)+(5*(j+2))] = (double)(point(x+i,y+j));
                }
        }
        invector[12] = 0.0;
        sum = 0.0;
        for(i = 0;i<25;i++)
        {
                sum += fabs(invector[i]);
        }
        if(sum > 0) contact++;

        /* feed forward - input to hidden */
        for(j=0;j<HUNITS;j++)
        {
                netih[j] = 0;
                for(i=0;i<IUNITS;i++)
                {
                        netih[j] = netih[j] + itohweight[i][j]*invector[i];
                }
                hout[j] = tanh(netih[j]);
        }
        /* feed forward - hidden to output */
        oout[0] = (double)vx;
        oout[1] = (double)vy;
        for(k=0;k<OUNITS;k++)
        {
                for(j=0;j<HUNITS;j++)
                {
                        oout[k] = oout[k] + htooweight[j][k]*hout[j];
                }
        }

        vx = (int)oout[0];
        vy = (int)oout[1];
        /* vx and vy must be in the range -2 to +2 */
        if(vx < -2) vx = -2;
        if(vx > 2) vx = 2;
        if(vy < -2) vy = -2;
        if(vy > 2) vy = 2;
        badflag =0;

        select_new:
        newx = x + vx;
        newy = y + vy;


        /* check if contact with object */
        if(point(newx,newy) > 0)
        {
                badflag = 1;
```

```
                        vx = newcoord();
                        vy = newcoord();
                        /* check again */
                        goto select_new;
                }

                if( badflag)
                {
                        badmove++;
                }
                if(contact % 100 == 0)
                {
                        fprintf(fd,"%ld\t%ld\n",contact,badmove);
                }
                error[0] = (double)(vx - (int)oout[0]);
                error[1] = (double)(vy - (int)oout[1]);

                /* compute the error correction for htooweight matrix elements */
                for(k=0;k<OUNITS;k++)
                {
                        for(j=0;j<HUNITS;j++)
                        {
                                delta = error[k]*hout[j]*eta;
                                htooweight[j][k] = htooweight[j][k] + delta;
                        }
                }

                /* compute the error correction for itohweitht matrix elements */
                for(j=0;j<HUNITS;j++)    /* first find sigma(e(k)u(kj)) */
                {
                        sigma[j] = 0;
                        for(k=0;k<OUNITS;k++)
                        {
                                sigma[j] = sigma[j] + error[k]*htooweight[j][k];
                        }
                        temp = (double)(1.0/cosh(netih[j]));
                        fprime[j] = temp*temp;
                }


                for(i=0;i<IUNITS;i++)
                {
                        for(j=0;j<HUNITS;j++)
                        {
                                delta = eta*fprime[j]*sigma[j]*invector[i];
                                itohweight[i][j] = itohweight[i][j] + delta;
                        }
                }

                /* move */
                _setcolor(0);
                _setpixel(x,y);
                _setcolor(1);
                _setpixel(newx,newy);
                x = newx;
                y = newy;

                if(kbhit())
                {
                        break;
                }
        }

}
/* clear screen & return control hit enter */
_clearscreen(_GCLEARSCREEN);
_setvideomode(_DEFAULTMODE);
printf("%ld\t%ld\n",contact,badmove);
```

```
        fclose(fd);}

int newcoord()
{
        int new;
        new = (int)(5*(rand()/32768.0))-2;
        return(new);
}

int point(x,y)
int x,y;
{
        union REGS r, *inregs, *outregs;

        inregs = &r;
        outregs = &r;

        r.x.cx = x;
        r.x.dx = y;
        r.h.ah = 13;

        int86(0x10, inregs, outregs);

        return((int) r.h.al);

}

double rnd()
{
        double result;
        result = (double)(rand())/(double)(32767.0);

        return (result);
}
```

4

CHAPTER

Mathematical bioforms

L-systems, fractals, & constrained morphologies

In this chapter, we will examine the simple iterative mathematics that generate life-like morphologies.

These simple forms represent much of the biological life around us. We will see structures resembling strange aquatic and marine life-forms. We will also examine structures resembling plants and animals. By changing only one parameter in a genetic code, we can get trees or insects. The chapter starts by examining L-systems, which are analogous to cellular automata and abstract languages. The next part of the chapter will focus on iterated function systems (IFS) and the connection between IFS and L-systems. The iterated function systems will lead into a discussion of Dawkins-like constrained morphologies. The chapter will end with fractal animals and Pickover Biomorphs.

L-systems

Much of the following is after Prusinkiewicz and Lindenmayer (1990) and Prusinkiewicz and Hanan (1980). L-systems are a simplified abstract language (grammar) that, when represented in graphical form, resembles plants. These systems have been developed by Lindenmayer, a theoretical biologist, who was studying plant growth and morphology. The entire subject of L-systems is linked with fractals (although they aren't necessarily fractals) and other graph structures described by Smith (1971, 1984, 1985, 1987). I should also mention the collection of papers on L-systems edited by Rosenberg and Solomaa (1986). Many of these are excellent. The subject of L-systems can be quite complicated, so here I will give only an overview; then we'll examine some simple programs and computer-generated pictures.

An example of L-systems will suffice to get us started. The rules or genotype for a simple system might be written as follows:

$$
\begin{array}{lcl}
CB & \leftarrow & A \\
A & \leftarrow & B \\
DA & \leftarrow & C \\
C & \leftarrow & D
\end{array}
$$

This represents a simple context-free (non-Markovian) deterministic L-system often called a DOL-system. These rules indicate the iteration procedure starting from any allowed seed within the alphabet of the language. For example, if the seed A is given at time $t = 0$, then CB is produced at time $t = 1$. This iteration procedure leads to larger structures, as shown in Fig. 4-1. Looking at this figure, you can see that an L-system is a simple rewriting system. A yields CB, C yields DA, and B yields A. The four rules of grammar are iterated each time step building up an evermore complex word.

```
       A     t=0
      CB     t=1
     DAA     t=2
   CCBCB     t=3
DADAADAA     t=4
```

4-1 Simple DOL-system example.

By extending our rules of grammar, we can develop tree-like graphs. For example, by adding brackets to the alphabet, we can represent the words as graphical structures.

$$
\begin{array}{ccc}
C[B]D & \leftarrow & DA \\
A & \leftarrow & B \\
C & \leftarrow & C \\
C(E)A & \leftarrow & D \\
D & \leftarrow & E
\end{array}
$$

With this set of rules, if the initial seed A is planted at time $t = 0$, then the structure shown in Fig. 4-2a will be generated. By assigning the brackets to mean left and

4-2a Bracketed DOL-system.	A	t=0
	C[B]D	t=1
	C[A]C(E)A	t=2
	C[C[B]D]C(D)C[B]D	t=3

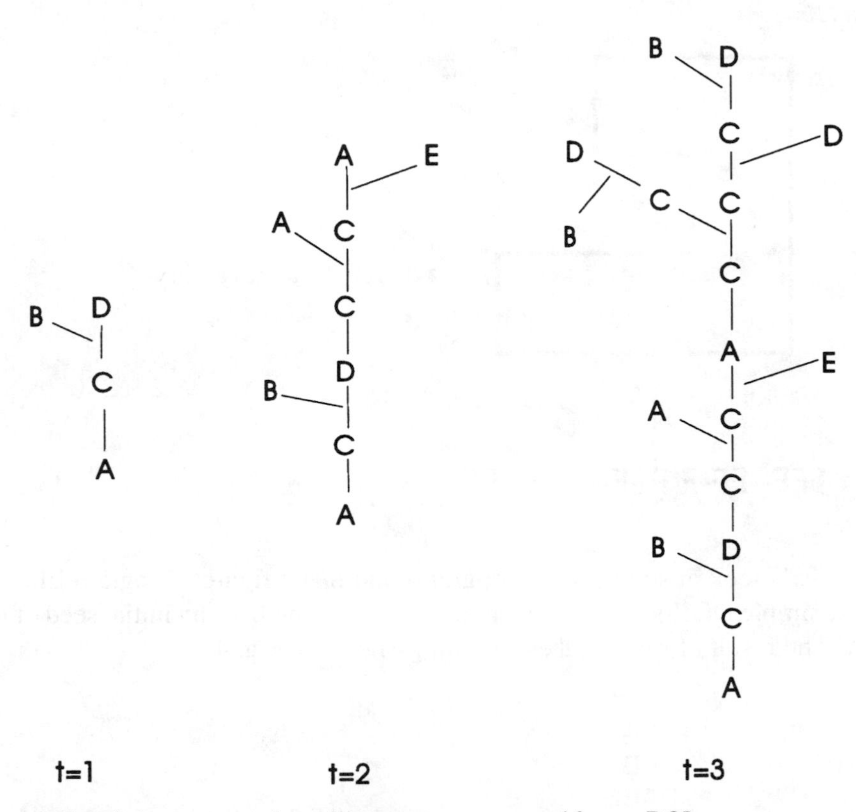

4-2b Time sequence of tree-like structure generated from a DOL-system.

parentheses to mean right, a tree-like structure is generated like the one shown in Fig. 4-2b.

By using Turbo Pascal Turtle graphics, one can give a more graphic interpretation to the words generated by the L-system grammars. With the following alphabet, many complex structures can be drawn—including Penrose tiles, Koch curves, and Hilbert curves. The alphabet is

- F Move forward step length *d* while drawing a line segment.
- f Move forward step length *d* while not drawing a line.
- \+ Turn left by angle α.
- – Turn right by angle α.

Using this alphabet, the curve shown in Fig. 4-3 can be drawn. The generating rules for the entire structure is given by

FFF–FF–F–F+F+FF–F–FFF.

This is the basic idea of L-systems. The references cited here should be consulted for more extensive details.

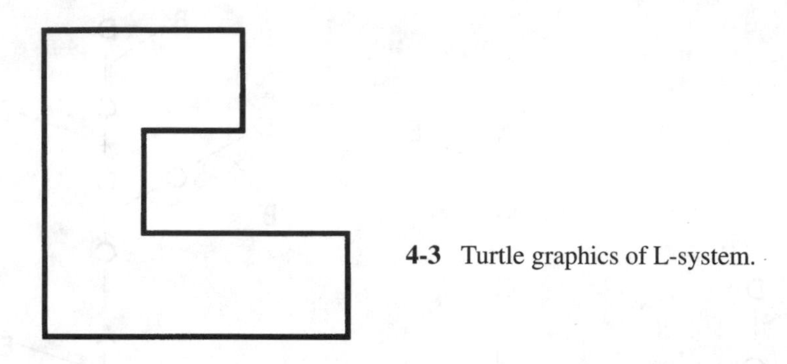

4-3 Turtle graphics of L-system.

FFF--FF-F-F+F+FF-F-FFF

Now let's look at some simple programs and many figures. Program Ll.c extends our simple DOL-system to seven iterations starting from an initial seed of the word A. The result of running the program is shown in Fig. 4-4.

```
                A
               CB
              DACB
            CCBDACB
          DADAACCBDAA
       CCBCCBCBDADAACCBCB
DADAADADAADAACCBCCBCBDADAADAA
```

4-4 L-system organism drawn from the program L1.c.

A more complex demonstration is generated by the program L5.c. In the declarations section, several variables are assigned integer values in the range [0,3]; these represent colors for pixels. The basic algorithm is as follows: An initial seed is planted in the upper-left corner of the display. Then the pixels are read in that line while drawing pixels in the line below. Thus, at time $t = 0$, the first line of pixels is drawn. At time $t = 1$, the line generated at $t = 0$ is read while drawing the pixels in a new line. The procedure is continued until the screen is filled.

These figures are strikingly similar to the one-dimensional cellular automata studied by Wolfram (1983). By experimenting with different colors for the symbols, you can generate pictures similar to those shown in Fig. 4-5.

My program L6.c is an extension, designed to draw a very simple tree structure (Fig. 4-6). The program prompts the user to input trunk height, branch angle, and branch length. The tree is then drawn on the screen.

My primary goal in this program was to explore computer graphics that would be extended into a more advanced program, to be explored later in this chapter. I didn't see any point in writing my own L-system program because two good programs are already available for L-system studies. For those interested in writing their own L-system program, I should point out that Turbo Pascal turtle graphics is in the same language as L-systems, so it should be very easy to write a program.

Two commercial C programs are available for drawing L-systems. Stevens (1989) has written a book that includes a floppy disk of fractal and L-system programs. Figure 4-7 and Fig. 4-8 are screen dumps of L-system trees drawn by the Stevens program. Another book with a floppy disk is by Wegner and Peterson (1991). Their book includes extensive L-system software. Figures 4-9 and 4-10

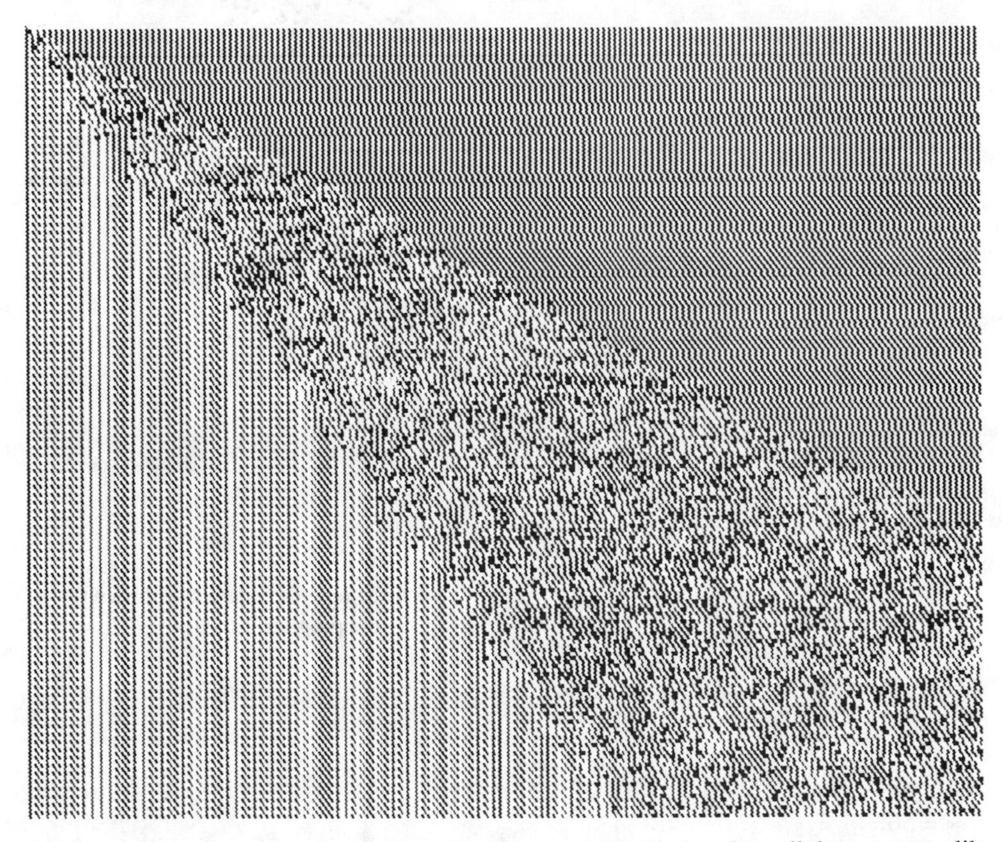

4-5a L-system organisms drawn from the program L5.c. Notice the cellular-automata-like nature of the figures.

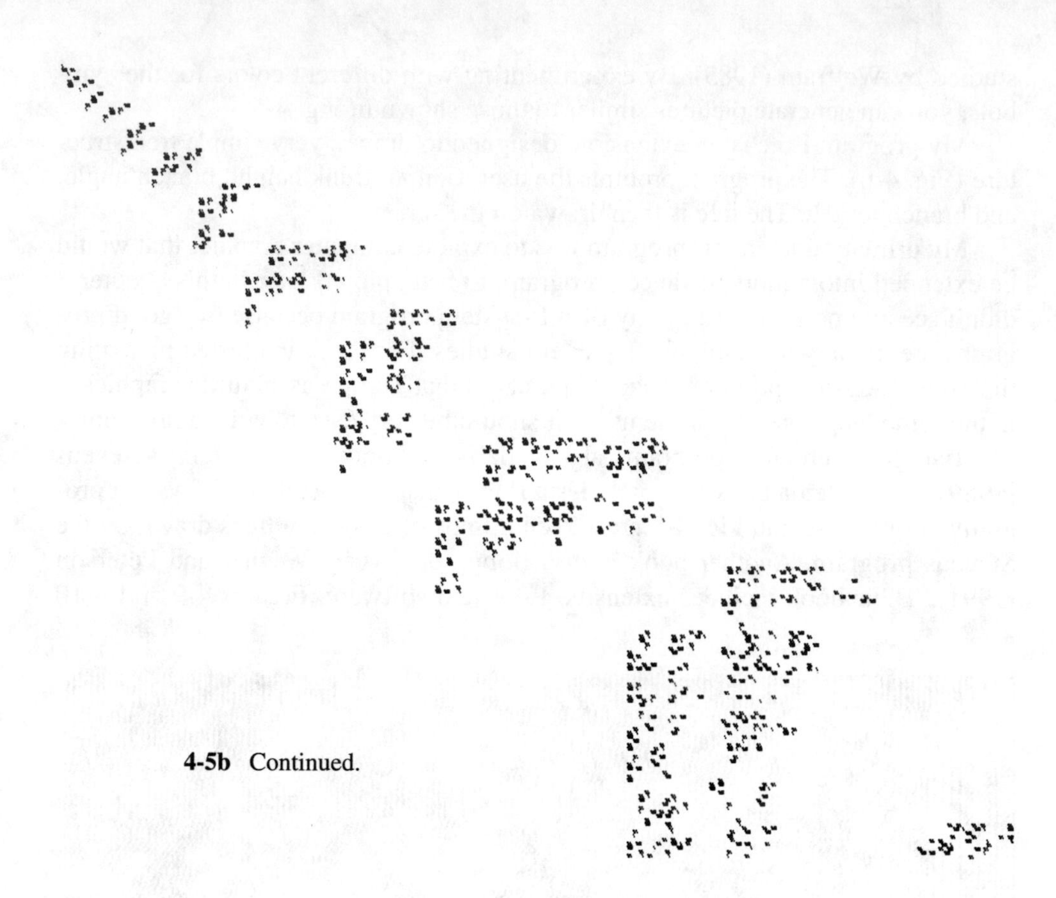

4-5b Continued.

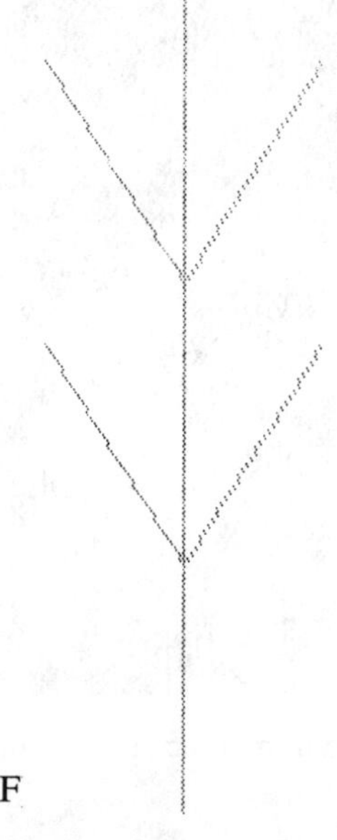

4-6 Simple L-system tree drawn from the program L6.c.

n = 50
∝ = 50°
x
x → F[+F] [-F] F [+F

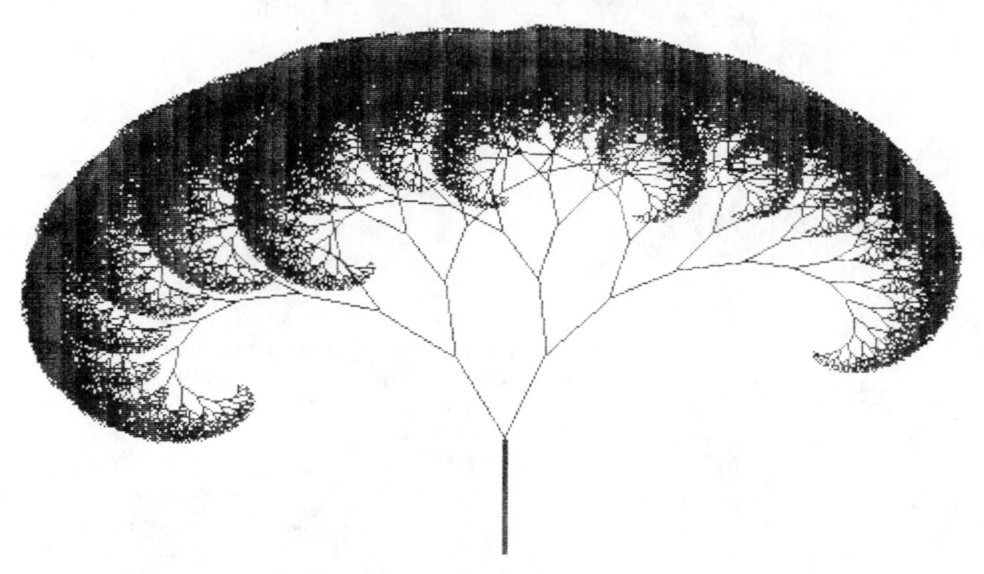

4-7 L-system tree drawn from the Stevens (1989) program.

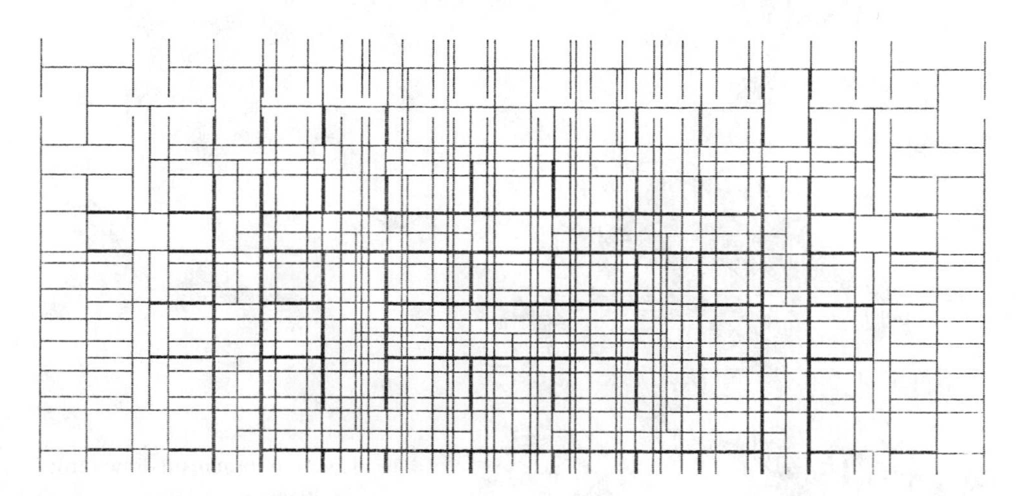

4-8 L-system tree drawn from the Stevens (1989) program.

are screen dumps of L-systems that resemble bushes. Figure 4-11 resembles a branch of a tree, and Fig. 4-12 is a Penrose tilling. All of these figures are screen dumps of L-systems generated by the Wegner and Peterson program.

Iterated function systems

Mandelbrot (1982) gives examples of plant-like structures generated from fractal algorithms, which suggests a relationship between fractals and L-systems. Fractal

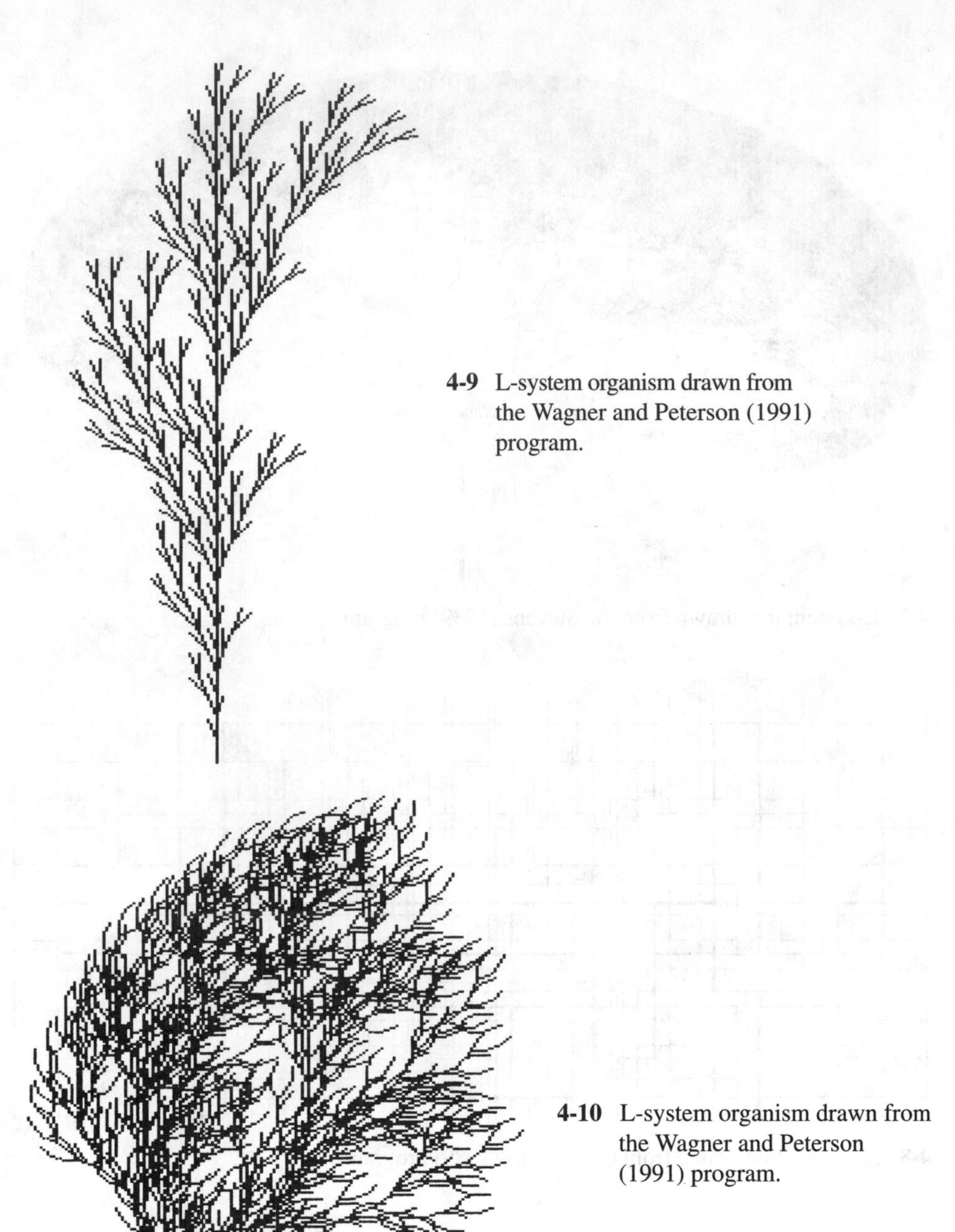

4-9 L-system organism drawn from the Wagner and Peterson (1991) program.

4-10 L-system organism drawn from the Wagner and Peterson (1991) program.

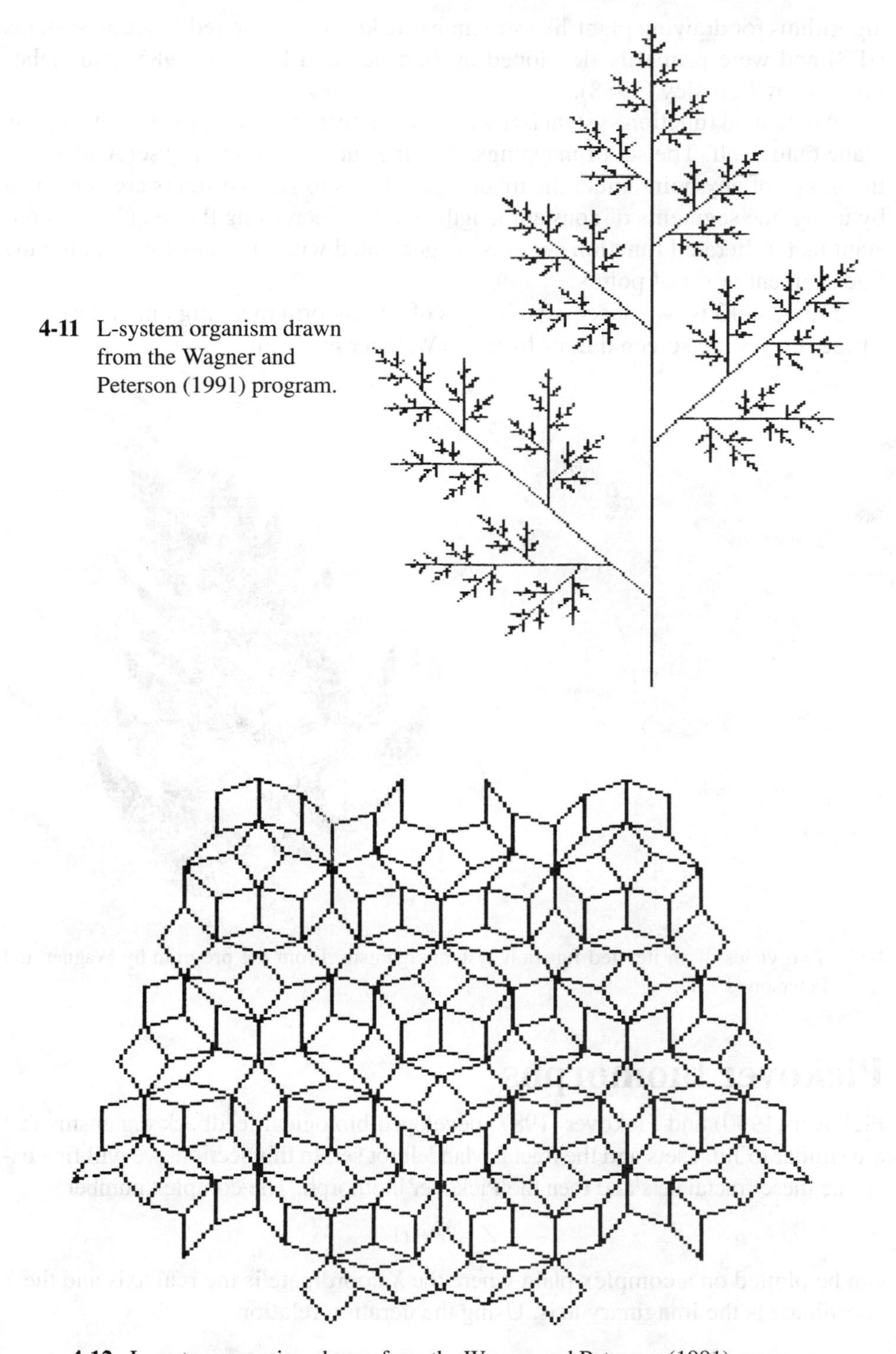

4-11 L-system organism drawn from the Wagner and Peterson (1991) program.

4-12 L-system organism drawn from the Wagner and Peterson (1991) program.

algorithms for drawing plant-like structures are known as iterated function systems (IFS) and were primarily developed by Barnsley and Demko (1985) and elaborated on by Barnsley (1988).

An iterated function system is a set of contractive affine mappings that map the plane onto itself. The set of mappings, T, is the smallest nonempty set A such that the image of any point under the mappings belongs to A. L-systems are generated by using line segments of constant length and then increasing the length by a constant factor. Iterated function systems are generated with matrices for mapping the line segment or set of points.

Figure 4-13 is two successive images of a transform mapping onto a fern leaf. These images are screen dumps from the Weigner program.

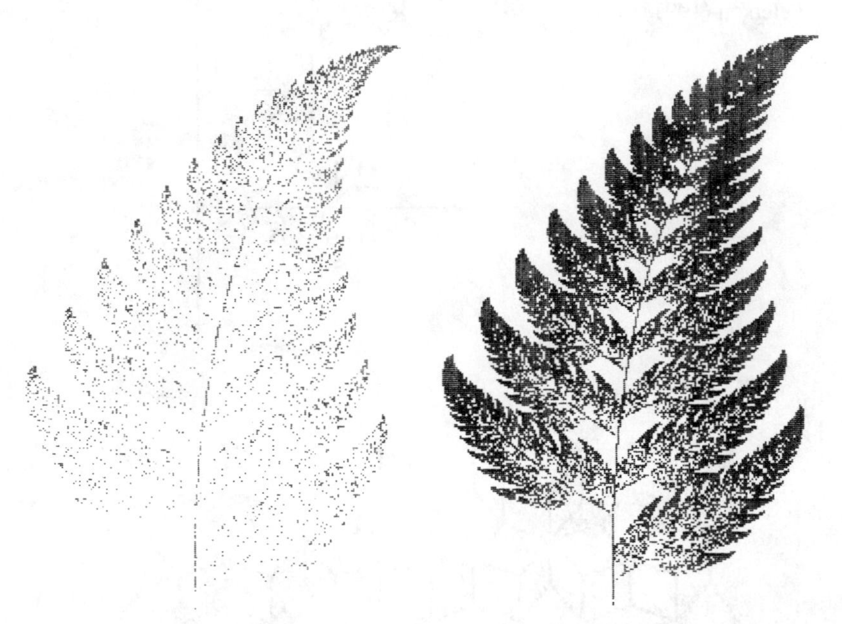

4-13 Two views of an iterated function system organism. From the program by Wagner and Peterson (1991).

Pickover biomorphs

Pickover (1990) and Pickover (1987) developed biological feedback organisms that are similar to Julia sets and the fractal Mandelbrot set. In this section, we will first examine these fractal sets and then the Pickover biomorph. The complex number

$$Z = X + iY$$

can be plotted on a complex plain where the X coordinate is the real axis and the Y coordinate is the imaginary axis. Using the iteration relation

$$Z_{n+1} = Z_n^2$$

there are only two attracting points. For an initial $Z < 1$, the attractor point is 0. For an initial $Z > 1$, the attractor point is infinity. If a complex constant is added at each iteration such that

$$Z_{n+1} = Z_n^2 + c$$

then the attracting points map out Julia sets and fractals. When the iterated points are plotted in Z-space and the parameter c is held fixed, the function maps out Julia sets. When the iterated points are started at $Z = 0$ and iterated for various values of the parameter c and plotted in c-space, the function maps out the Mandelbrot set. This can be made clearer with some algebra. The basic equation is

$$Z_{n+1} = Z_n^2 + c$$

where Z and c are complex numbers given as follows:

$$Z = X + iY$$
$$c = p + iq$$

The function is then given by the following:

$$X_{n+1} = X^2_n + Y_n^2 + p$$
$$Y_{n+1} = 2X_nY_n + q$$

For Julia sets, you hold p and q constant for the entire region of space, select an initial point (X,Y), and iterate this to an attractor point. The number of iterations required to reach the attractor point is recorded and assigned a color. The point (X,Y) is then assigned this color and plotted. The resulting Julia set is then a map of the number of iterations to reach an attractor point.

Many points do not reach an attractor point other than infinity. In order to prevent this, the modulus of the complex number must be calculated. In general, if the modulus is small, then the iterates will not escape to infinity. The modulus is calculated using

$$\text{mod}\ (z) = |\sqrt{X^2 + Y^2}|$$

or something like the following will do:

$$X^2 + Y^2 \geq 4$$

The Julia sets are an infinite number of mappings for a whole range of constant values for p and q.

We spent time on the Julia sets because they are analogs of the Pickover biomorphs. Pickover (1990) refers to the biological forms we are about to model as feedback forms. Just like the Julia sets, we start with an initial Z and constant c. The iteration is then performed until the magnitude of Z reaches a threshold value. This is the usual procedure in Julia set studies.

If, after a number of iterations, the magnitude of the Z value is less than the threshold, a point in the Z-space is plotted; otherwise, a new Z is selected without printing to the screen. Instead of checking the magnitude of Z, we will check the magnitude of the components. If either the real or the imaginary components are small after n iterations, then the point is plotted.

Now let's look at two specific systems. I call the first one the z-cubed system, which is given by this equation:

$$Z_{t+1} = Z_t^3 + c$$

The second one is the z-fifth system:

$$Z_{t+1} = Z_t^5 + c$$

Because Z and c are complex numbers, these equations can be expanded and written in terms of the components. For the z-cubed system, we get this:

$$(a + ib)^3 = a^3 - 3ab^2 + (3a^2b - b^3)$$

The z-fifth system is given by

$$(a + ib)^5 = a^5 + 5a^4bi - 10a^3b^2 - 10a^2b^3i + 5ab^4 + b^5i$$

which can be grouped as follows:

$$(a + ib)^5 = (a^5 - 10a^3b^2 + 5ab^4) + (5a^4b - 10a^2b^3 + b^5)i$$

These two relations are used in the programs pick1.c and pick2.c, which are designed for modeling Pickover type biomorphs (Pickover, 1990). The program pick1.c is designed to model the z-cubed type biomorphs, and the pick2.c is designed to model z-fifth biomorphs. Both programs are well commented, so I'll just discuss the algorithm.

Two parameters, x and y, are set equal to 0; these are the coordinates for plotting the attractors. The y direction is the imaginary axis, and the x direction is the real axis of the z-space. The space consists of 320x200 pixels. The real axis is the outer loop, and the imaginary axis is the inner loop. Each time through these loops, the x,y counters are incremented. This will be the actual coordinate plotted on the screen.

The loops increment by a factor of 2*rmax/320 or 2*imax/200, where rmax and imax are the real and imaginary maximum for the space. These parameters determine the magnification for viewing the biomorph and are input by the user at the start of the program. Like Julia sets, the C is a constant and complex. The real and imaginary parts of this constant are input by the user and determine the "species" of the biomorph. The choice of whether or not to draw a pixel is made 30 times due to the loop. If the modulus is > 100, the loop is broken out of and the next pixel is selected. If the iteration reaches 30, the loop ends and the real and imaginary parts of the solution are checked for their absolute value to be less than 10. If either parameter has an absolute value less than 10, then the pixel is drawn and the new coordinate is selected.

Now let's examine the morphological changes produced by simple changes in the parameters (genetic code). The program prompts the user for the genetic code of the organism. This is the real and imaginary maxima of *Z* and the real and imaginary part of the constant *c*. With these four parameters as input, we can see their effect on the organism morphology. You can view changes in figures in two ways: you can let each change in any parameter represent a new species of organism, or you can let the first two parameters (the real and imaginary maximum of *Z*) represent a microscope and let the real and imaginary parts of the constant be codes that generate new species. I will take the second approach when drawing and explaining the figures.

Figure 4-14 is a large z-cubed biomorph with parameters (10,10,10,0.1). Figure 4-15 through Fig. 4-27 are images of organisms at the same microscopic level of (2.5,2.5). The changes in the different species can be quite significant, varying from Julia set-like organisms to starfish-like forms (Fig. 4-26). The z-fifth program prompts the user for the same type of inputs. Fig. 4-28 is a z-fifth Pickover organism at a very high magnification.

Figures 4-29 through 4-33 each has a different gene code.

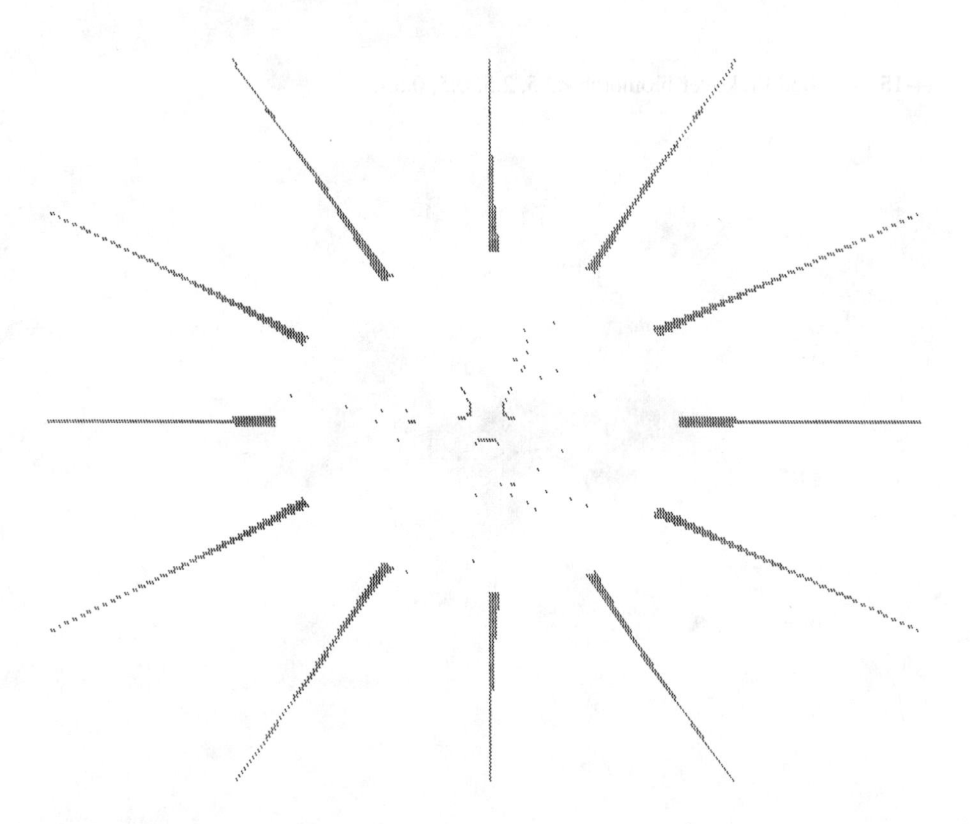

4-14 z-cubed Pickover biomorph <10, 10, 10, 0.1>.

4-15 z-cubed Pickover biomorph <2.5, 2.5, 0.5, 0.5>.

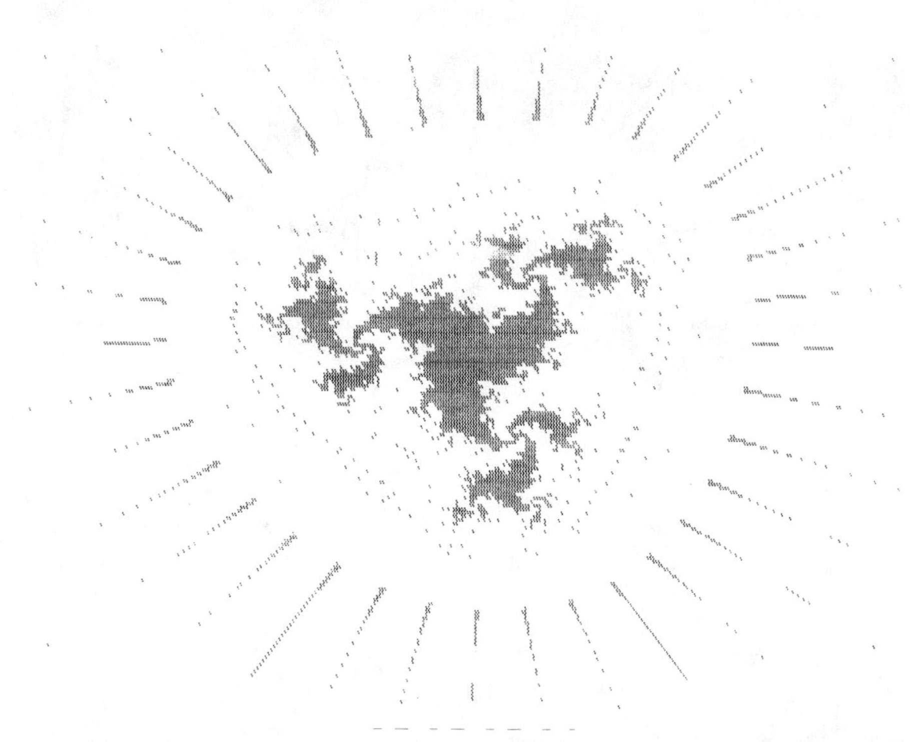

4-16 z-cubed Pickover biomorph <2.5, 2.5, 0.5, 0.6>.

4-17 z-cubed Pickover biomorph <2.5, 2.5, 0.4, 0.6>.

4-18 z-cubed Pickover biomorph <2.5, 2.5, 0.5, 0.7>.

4-19 z-cubed Pickover biomorph <2.5, 2.5, –1, 1>.

4-20 z-cubed Pickover biomorph <2.5, 2.5, 0, 1>.

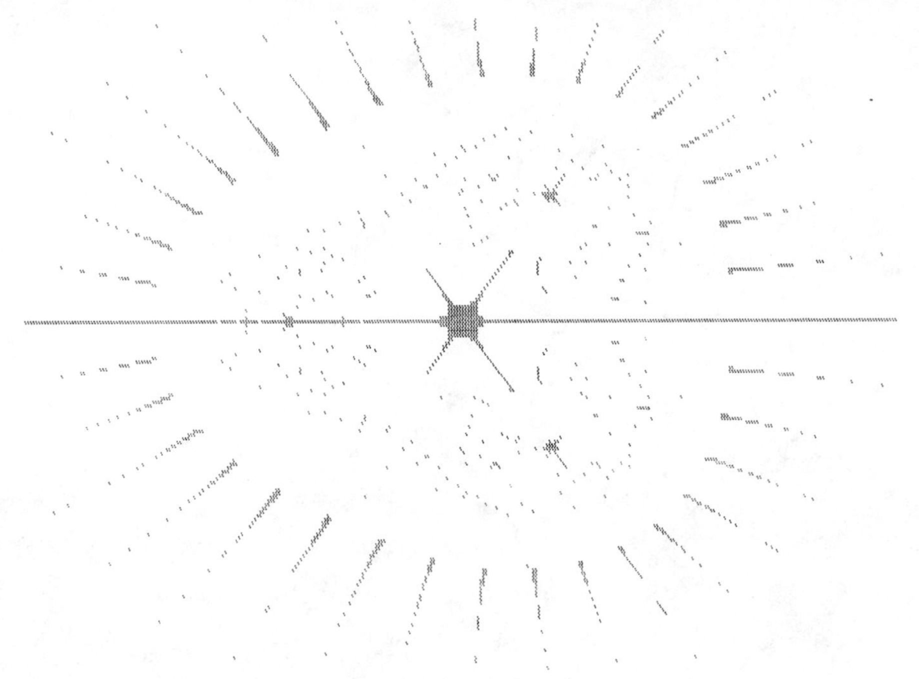

4-21 z-cubed Pickover biomorph <2.5, 2.5, 1, 0>.

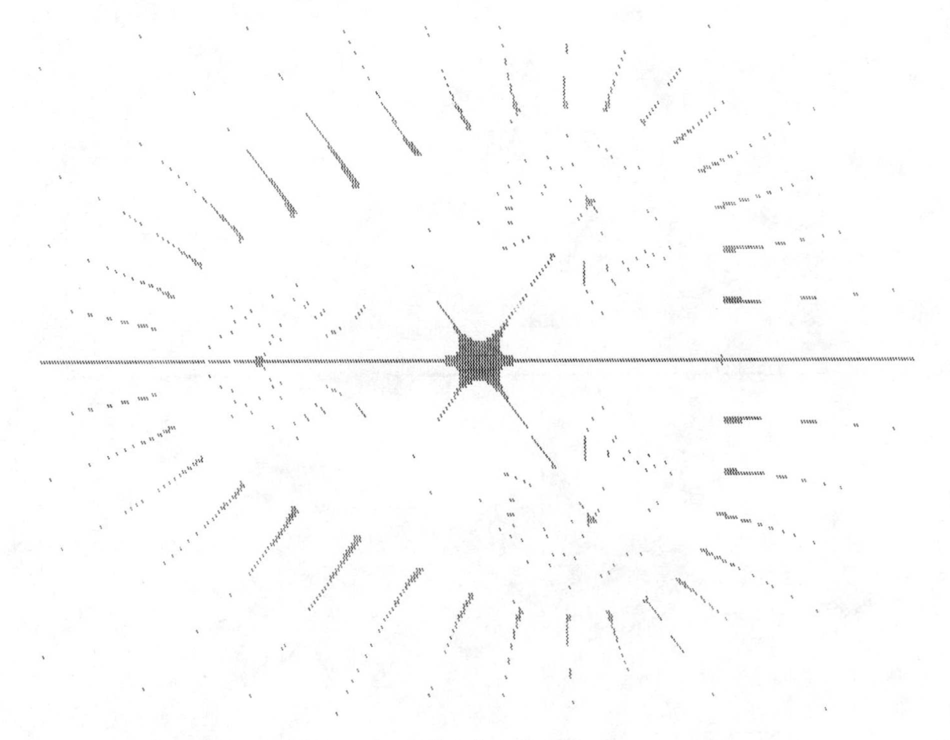

4-22 z-cubed Pickover biomorph <2.5, 2.5, 2, 0>.

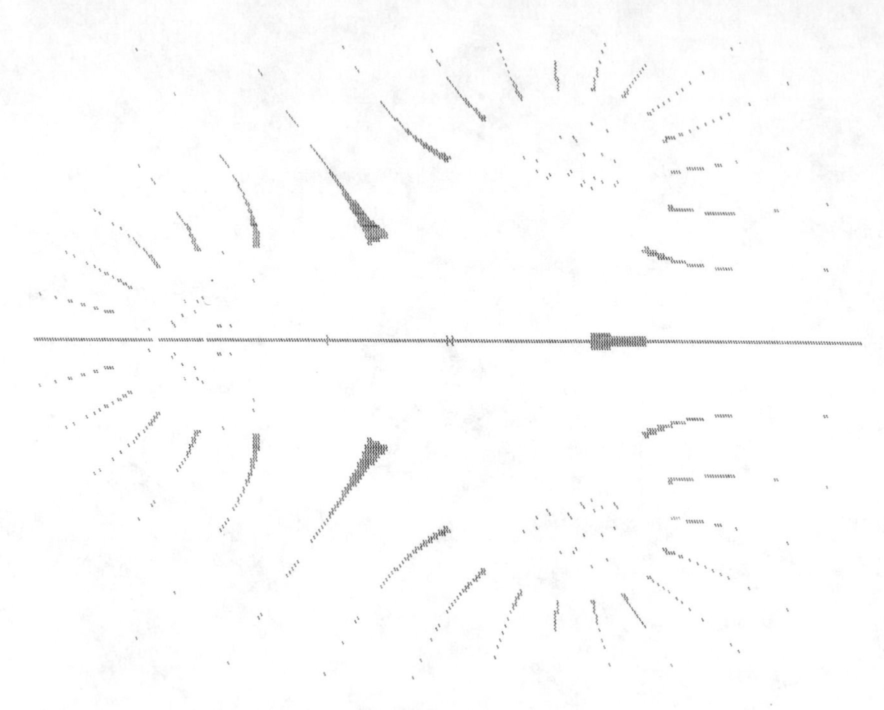

4-23 z-cubed Pickover biomorph <2.5, 2.5, 4, 0>.

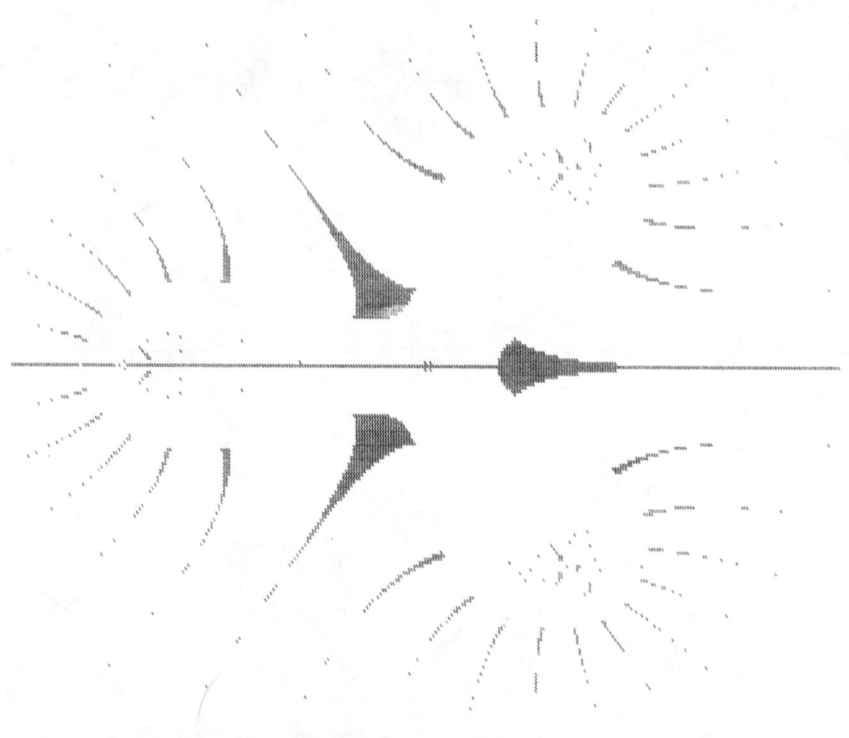

4-24 z-cubed Pickover biomorph <2.5, 2.5, 4.5, 0>.

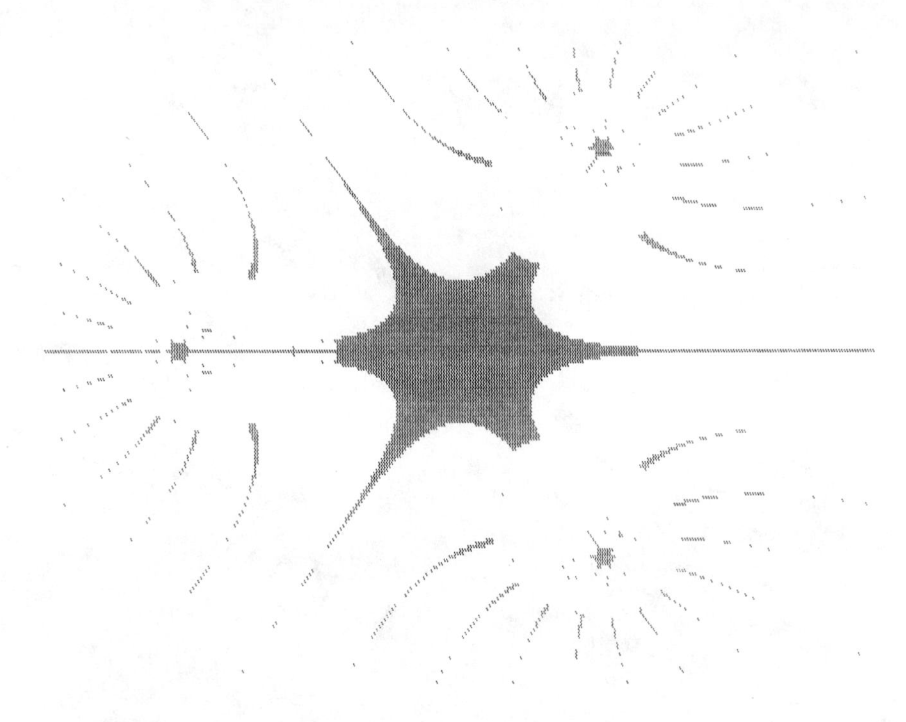

4-25 z-cubed Pickover biomorph <2.5, 2.5, 5, 0>.

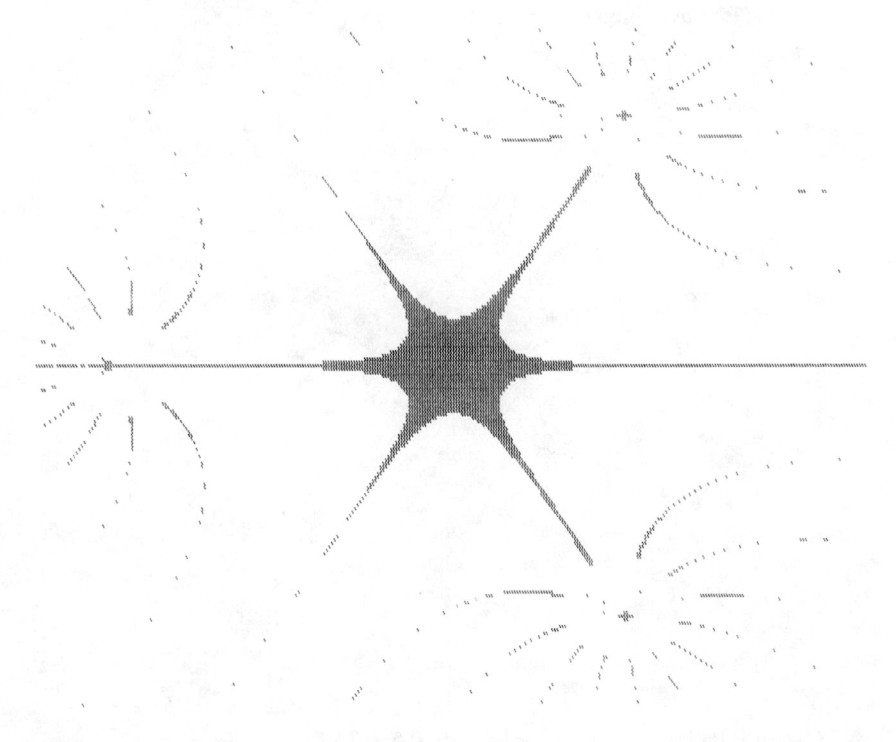

4-26 z-cubed Pickover biomorph <2.5, 2.5, 9, 0>.

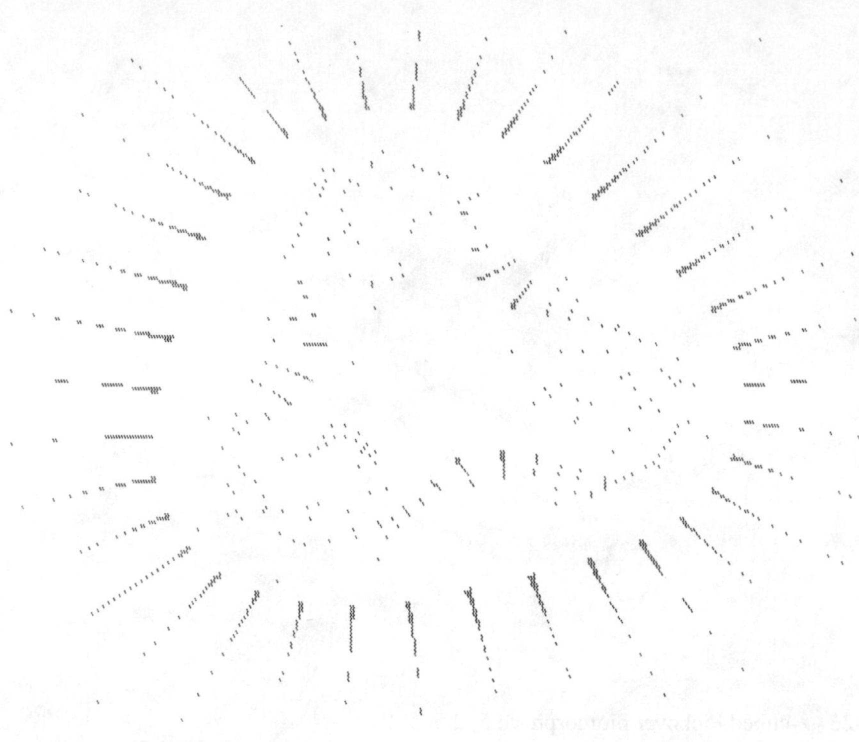

4-27 z-cubed Pickover biomorph <2.5, 2.5, –1, –1>.

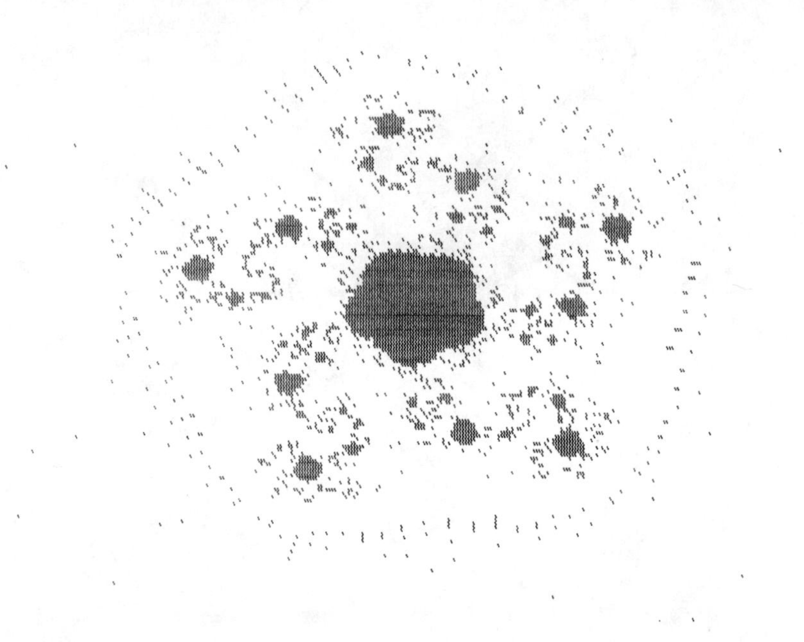

4-28 z-fifth Pickover biomorph <1.9, 1.9, 0.5, 0.7>.

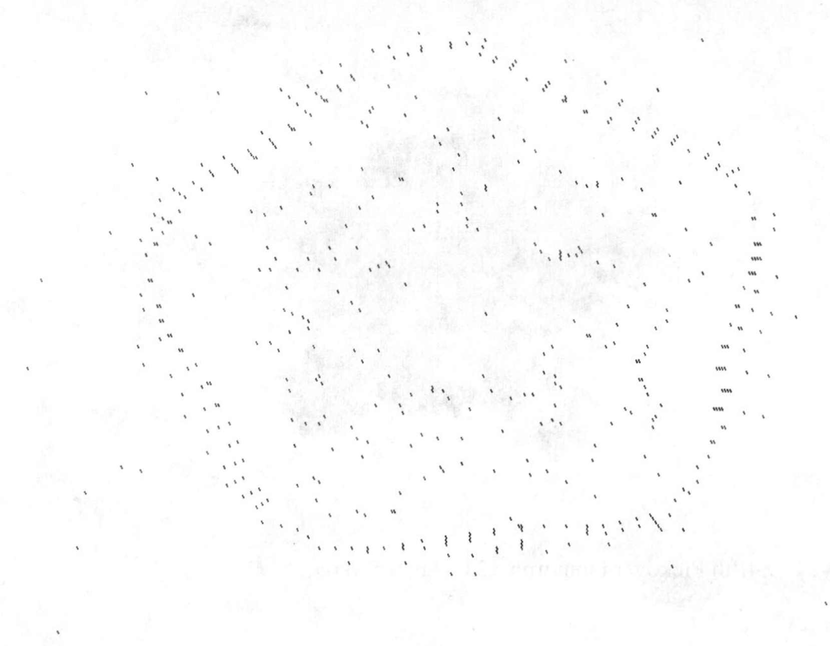

4-29 z-fifth Pickover biomorph <1.8, 1.8, 0.5, 0.8>.

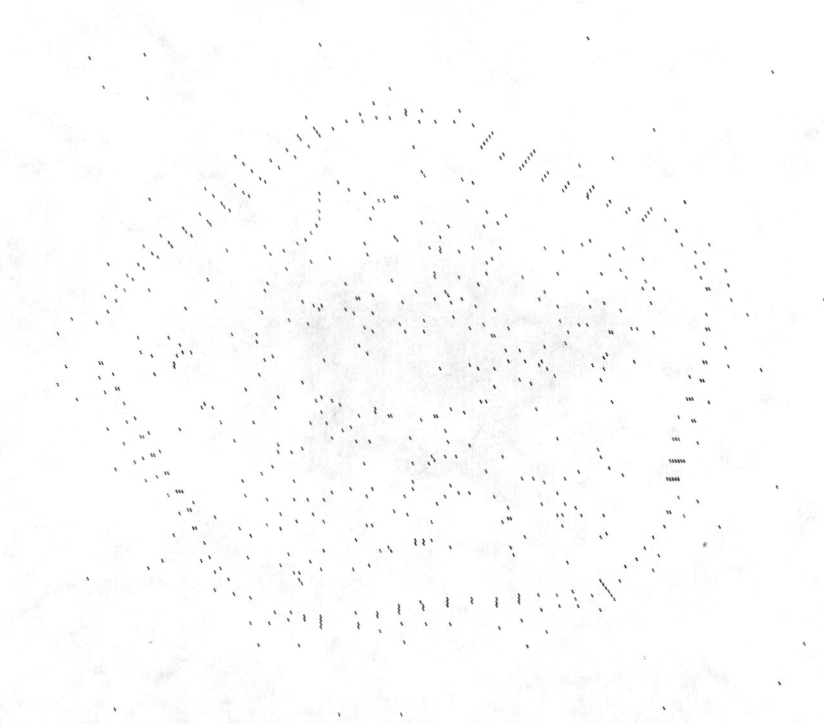

4-30 z-fifth Pickover biomorph <1.8, 1.8, 0.5, 0.75>.

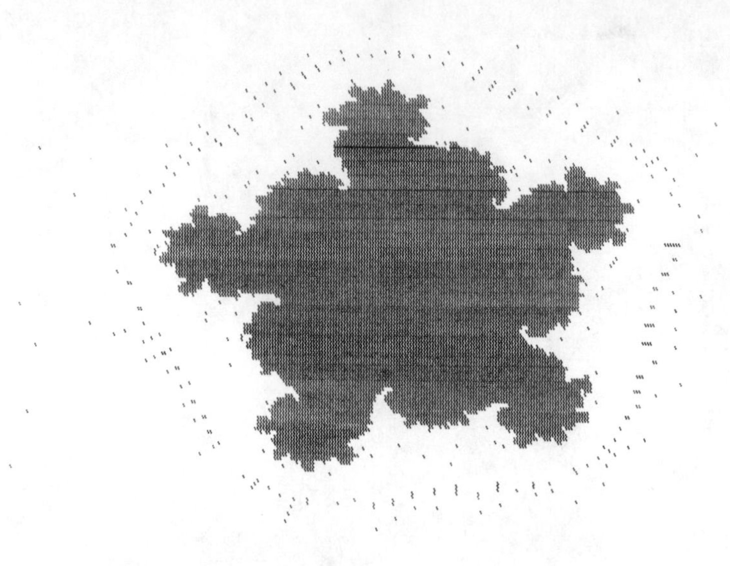

4-31 z-fifth Pickover biomorph <2.0, 2.0, 0.5, 0.6>.

4-32 z-fifth Pickover biomorph <2.5, 2.5, 0.5, 0.5>.

4-33 z-fifth Pickover biomorph <5, 5, 1, 1>.

You could examine huge numbers of species at many powers of magnification in these two biomorph families. After you finish that study, you might examine the following organisms also suggested by Pickover (1990).

$$Z_{t+1} = Z_t^{Z_t} + Z^5 + C$$
$$Z_{t+1} = \sin(Z_t) + Z^2{}_t + C$$
$$Z_{t+1} = Z_t^{Z_t} + Z^6 + C$$
$$Z_{t+1} = \sin(Z_t) + e^{Z_t} + C$$

Remember, in each of these cases, Z is a complex number.

Ural (1990) gives Pascal code for most of these operations on complex numbers. These fractal animals must be related to iterated function systems, which are also fractal. There probably exists a mathematical mapping from one relation to the other.

Dawkins biomorphs

Dawkins (1989) elaborated on Dawkins (1986) on another type of biomorph. These organisms are very similar to L-systems in that they are produced by line segments. By the recursive drawing of these organisms, they also resemble iterated-function systems. Dawkins' objective was to use a computer display to show organisms of constrained morphology and to model the phenotopic changes by small mutations in the genotype. The resulting program has a gene code with a few parameters. By changing one of these parameters, one can generate structures resembling trees or insects.

My program dawkins8.pas is a spinoff of l6.c. I shifted to Pascal as the programing language because complex data structures, like stacks and linked lists, are more easily implemented in Pascal.

The program dawkins8.pas has been designed to draw Dawkins type biomorphs as described in The Blind Watch Maker (Dawkins, 1987).

The program uses the Turbo Pascal units, crt, and graphics. It also uses the custom unit, genecode, which will be examined first.

The unit genecode contains software code for a stack structure, as explained in most data structures books (Stubbs and Webre, 1989). The unit contains procedures that allow the creation, manipulation, and destruction of stack-nodes consisting of records of genecode parameters. Each Pascal record is called a chromosome record and contains the following parameters:

direction:	integer	Allowed values +1 and –1
x:	integer	x coordinate on screen
y:	integer	y coordinate on screen
newx:	integer	New x value computed
newy:	integer	New y value computed
height:	integer	Initial height of the organism
angle:	integer	Initial angle for branches
length:	integer	Branch length change
blength:	integer	A computed variable

Each node in the stack contains pointers to a chromosome record and a link to another stack node.

The procedure create( ) is used to create the first node for the stack. The bottom of the stack points to nil. The function empty( ) returns a boolean value for the stack being empty or not.

The procedure push is used to push a chromosome record onto the stack. This is done by using the new( ) command to create a new stack node. The element of the node is assigned to point to a chromosome record, and then the node link is connected to the stack and set as the top of the stack. When a node is removed from the stack, it is popped by reading the element pointer and then moving the link pointer and disposing of the node.

The program dawkins8.pas declares a constant called rad_factor that equals 57.2957795. This factor is used in conversion from angular degrees to radians. The variables declared are gene (of type chromosome), alpha (an integer used in angle computations), sleft and sright (two stacks), temp (an integer variable used in a swap calculation), heaptop (a pointer to word and a marker in the physical memory of the computer), done (a boolean type), and d (a simple integer).

Main starts in a repeat loop. The first procedure called is user_input. The procedure starts by allowing the user to enter a boolean value that determines whether or not this is the last run of the program. The user then enters the initial organism height (here called a tree, but insects are also possible), the angle of branches (legs, etc.), and the change in branch length.

The next procedure called is setup_graphics. As the name implies, this procedure sets up the graphics driver. The procedure mark(heaptop) is use to allocate memory by recording a pointer at the top of the memory heap. By a call to release(heaptop), the entire block of memory can be set free for other use.

A call to seed plants the first pixel on the screen and computes the next pixel position. The embryonic organism is then started by first selecting a +1 direction and creating a stack called sright. The computed gene record is then pushed onto the stack, and morphogenic development of the organism begins in the first while loop.

While the length of the parameter gene.blength is > 0, the loop is executed. The first procedure called in this while loop is branch( ) and requires four arguments—x, y, newx, newy—and defines a line. The procedure itself just draws the line segment. Still in the while loop, the new branch is now computed. Gene.direction is still +1, gene.blength is decreased by gene.length, alpha is an angle and is increased by gene.angle. Gene.direction will determine if this angle increase is positive or negative. The current (x,y) position is assigned to the (newx,newy) position, and newx and newy positions are computed by calls to the procedures x_position and y_position. These new gene parameters are now pushed into the stack by a call to push( ), thus ending the while loop.

So what is going on in this loop is that a branch is drawn from (x,y) to (newx,newy). The coordinates (newx,newy) are now assigned to the coordinates (x,y), and newx and newy coordinates are computed. This new branch length will be decreased in length by the amount set in the gene code, and this position and code is remembered by pushing into the stack. The next branch is then computed and the process repeated until the branch length reaches 0. The top of the stack is then popped to represent the next starting pixel for the organism development. This might result in several pushings until the branch length is again 0 and a new gene node is popped from the stack.

This computing, drawing, pushing and popping is repeated until the stack is empty. At this time, one fifth of the organism has grown. Four more such stacks are operated on until the organism is fully grown. The only difference in the five growth functions are the direction and angle computation. In order to see the individual parts and the effects they have on the organism morphogenesis, I recommend commenting out one or more of these big while loops and compiling and running the program. The program then calls the procedure pause to hold for display until a key is pressed when the memory is released and graphics cleared.

This program could be modified for automatic modification of individual genes. These modifications could be selected by the user to allow selective breeding, just as Dawkins did in his programs. The structures could be drawn using recursive functions calls rather than stacks. The two approaches are equivalent.

The photos of constrained morphology biomorphs are shown in Photos 4-1 through 4-35. These photos were made from the program dawkins8.pas. The program prompts the user to input the initial height of the organism (tree height), the angle of the branches, and the change in branch length. In all these photos, the initial height was set at 60, the change in branch length was set at 10, and the angle was changed in 10 degree increments starting at 10 and ending at 350 degrees. By using this approach, you'll certainly constrain the morphology more than necessary, but you also can clearly observe the effects of changing one gene. You might think of these photos as representing a portion of gene space.

Relevance to artificial life

How do these biomorphs fit into the big picture of ALife? These biomorphs are all very simple algorithms or rule sets that give rise to complex morphologies. One of these biomorph algorithms is already being used in design of robot hardware.

Holland and Snath (1991) describe using the Dawkins constrained morphologies to develop a neural network controller for a legged robot. The quadrapedal walking robot and its neural system is described in Snath and Holland (1991).

The iterated function systems are a computer-bound problem. Stark (1991) has shown that neural networks can be mapped into IFS, thus allowing IFS generation by a massively parallel processor such as a neural network. This algorithm might have relevance to genetic evolving hardware robots with neural network controllers. The algorithms could generate different robot morphologies.

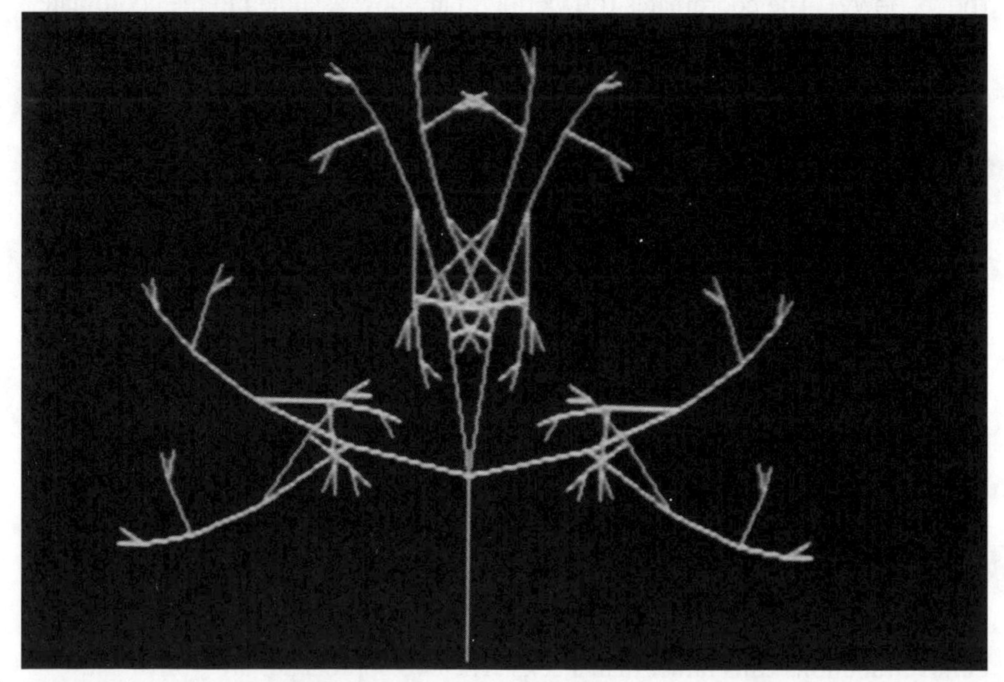

Photo 4-1 Dawkins biomorph <60, 10, 10>.

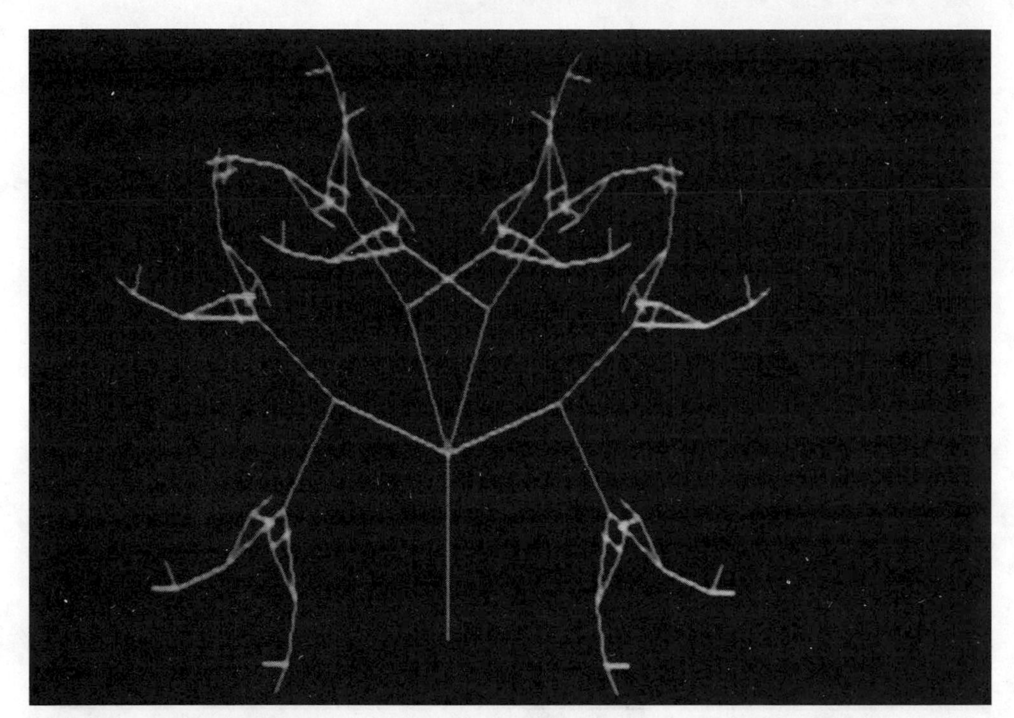

Photo 4-2 Dawkins biomorph <60, 20, 10>.

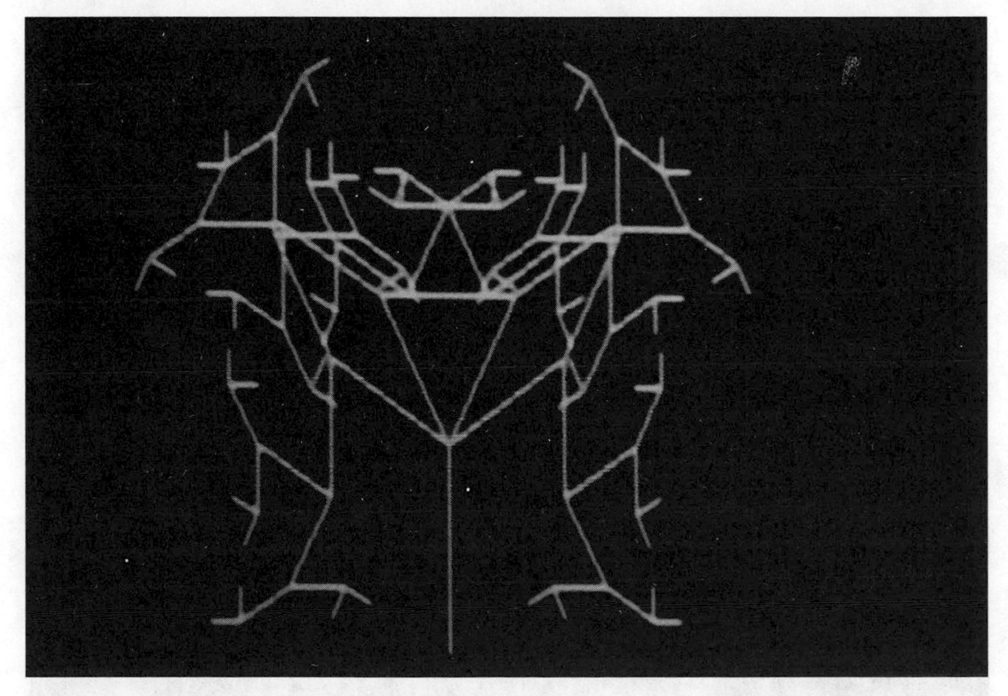

Photo 4-3 Dawkins biomorph <60, 30, 10>.

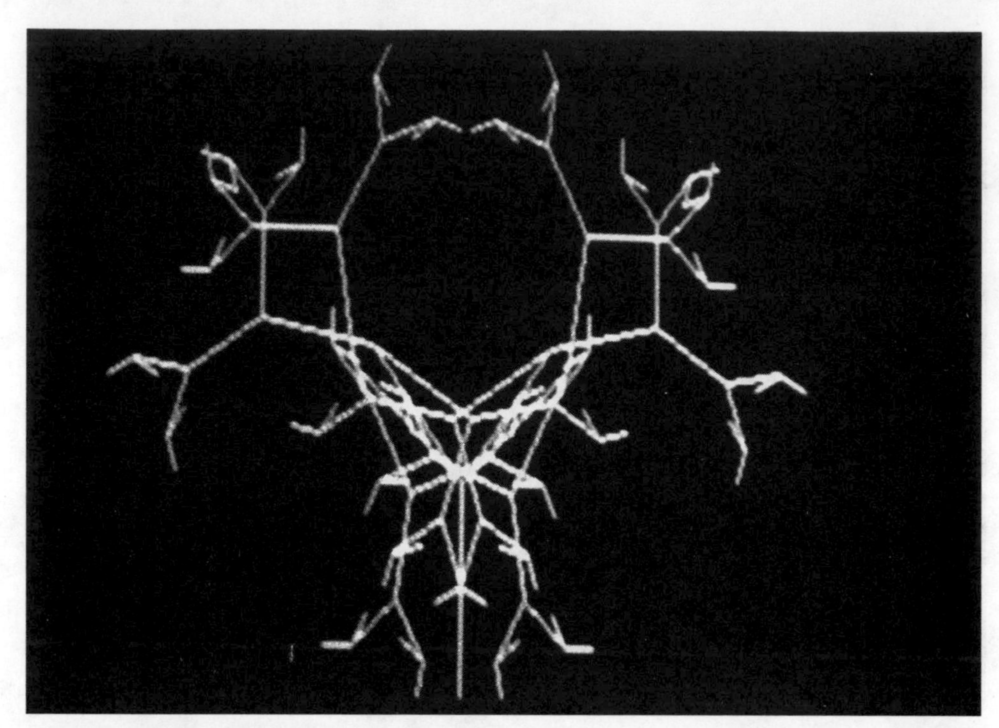

Photo 4-4 Dawkins biomorph <60, 40, 10>.

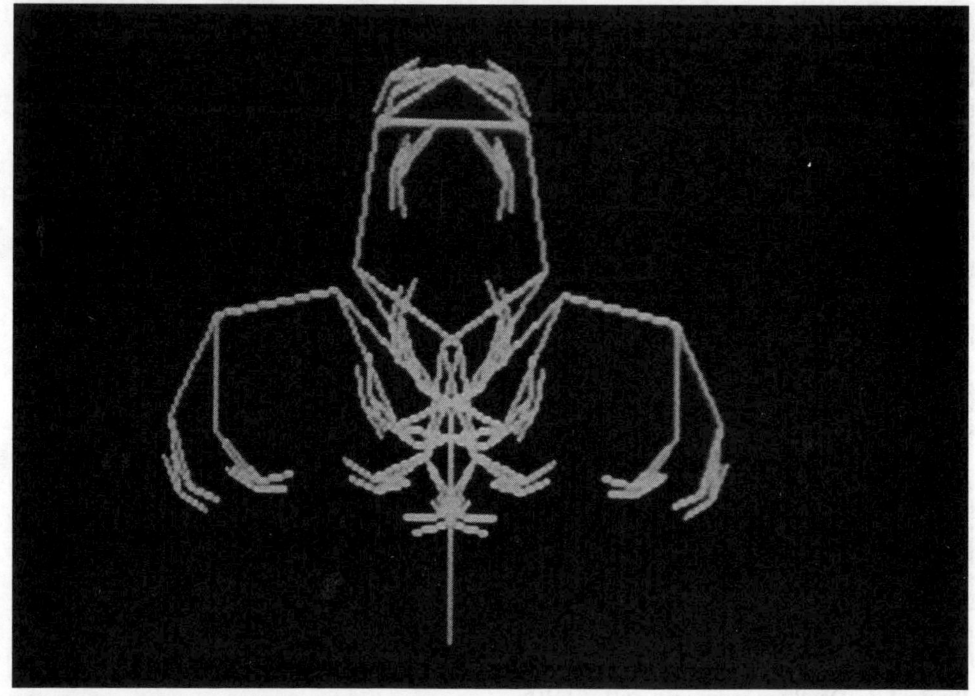

Photo 4-5 Dawkins biomorph <60, 50, 10>.

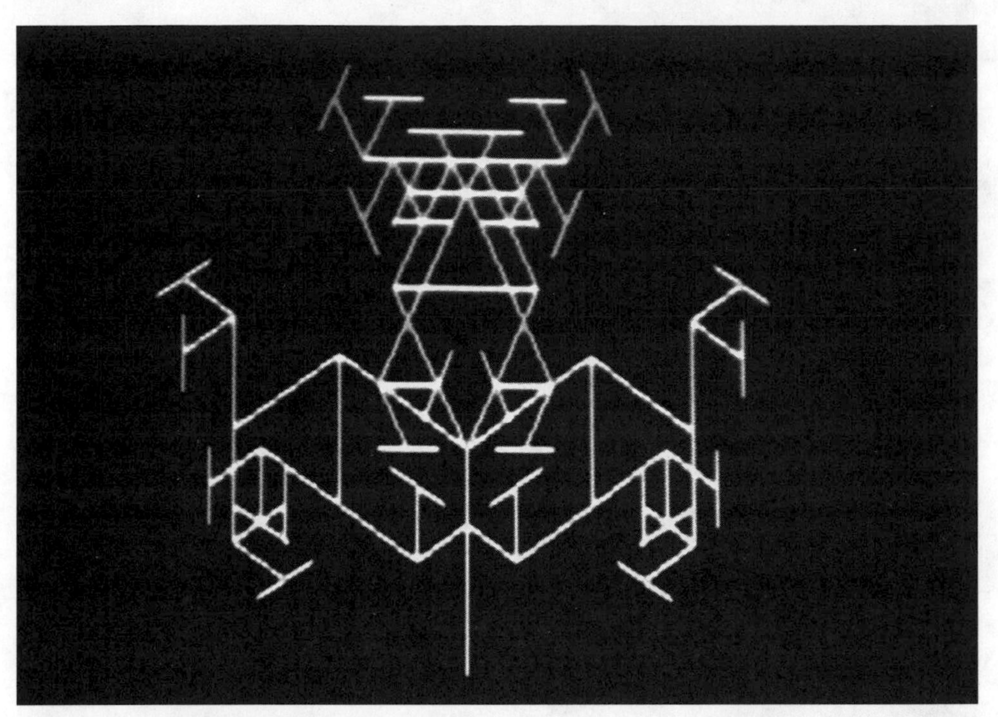

Photo 4-6 Dawkins biomorph <60, 60, 10>.

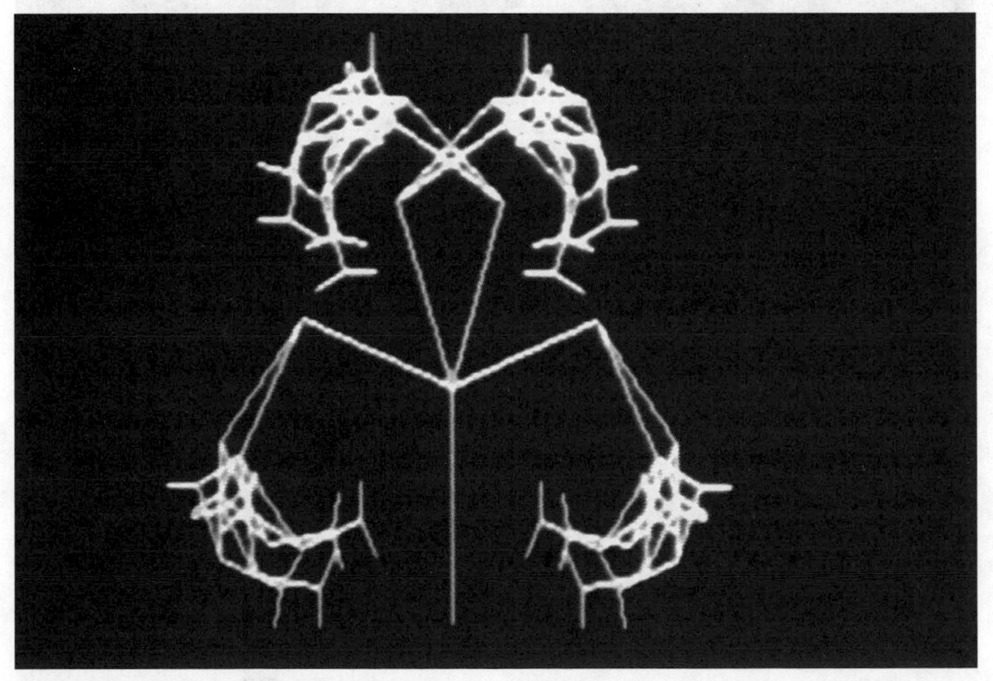

Photo 4-7 Dawkins biomorph <60, 70, 10>.

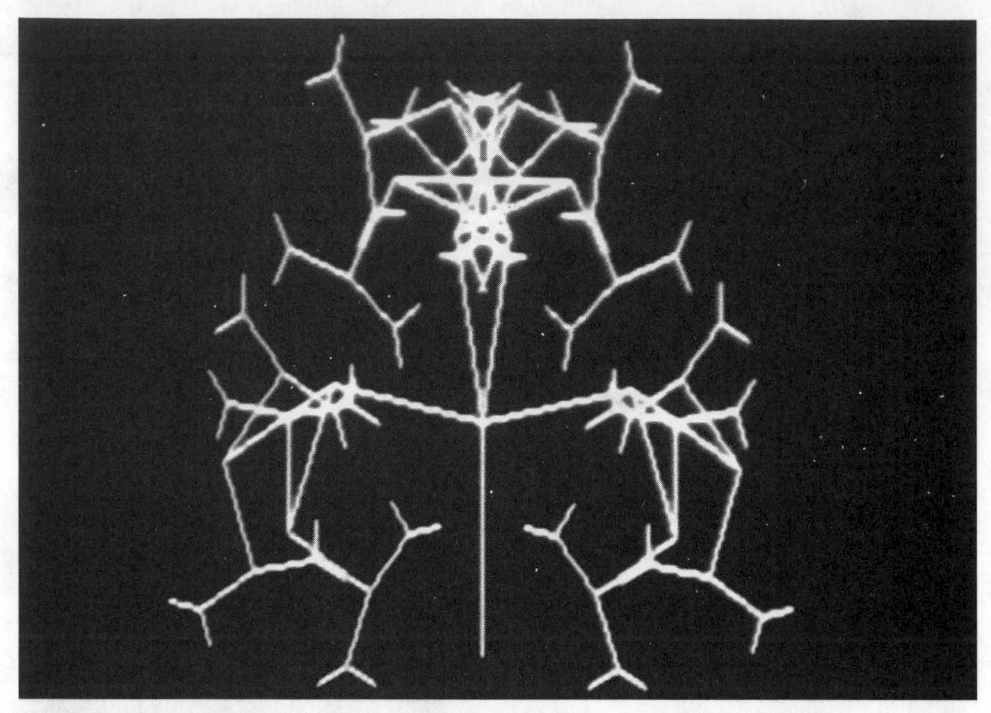

Photo 4-8 Dawkins biomorph <60, 80, 10>.

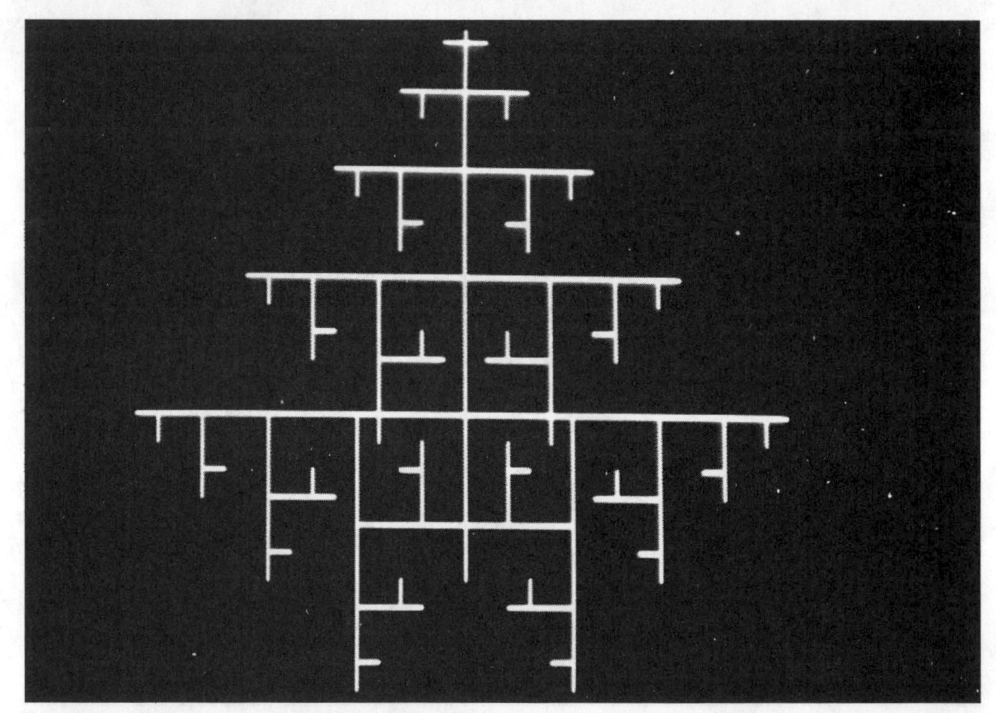

Photo 4-9 Dawkins biomorph <60, 90, 10>.

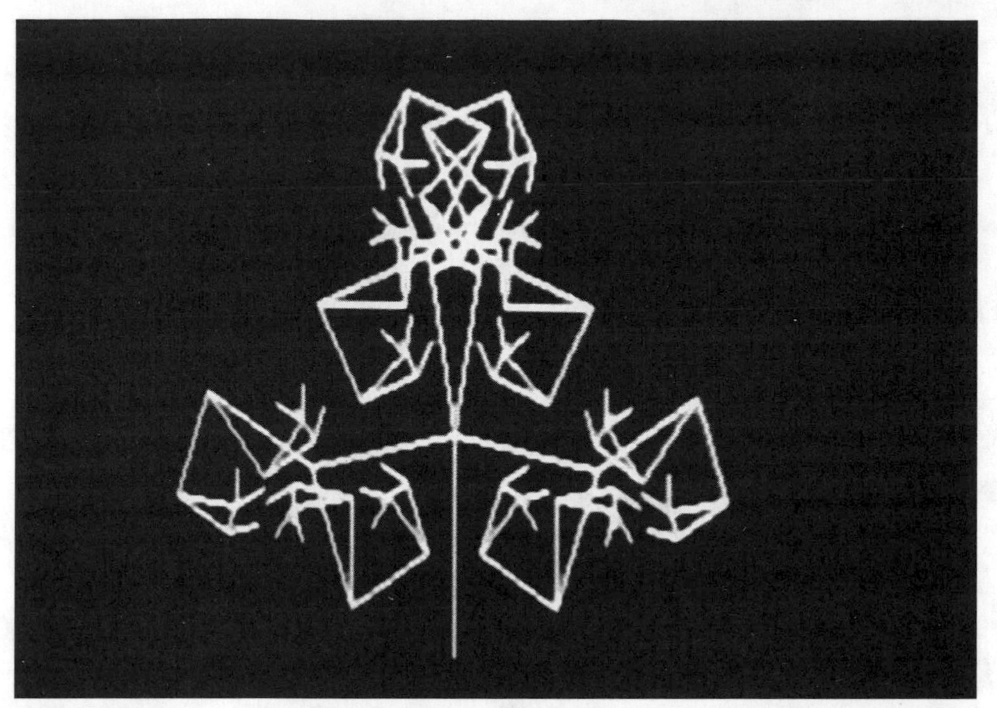

Photo 4-10 Dawkins biomorph <60, 100, 10>.

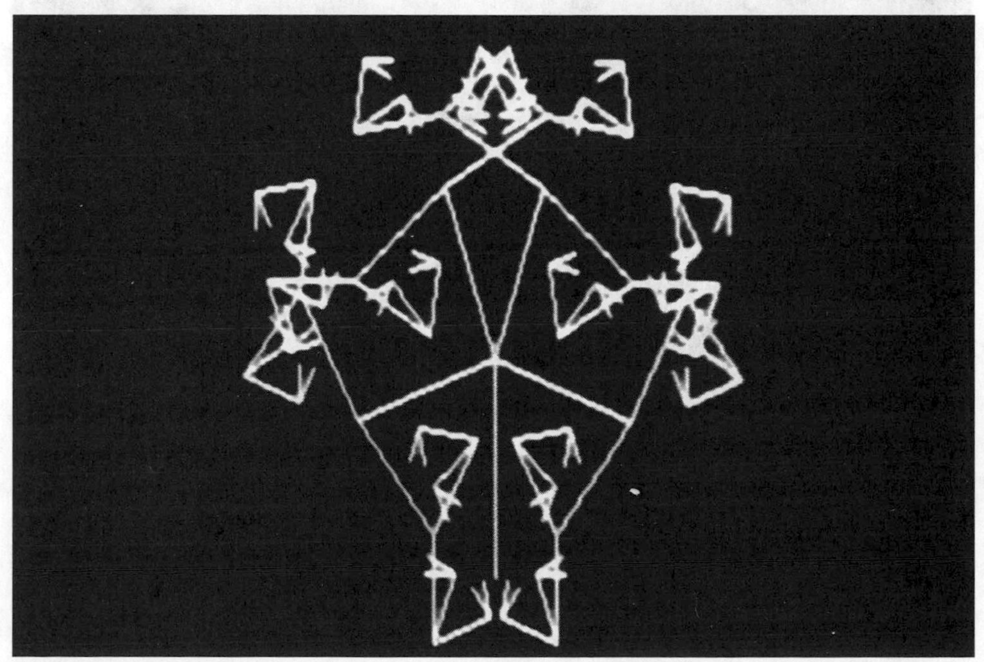

Photo 4-11 Dawkins biomorph <60, 110, 10>.

Photo 4-12 Dawkins biomorph <60, 120, 10>.

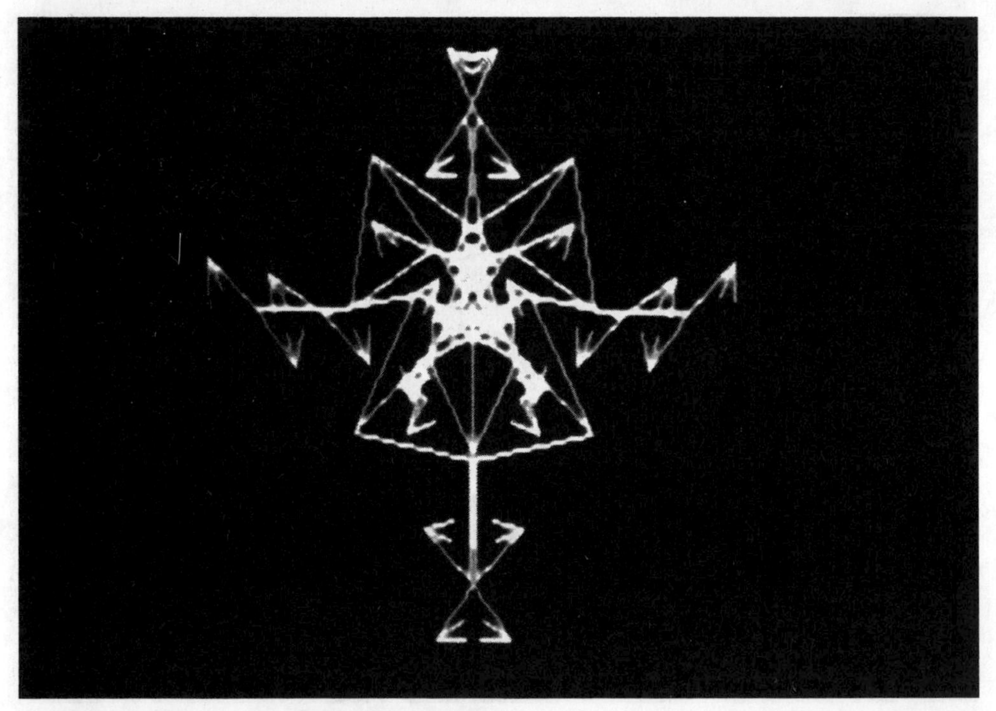

Photo 4-13 Dawkins biomorph <60, 130, 10>.

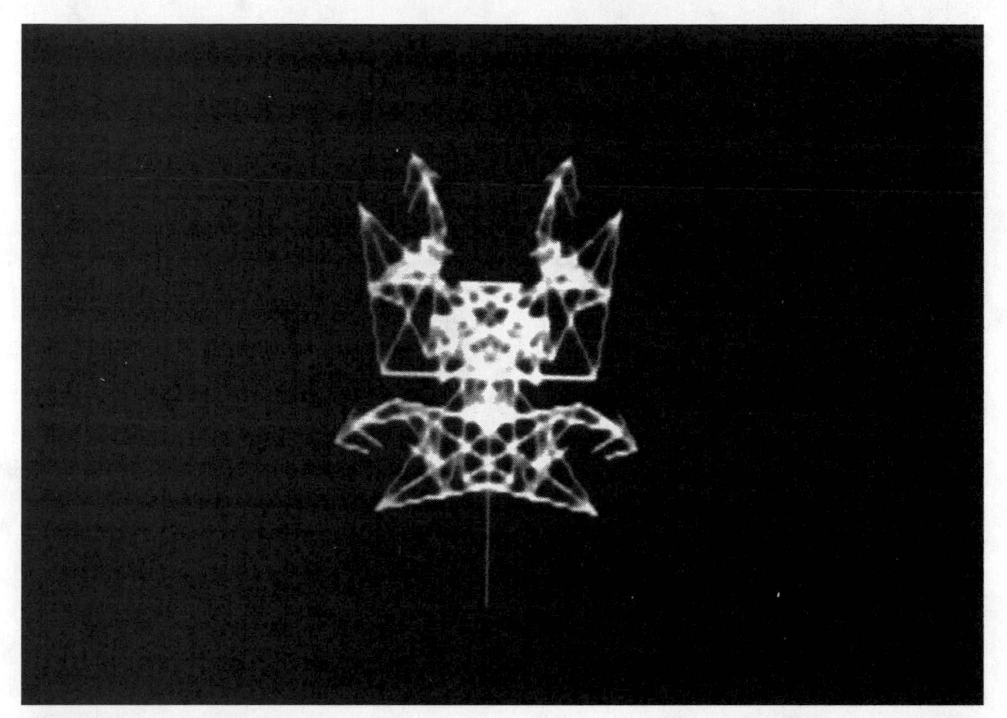

Photo 4-14 Dawkins biomorph <60, 140, 10>.

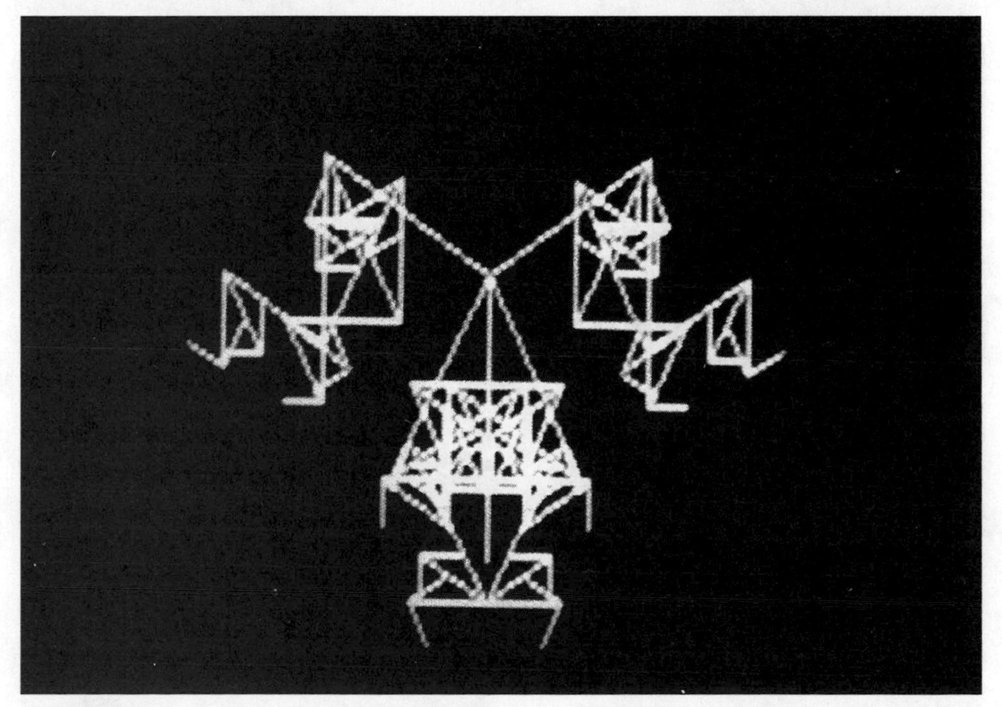

Photo 4-15 Dawkins biomorph <60, 150, 10>.

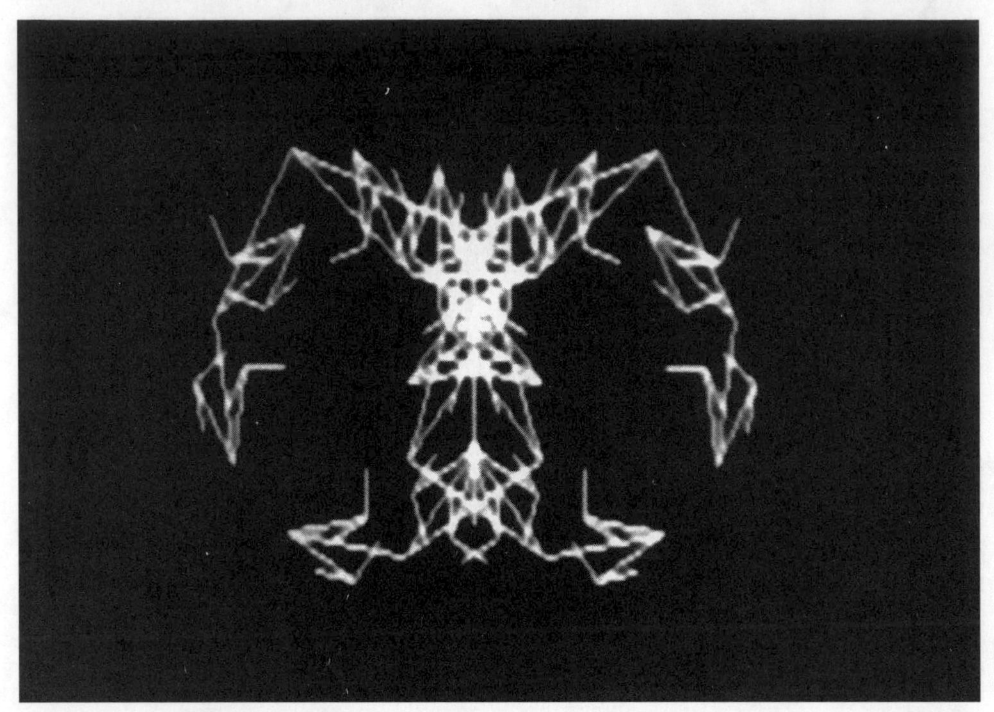

Photo 4-16 Dawkins biomorph <60, 160, 10>.

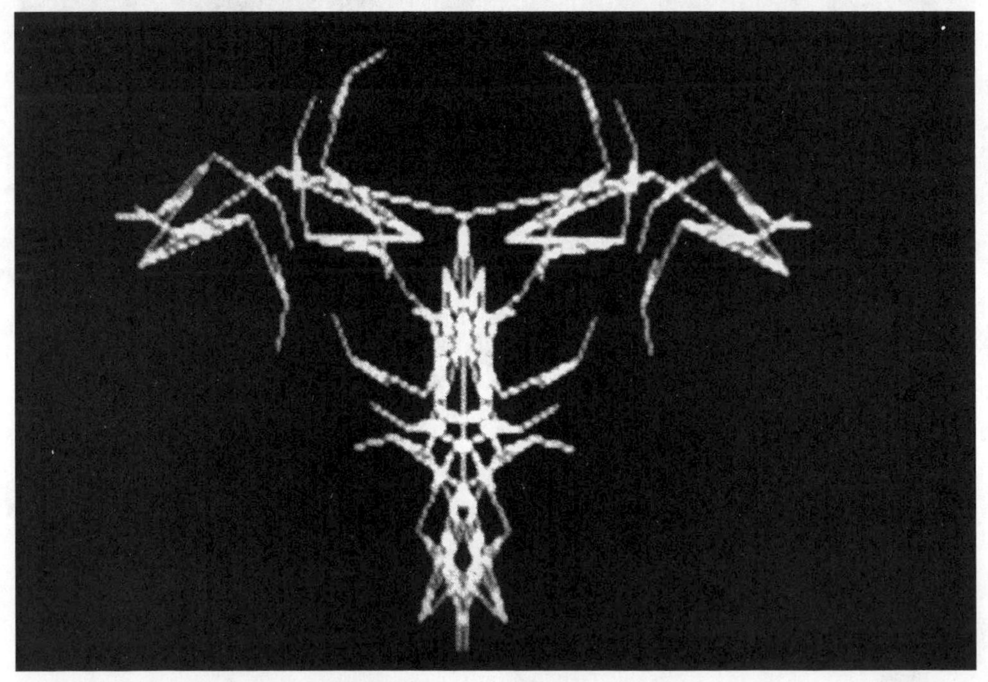

Photo 4-17 Dawkins biomorph <60, 170, 10>.

Photo 4-18 Dawkins biomorph <60, 180, 10>.

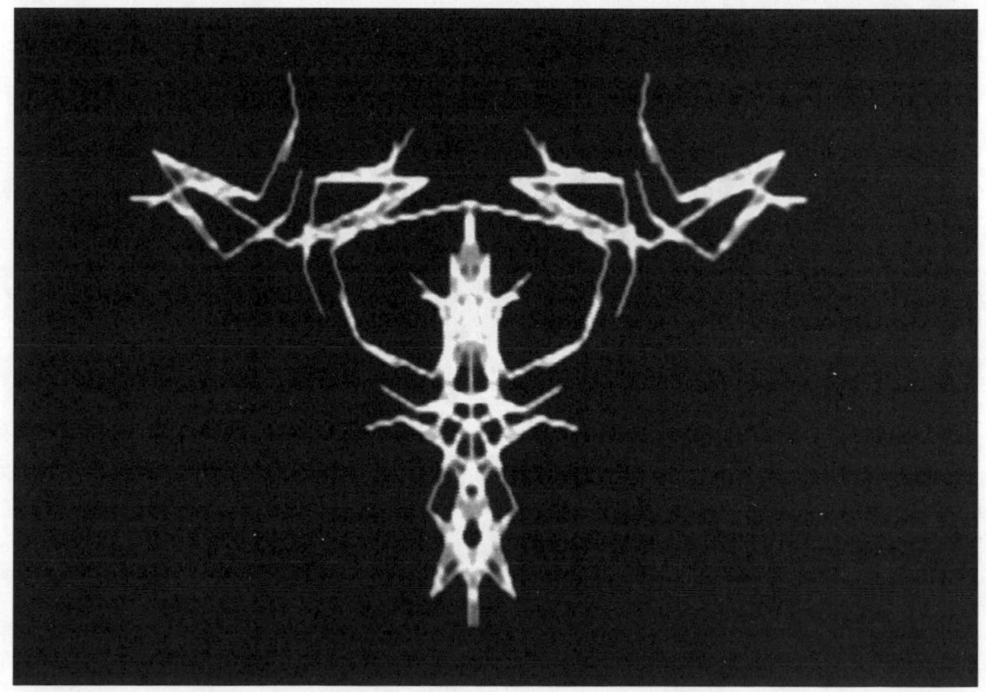

Photo 4-19 Dawkins biomorph <60, 190, 10>.

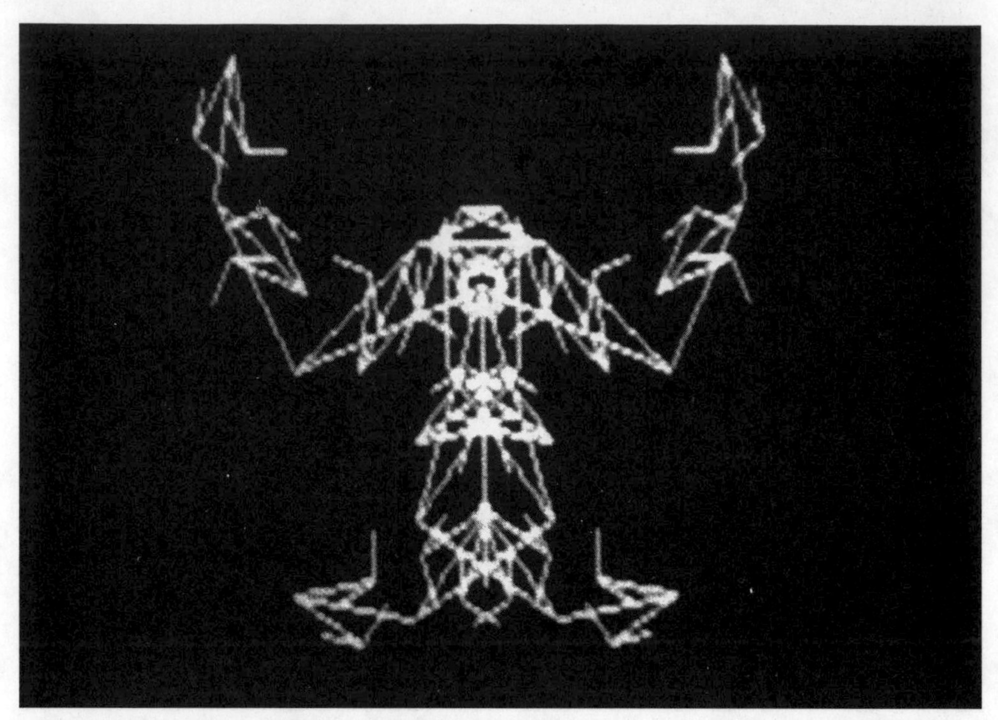

Photo 4-20 Dawkins biomorph <60, 200, 10>.

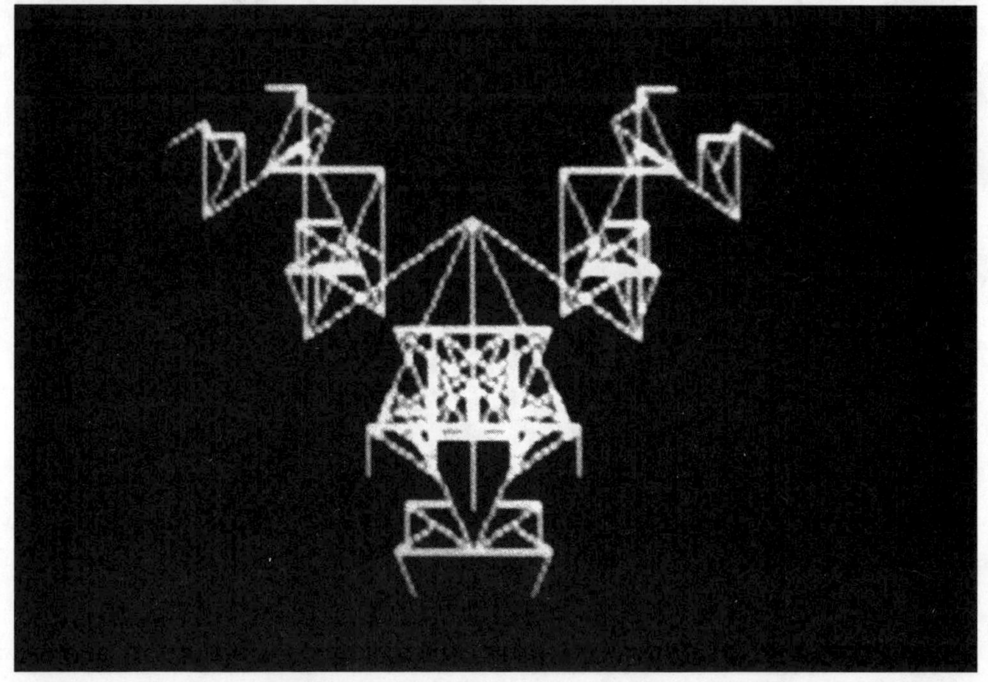

Photo 4-21 Dawkins biomorph <60, 210, 10>.

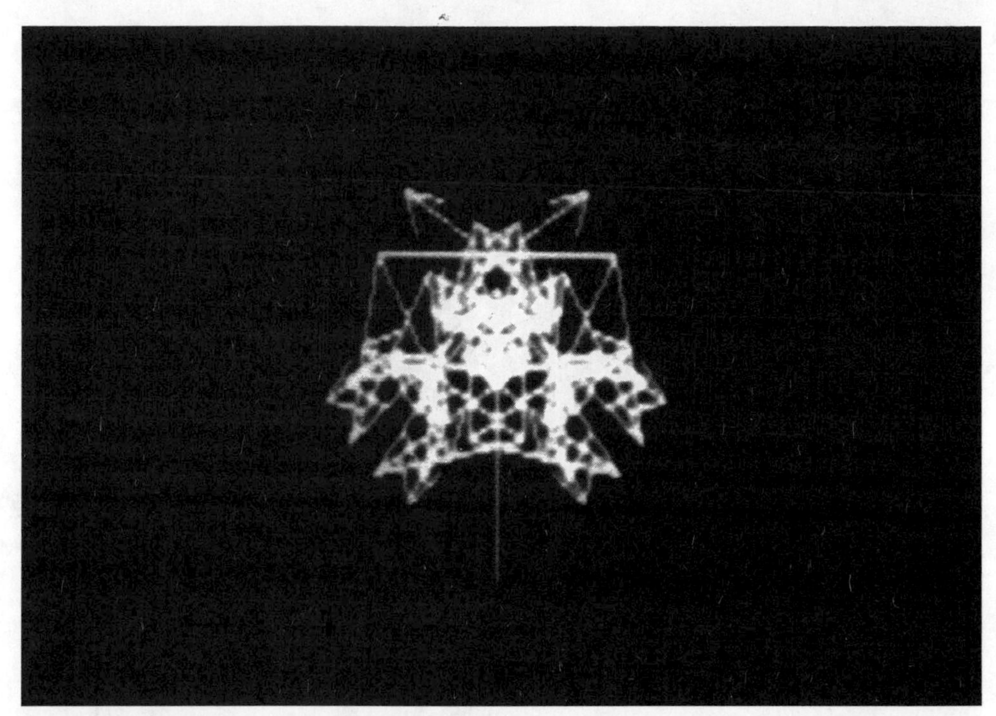

Photo 4-22 Dawkins biomorph <60, 220, 10>.

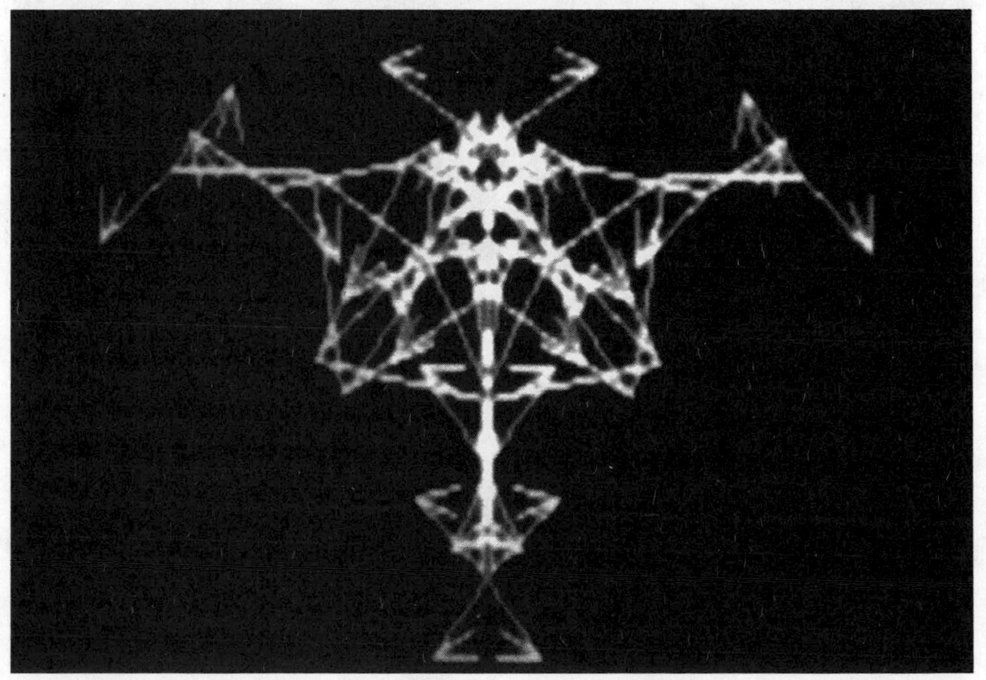

Photo 4-23 Dawkins biomorph <60, 230, 10>.

Photo 4-24 Dawkins biomorph <60, 240, 10>.

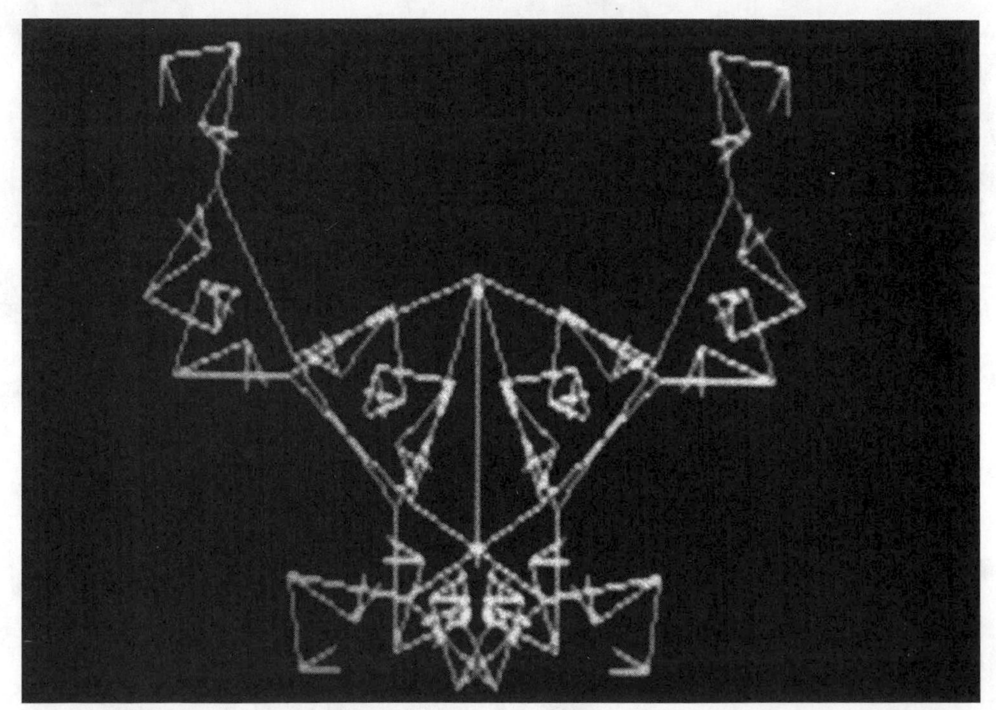

Photo 4-25 Dawkins biomorph <60, 250, 10>.

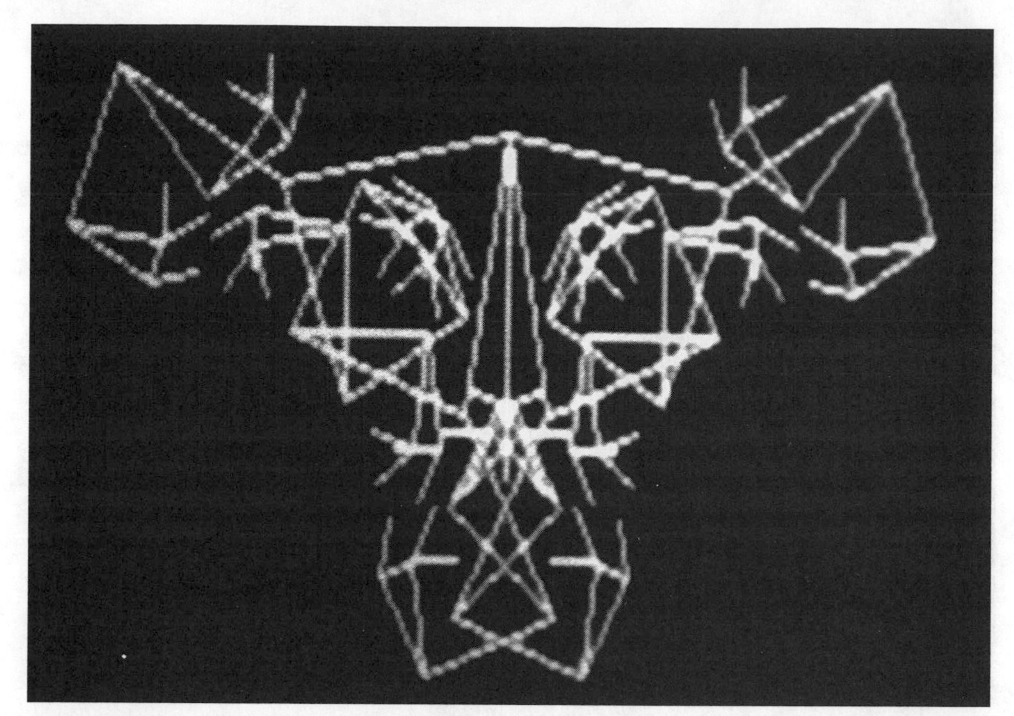

Photo 4-26 Dawkins biomorph <60, 260, 10>.

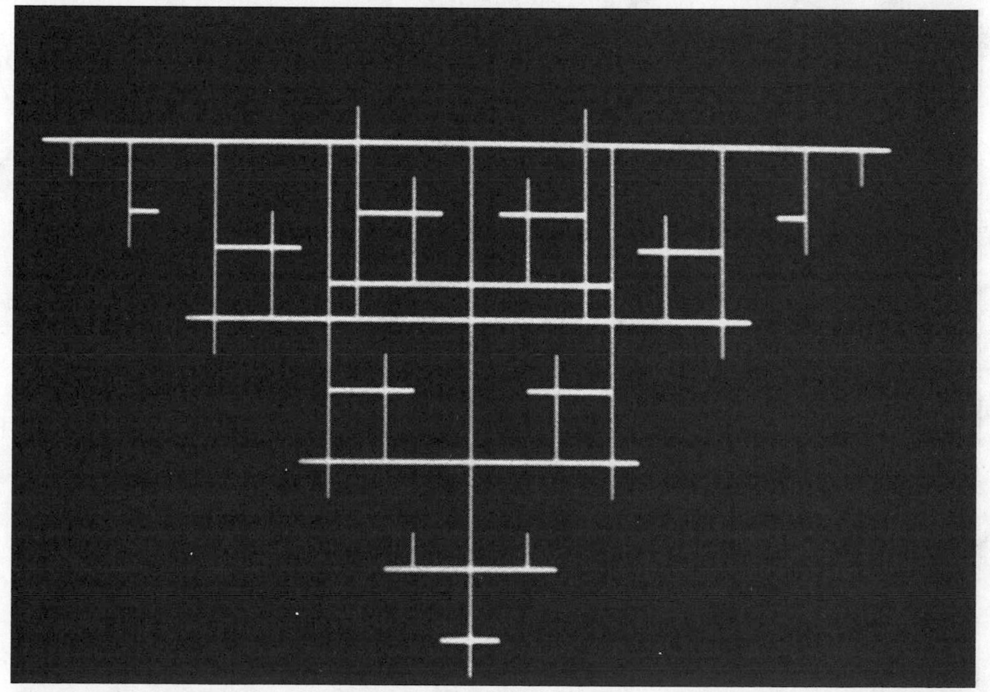

Photo 4-27 Dawkins biomorph <60, 270, 10>.

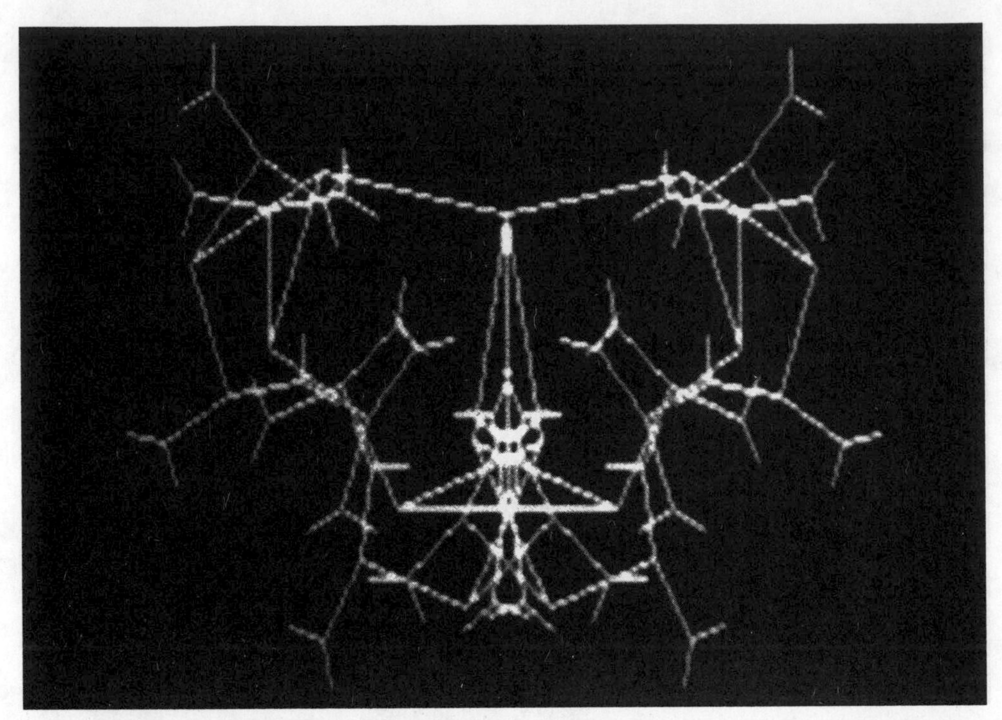

Photo 4-28 Dawkins biomorph <60, 280, 10>.

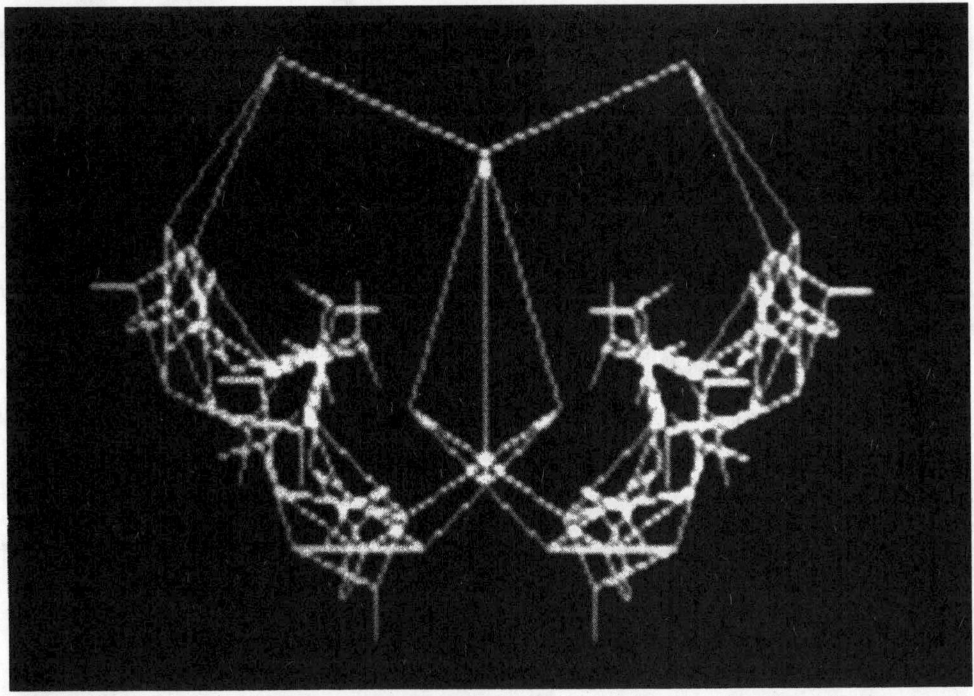

Photo 4-29 Dawkins biomorph <60, 290, 10>.

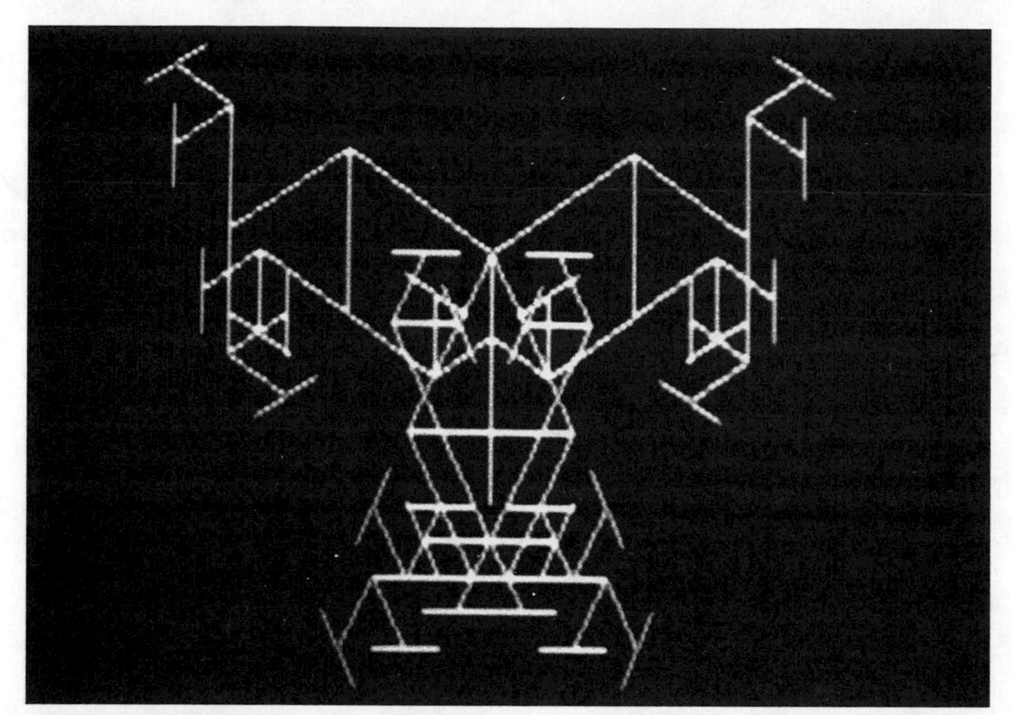

Photo 4-30 Dawkins biomorph <60, 300, 10>.

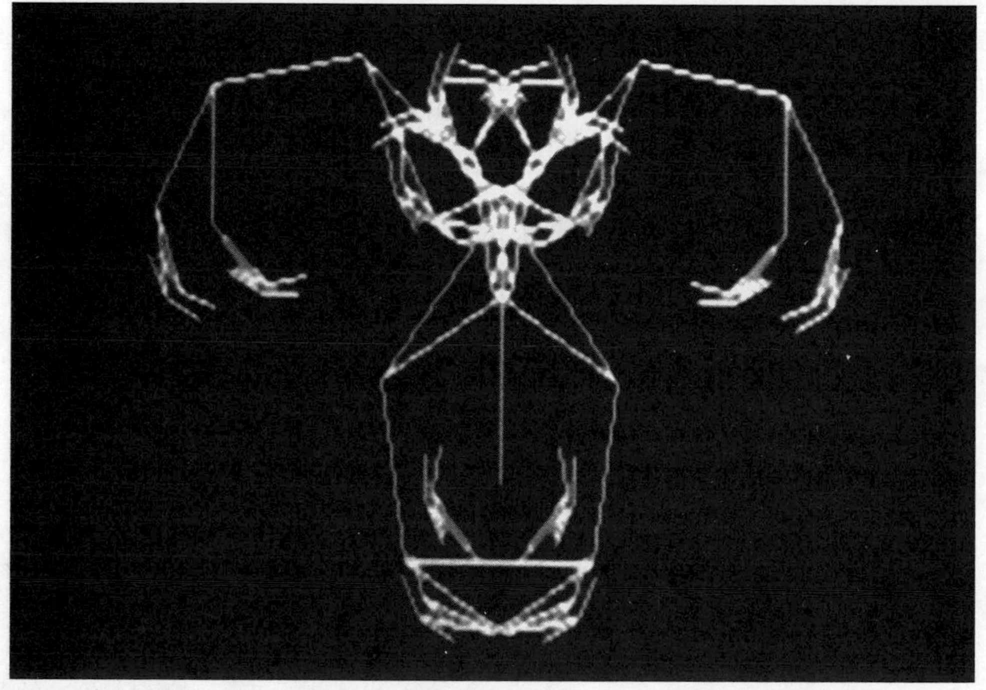

Photo 4-31 Dawkins biomorph <60, 310, 10>.

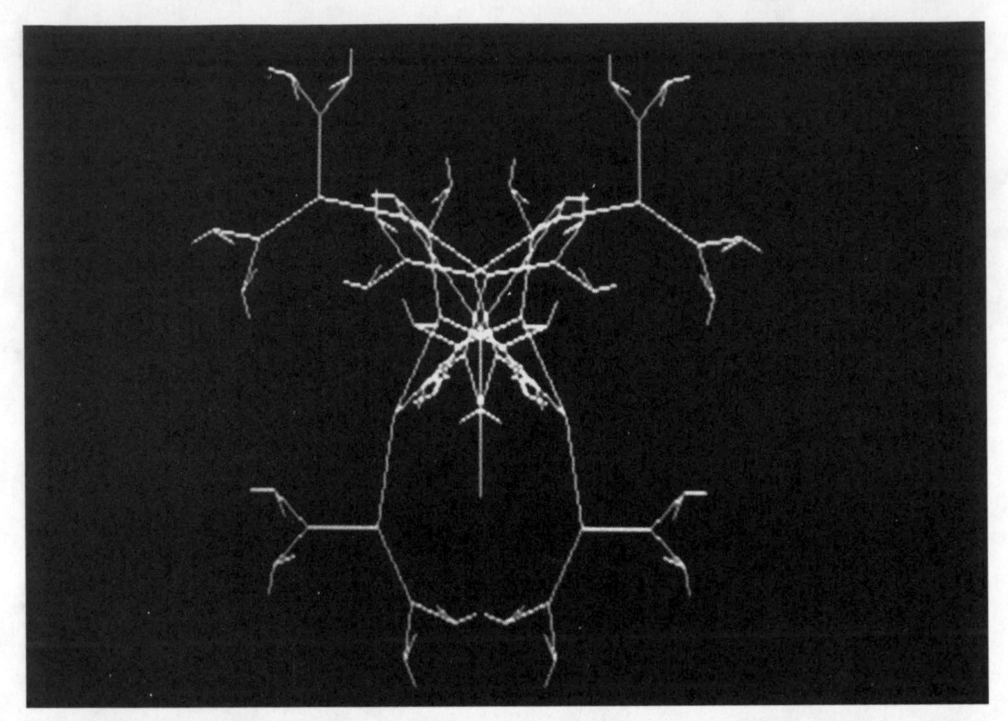

Photo 4-32 Dawkins biomorph <60, 320, 10>.

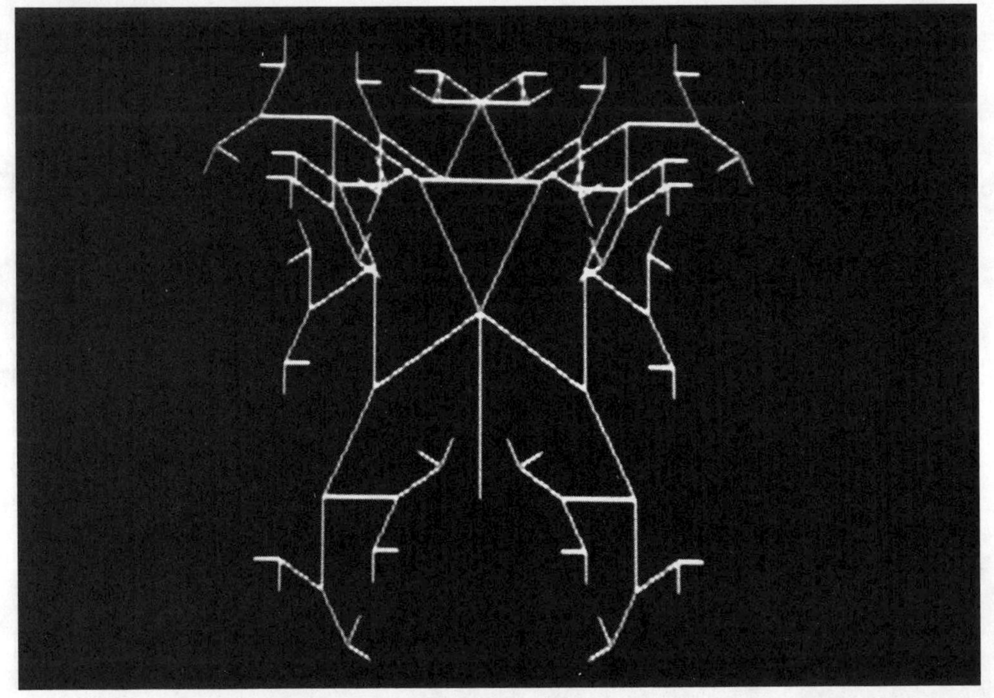

Photo 4-33 Dawkins biomorph <60, 330, 10>.

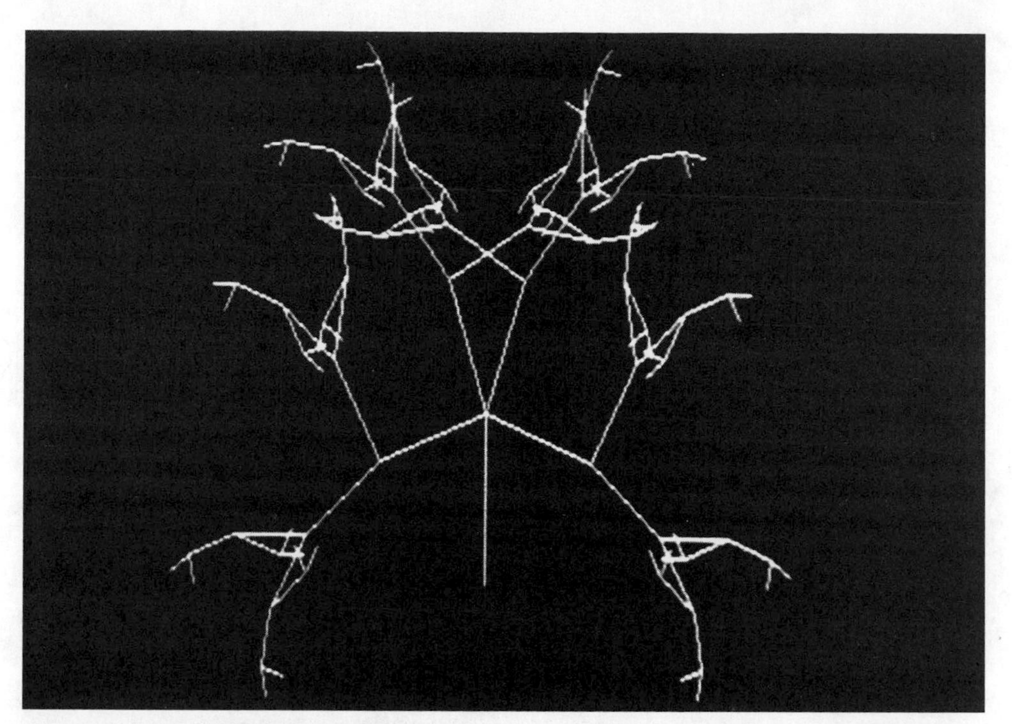

Photo 4-34 Dawkins biomorph <60, 340, 10>.

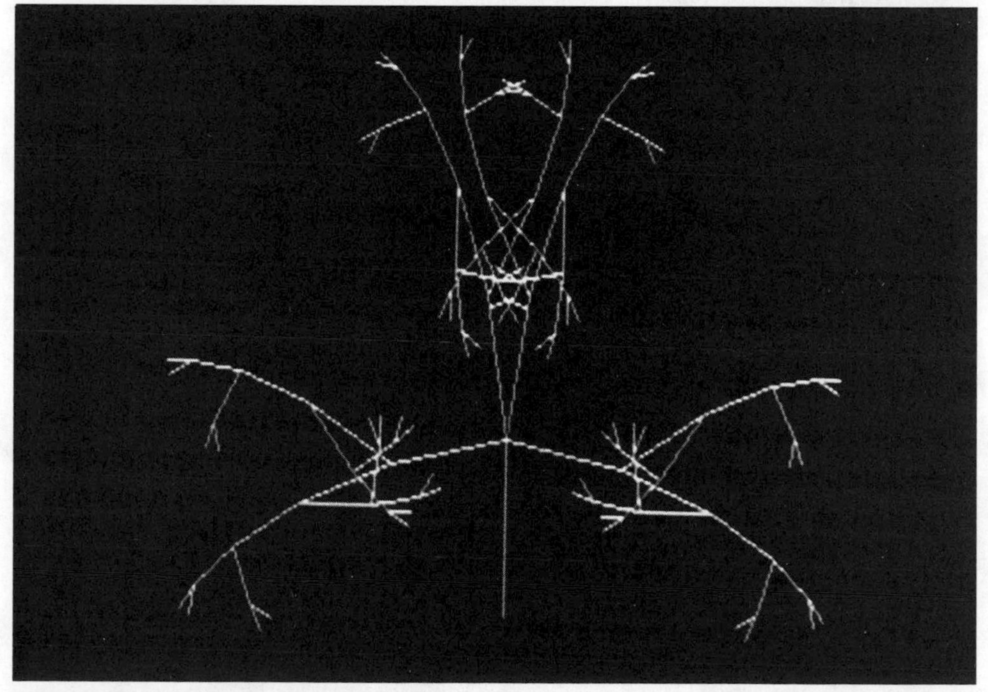

Photo 4-35 Dawkins biomorph <60, 350, 10>.

dawkins8.pas

```
(* program designed to study
   Dawkins' biomorphs.
   See: The Blind Watch Maker *)

{$N+} (* numeric processing *)
{$R+} (* range checking *)
{$S+} (* stack checking *)

program dawkins8;

uses genecode,crt,graph;

const
     rad_factor = 57.2957795;


var gene:chromosome; (* in genecode unit *)
    alpha:integer;
    sleft:node_type; (* left stack *)
    sright:node_type; (* right stack *)
    temp:integer;
    heaptop:^word;
    done:boolean;
    d:integer;


procedure user_input;

begin
     writeln('last run? enter 0=yes, 1=no');
     readln(d);
     done := true;
     if(d=1)then done:=false;
     writeln('enter the desired height of tree ');
     readln(gene.height);
     writeln('enter the angle of branches ');
     readln(gene.angle);
     writeln('enter the desired change in branch length ');
     readln(gene.length)
end; (* user input *)


procedure setup_graphics;

var
   grDriver:integer;
   grMode:integer;
   ErrCode:integer;

begin
     grDriver := Detect;
     InitGraph(grDriver,grMode,'');
     ErrCode := GraphResult;
     if ErrCode <> 0 then
         writeln('graphics error');
end; (* end graphics setup *)


procedure x_position(var x,length,angle:integer);
begin
     gene.newx := x - round(length*cos(angle/rad_factor));
end;


procedure y_position(var y,length,angle:integer);
begin
```

```
     gene.newy := y - round(length*sin(angle/rad_factor));
end;


procedure branch(x,y,newx,newy:integer);
begin

     x := gene.x;
     y := gene.y;
     newx := gene.newx;
     newy := gene.newy;

     line(x,y,newx,newy);

end; (* end tree *)


procedure pause;
begin
     readln;
     release(heaptop);
     closegraph;
end; (* end pause *)


procedure seed;
begin

      gene.x := 320;
      gene.y := 175 + round(gene.height/2);
      gene.newx := gene.x;
      gene.newy := 175 - round(gene.height/2);

      alpha := 90;
      gene.blength := gene.height;

end;


(* procedure main *)
begin


repeat


      user_input;

      setup_graphics;

      mark(heaptop);



      seed;


      gene.direction := 1;
      create(sright); (* stack for right branches *)
      push(sright,gene);


      (* while stack not empty *)
      while not empty(sright) do
      begin
```

```
     (* draw tree *)
     while(gene.blength > 0) do
     begin

          (* draw branch *)
          branch(gene.x,gene.y,gene.newx,gene.newy);


          (* compute next branch *)
          gene.direction := 1;
          gene.blength := gene.blength - gene.length;

          alpha := alpha + gene.direction*gene.angle;
          alpha := alpha mod 360;

          gene.x := gene.newx;
          gene.y := gene.newy;

          x_position(gene.newx,gene.blength,alpha);
          y_position(gene.newy,gene.blength,alpha);


          (* push gene into right chromosome *)
          push(sright,gene);

     end; (* end while to draw tree *)

     pop(sright,gene);

end; (* end while stack *)


seed;


gene.direction := -1;
create(sleft); (* stack for left branches *)
push(sleft,gene);


(* while stack not empty *)
while not empty(sleft) do
begin

     (* draw tree *)
     while(gene.blength > 0) do
     begin

          (* draw branch *)
          branch(gene.x,gene.y,gene.newx,gene.newy);


          (* compute next branch *)
          gene.direction := -1;
          gene.blength := gene.blength - gene.length;

          alpha := alpha + gene.direction*gene.angle;
          alpha := alpha mod 360;

          gene.x := gene.newx;
          gene.y := gene.newy;

          x_position(gene.newx,gene.blength,alpha);
          y_position(gene.newy,gene.blength,alpha);
```

```
          (* push gene into right chromosome *)
          push(sleft,gene);


     end; (* end while to draw tree *)

     pop(sleft,gene);

end; (* end while stack *)



seed;
alpha := 0; (* reset angle *)


gene.direction := 1;
create(sright); (* stack for right branches *)
push(sright,gene);

(* while stack not empty *)
while not empty(sright) do
begin

     (* draw tree *)
     while(gene.blength > 0) do
     begin


          (* draw branch *)
          branch(gene.x,gene.y,gene.newx,gene.newy);


          (* compute next branch *)
          gene.direction := 1;
          gene.blength := gene.blength - gene.length;

          alpha := alpha + gene.direction*gene.angle;
          alpha := alpha mod 360;

          gene.x := gene.newx;
          gene.y := gene.newy;

          x_position(gene.newx,gene.blength,alpha);
          y_position(gene.newy,gene.blength,alpha);


          (* push gene into right chromosome *)
          push(sright,gene);


     end; (* end while to draw tree *)

     pop(sright,gene);

end; (* end while stack *)




seed;
alpha := 180; (* reset alpha *)
gene.direction := -1;
create(sleft); (* stack for left branches *)
push(sleft,gene);


(* while stack not empty *)
while not empty(sleft) do
```

```
begin
     (* draw tree *)
     while(gene.blength > 0) do
     begin

          (* draw branch *)
          branch(gene.x,gene.y,gene.newx,gene.newy);


          (* compute next branch *)
          gene.direction := -1;
          gene.blength := gene.blength - gene.length;

          alpha := alpha + gene.direction*gene.angle;
          alpha := alpha mod 360;

          gene.x := gene.newx;
          gene.y := gene.newy;

          x_position(gene.newx,gene.blength,alpha);
          y_position(gene.newy,gene.blength,alpha);


          (* push gene into right chromosome *)
          push(sleft,gene);


     end; (* end while to draw tree *)

     pop(sleft,gene);
end; (* end while stack *)


(* By commenting out these next two sections
   it is possible to remove quadrilateral symmetry
   in the biomorphs.  The result is bilatrial symmetry
   with striking resemblances to real life forms. *)
{
seed;
alpha := 270; (* reset angle *)


gene.direction := 1;
create(sright); (* stack for right branches *)
push(sright,gene);

(* while stack not empty *)
while not empty(sright) do
begin

     (* draw tree *)
     while(gene.blength > 0) do
     begin


          (* draw branch *)
          branch(gene.x,gene.y,gene.newx,gene.newy);


          (* compute next branch *)
          gene.direction := 1;
          gene.blength := gene.blength - gene.length;

          alpha := alpha + gene.direction*gene.angle;
          alpha := alpha mod 360;

          gene.x := gene.newx;
          gene.y := gene.newy;
```

```
                x_position(gene.newx,gene.blength,alpha);
                y_position(gene.newy,gene.blength,alpha);

                (* push gene into right chromosome *)
                push(sright,gene);

            end; (* end while to draw tree *)

            pop(sright,gene);

        end; (* end while stack *)

        seed;
        alpha := 270; (* reset alpha *)

        gene.direction := -1;
        create(sleft); (* stack for left branches *)
        push(sleft,gene);

        (* while stack not empty *)
        while not empty(sleft) do
        begin

            (* draw tree *)
            while(gene.blength > 0) do
            begin

                (* draw branch *)
                branch(gene.x,gene.y,gene.newx,gene.newy);

                (* compute next branch *)
                gene.direction := -1;
                gene.blength := gene.blength - gene.length;

                alpha := alpha + gene.direction*gene.angle;
                alpha := alpha mod 360;

                gene.x := gene.newx;
                gene.y := gene.newy;

                x_position(gene.newx,gene.blength,alpha);
                y_position(gene.newy,gene.blength,alpha);

                (* push gene into right chromosome *)
                push(sleft,gene);

            end; (* end while to draw tree *)

            pop(sleft,gene);

        end; (* end while stack *)

    }

        pause; (* pause for display *)

    until done;

    end.  (* end of main *)
```

l1.c

```
#include "quickc\include\dos.h"
#include "quickc\include\float.h"
#include "quickc\include\stdio.h"
#include "quickc\include\math.h

void main()
{
        /* declarations */

        /* setup L system rules */
        /* A -> CB                */
        /* B -> A                 */
        /* C -> DA                */
        /* D -> C                 */


        /* write the words */
        printf("              A              \n");
        printf("             CB              \n");
        printf("            DACB             \n");
        printf("          CCBDACB            \n");
        printf("        DADAACCBDAA          \n");
        printf("    CCBCCBCBDADAACCBCB       \n");
        printf(" DADAADADAADAACCBCCBCBDADAADAA \n");

} /* end main */
```

l5.c

```
/* program to draw L system organisms */
/*            100691                  */

#include "\quickc\include\dos.h"
#include "\quickc\include\float.h"
#include "\quickc\include\stdio.h"
#include "\quickc\include\graph.h"
#include "\quickc\include\math.h"

#define BLACK 0


struct videoconfig vc;
char error_message[] = "this video mode is not suported";


void main()
{

        /* declarations */
        int point();
        int row,element;
        int color;
        int counter;
        int Ar = 1;   /* blue */
        int Al = 2;   /* red */
        int Br = 3;   /* white */
        int Bl = 0;   /* white */


        /* set up graphics driver */
        if (_setvideomode(_MRES4COLOR) == 0)
        {
                printf("%s\n",error_message);
                exit(0);
```

```
}
_getvideoconfig(&vc);
_clearscreen(_GCLEARSCREEN);
_selectpalette(3);
_setviewport(0,0,320,200);


/*        setup L system rules        */
/* Ar -> AlBr  */
/* Al -> BlAr  */
/* Br -> Ar    */
/* Bl -> Al    */

/* plant initial seed */
_setcolor(Ar);
_setpixel(0,0);

/* generate and display the organism */
for(row=0;row<200;row++)
{

        counter = 0;
        for(element=0;element<320;element++)
        {
                /* point to each element */
                color = point(row,element);

                if(color == Ar)
                {
                        if(element+counter>320) break;
                        _setcolor(Al);
                        _setpixel(element+counter,row+1);
                        counter++;
                        if(element+counter>320) break;
                        _setcolor(Br);
                        _setpixel(element+counter,row+1);
                        counter++;
                 }
                 else
                 if(color == Al)
                 {
                        if(element+counter>320) break;
                        _setcolor(Bl);
                        _setpixel(element+counter,row+1);
                        counter++;
                        if(element+counter>320) break;
                        _setcolor(Ar);
                        _setpixel(element+counter,row+1);
                        counter++;
                 }
                 else
                 if(color == Br)
                 {
                        if(element+counter>320) break;
                        _setcolor(Ar);
                        _setpixel(element+counter,row+1);
                        counter++;
                 }
                 else
                 if(color == Bl)
                 {
                        if(element+counter>320) break;
                        _setcolor(Al);
                        _setpixel(element+counter,row+1);
                        counter++;
                 }
                 else
                 {
                        if(element+counter>320) break;
```

```
                        _setcolor(BLACK);
                        _setpixel(element+counter,row+1);
                        counter++;
                    }

                if(kbhit())
                {
                        goto escape;
                }


            } /* next element */

        } /* next row */

        /* pause for display */
        for(;;)
        {
            if(kbhit())
            {
                goto escape;
            }
        }


        escape:
        /* clear screen & return control */
        _clearscreen(_GCLEARSCREEN);
        _setvideomode(_DEFAULTMODE);



} /* end main */


int point(x,y)
int x,y;
{
        union REGS r, *inregs, *outregs;

        inregs = &r;
        outregs = &r;

        r.x.cx = x;
        r.x.dx = y;
        r.h.ah = 13;

        int86(0x10, inregs, outregs);

        return((int) r.h.al);

} /* end point */
```

16.c

```
/*   program to study L system graphics    */
/*              102791                     */

#include "\quickc\include\dos.h"
#include "\quickc\include\float.h"
#include "\quickc\include\stdio.h"
#include "\quickc\include\graph.h"
#include "\quickc\include\math.h"

#define RAD_FACTOR 57.2957795 * 180/pi */
```

```
/* external structure */
struct gene_code
        {
                int height;
                int angle;
                int length;
        };


/* global variables */
struct gene_code organism;

struct videoconfig vc;
char error_message[] = "this video mode is not suported";


void main()
{

        /* declarations */
        void escape();
        void set_graphics();
        void draw_organism();
        void user();

        /* user input  of genetic code */
        user();


        /* set up the graphics screen */
        set_graphics();


        /* generate and display the organism */
        draw_organism();



        /* exit program */
        escape();


} /* end main */

void escape()
{
        /* pause for viewing */
        for(;;)
        {
                if(kbhit()) break;
        }

        /* clear screen & return to alpha */
        _clearscreen(_GCLEARSCREEN);
        _setvideomode(_DEFAULTMODE);
        exit(0);

} /* end escape */


void set_graphics()
{
        /* set up graphics display */
        if (_setvideomode(_MRES4COLOR) == 0)
        {
                printf("%s\n",error_message);
                exit(0);
        }
```

```
        _getvideoconfig(&vc);
        _clearscreen(_GCLEARSCREEN);
        _setviewport(0,0,320,200);
        _setcolor(1); /* light blue */

} /* end set_graphics */


void user()
{

        /* declarations */
        int dummy;

        /* user input of gene code */

        printf("input height to first branch point ");
        scanf("%d",&dummy);
        organism.height = dummy;

        printf("input angles of branches  ");
        scanf("%d",&dummy);
        organism.angle = dummy;

        printf("input length of branches  ");
        scanf("%d",&dummy);
        organism.length = dummy;

} /* end user input */


void draw_organism()
{
        /* declarations */
        double x_position();
        double y_position();

        /* draw trunk of tree */
        _moveto(160,199);
        _lineto(160,199-organism.height);

        /* first branch point */
        _moveto(160,199-organism.height);
        _lineto(160+x_position(),199-organism.height-y_position());
        _moveto(160,199-organism.height);
        _lineto(160-x_position(),199-organism.height-y_position());

        /* second branch point */
        _moveto(160,199-organism.height);
        _lineto(160,199-(2*organism.height));

        _moveto(160,199-(2*organism.height));
        _lineto(160+x_position(),199-2*organism.height-y_position());
        _moveto(160,199-(2*organism.height));
        _lineto(160-x_position(),199-2*organism.height-y_position());

        _moveto(160,199-2*organism.height);
        _lineto(160,199-3*organism.height);



} /* end draw_organism */
double x_position()
{
        double x;

        x = organism.length * cos(organism.angle/RAD_FACTOR);
```

```
        return(x);

} /* end x_position */

double y_position()
{
        double y;
        y = organism.length * sin(organism.angle/RAD_FACTOR);

        return(y);

} /* end y_position */
```

pick1.c

```
/* Pickover biomorph program
   see: Computers, patterns, chaos and beauty
                        112491                    */

#include "\quickc\include\stdio.h"
#include "\quickc\include\math.h"
#include "\quickc\include\float.h"
#include "\quickc\include\graph.h"

struct videoconfig vc;
char error_message[] = "this video mode is not suported";


void main()
{
        /* declatrations */
        float rz; /* counter for real part of z */
        float iz; /* counter for imag part of z */
        int iterate; /* iterate counter */
        int counter; /* loop counter */

        float rmax; /* real part of zmax entered by user */
        float imax; /* imag part of zmax entered by user */
        float realz; /* real part of z in computation loop */
        float imagz; /* imag part of z in computation loop */
        float real_zcube; /* real part of z cubed */
        float imag_zcube; /* imag part of z cubed */
        float real_const; /* real part of constant */
        float imag_const; /* imag part of constant */

        int x,y; /* coordinates for plotting */
        float dx,dy;


        /* user input of parameters */
        printf("input rmax of z  ");
        scanf("%f",&rmax);
        printf("input imax of z  ");
        scanf("%f",&imax);
        printf("input real part of constant  ");
        scanf("%f",&real_const);
        printf("input imag part of constant  ");
        scanf("%f",&imag_const);


        /* set up graphics */
        if (_setvideomode(_MRES4COLOR) == 0)
        {
                printf("%s\n",error_message);
                exit(0);
        }
        _getvideoconfig(&vc);
        _clearscreen(_GCLEARSCREEN);
```

```
_setviewport(0,0,320,200);
_setcolor(1);

/* check the entire z space */
x = 0;
y = 0;
for(rz = -rmax; rz < rmax; rz += 2*rmax/320)
{
        /* compute the x coordinate */
        x = x + 1;
        y = 0; /* reset y counter */
        for(iz = -imax; iz < imax; iz += 2*imax/200)
        {
                /* compute y coordinate */
                y = y + 1;

                /* find z from rz and iz: initialize z */
                realz = rz;
                imagz = iz;

                counter = 0;
                /* iterate each point in z space */
                for(iterate = 0; iterate < 30; iterate++)
                {
                        counter++;

                        /* find z cubed */
                        real_zcube = realz*realz*realz -
                                        3*realz*imagz*imagz;
                        imag_zcube = 3*realz*realz*imagz -
                                        imagz*imagz*imagz;


                        /* compute new z */
                        realz = real_zcube + real_const;
                        imagz = imag_zcube + imag_const;


                        /* test modulus */
                        if(sqrt(realz*realz + imagz*imagz) > 100.0)
                                break;

                } /* end iterate loop */

                if((abs(realz) < 10.0) || (abs(imagz) < 10.0))
                {

                        /* plot x,y */
                        /*
                        printf("%f  %f  %f  %f  %d %d\n",rz,iz,realz,imagz,x,y);
                        */
                        _setpixel(x,y);
                }

        } /* end imag loop */

} /* end real loop */


/* pause for display */
for(;;)
{
        if(kbhit()) break;
}

/* clear screen & return to alpha */

_clearscreen(_GCLEARSCREEN);
```

```
        _setvideomode(_DEFAULTMODE);
        exit(0);

} /* end main */
```

pick2.c

```
/* Pickover biomorph program
   see: Computers, patterns, chaos and beauty
                        112491                  */

#include "\quickc\include\stdio.h"
#include "\quickc\include\math.h"
#include "\quickc\include\float.h"
#include "\quickc\include\graph.h"

struct videoconfig vc;
char error_message[] = "this video mode is not suported";

void main()
{
        /* declatrations */
        double rz; /* counter for real part of z */
        double iz; /* counter for imag part of z */
        int iterate; /* iterate counter */
        int counter; /* loop counter */

        double rmax; /* real part of zmax entered by user */
        double imax; /* imag part of zmax entered by user */
        double realz; /* real part of z in computation loop */
        double imagz; /* imag part of z in computation loop */
        double real_zfifth; /* real part of z fifth */
        double imag_zfifth; /* imag part of z fifth */
        double real_const; /* real part of constant */
        double imag_const; /* imag part of constant */

        int x,y; /* coordinates for plotting */
        double dx,dy;


        /* user input of parameters */
        printf("input rmax of z  ");
        scanf("%lf",&rmax);
        printf("input imax of z  ");
        scanf("%lf",&imax);
        printf("input real part of constant  ");
        scanf("%lf",&real_const);
        printf("input imag part of constant  ");
        scanf("%lf",&imag_const);


        /* set up graphics */
        if (_setvideomode(_MRES4COLOR) == 0)
        {
                printf("%s\n",error_message);
                exit(0);
        }
        _getvideoconfig(&vc);
        _clearscreen(_GCLEARSCREEN);
        _setviewport(0,0,320,200);
        _setcolor(1);
        /* check the entire z space */
        x = 0;
        y = 0;
        for(rz = -rmax; rz < rmax; rz += 2*rmax/320)
        {
                /* compute the x coordinate */
```

```
                x = x + 1;
                y = 0; /* reset y counter */
                for(iz = -imax; iz < imax; iz += 2*imax/200)
                {
                        /* compute y coordinate */
                        y = y + 1;

                        /* find z from rz and iz: initialize z */
                        realz = rz;
                        imagz = iz;

                        counter = 0;
                        /* iterate each point in z space */
                        for(iterate = 0; iterate < 30; iterate++)
                        {
                                counter++;

                                /* find z fifth */
                                real_zfifth = pow(realz,5) -
                                         10*pow(realz,3)*pow(imagz,2) +
                                         5*realz*pow(imagz,4);


                                imag_zfifth = 5*pow(realz,4)*imagz -
                                         10*pow(realz,2)*pow(imagz,3) +
                                         pow(imagz,5);

                                /* compute new z */
                                realz = real_zfifth + real_const;
                                imagz = imag_zfifth + imag_const;


                                /* test modulus */
                                if(sqrt(realz*realz + imagz*imagz) > 100.0)
                                        break;

                        } /* end iterate loop */

                        if((abs(realz) < 10.0) || (abs(imagz) < 10.0))
                        {

                                /* plot x,y */
                                _setpixel(x,y);
                        }

                } /* end imag loop */

        } /* end real loop */


        /* pause for display */
        for(;;)
        {
                if(kbhit()) break;
        }

        /* clear screen & return to alpha */
        _clearscreen(_GCLEARSCREEN);
        _setvideomode(_DEFAULTMODE);
        exit(0);

} /* end main */
```

5
CHAPTER

Genetic algorithms and evolution

In my previous book, *Creating Artificial Life* (#3719), we examined the central dogma of molecular biology: the genetic algorithm of biology. Information flows from DNA in replication to an RNA intermediate in the transcription step. The final step is translation in which RNA codes for amino acids in the protein chains.

Here, in this chapter, we will examine in a little more detail the biological genetic algorithms of replication, transcription, and translation. The principles of a simplified computer genetic algorithm will then be presented, followed by some ideas for using genetic algorithms to breed neural networks. The last major section of the chapter will review some ideas on evolution and species formation. This should smoothly lead into the next chapter on ecosystems.

Biological genetic algorithms

Much of this section is from Sampson (1984). The first step in the biological genetic algorithm entails replicating the DNA. Enzymes called *polymerases* are used in replication and transcription and in the assembly of the nucleic acid polymers, RNA and DNA. Other enzymes are used in other aspects of the replication and transcription, but the polymerases are the primary ones used for the information transfer. Nucleuses are needed for the disassembly of the polymers. Exonucleases attack the free end of the polymer for decomposition, and endonucleases break internal bonds at specific sites. Many of these enzymes are the molecular scale tools of genetic and protein engineering.

Polymerases will replicate DNA in a test tube if a template strand is already present and a supply of monomers is available. Replication begins at several sites in parallel. The entire process takes place at about 200 nucleotides per second. In actual cells, untwisting and unwinding proteins are involved. The untwisting and unwinding takes place in parallel, with replication toward the end of the unwound strands.

The second major step in the biological genetic algorithm is transcription. This process, like the replication step, uses polymerases (in this case, RNA polymerases). The RNA differs from DNA by the ribose sugar base and uracil instead of thymine. Assembly takes place at about 50 nucleotides per second. An RNA of 10^6 base units will require about 5.5 hours to transcribe.

The last step in the biological genetic algorithm is translation. This step uses the RNA as a template for molecular recognition in the assembly of proteins, which are essentially the phenotype of the organism. Protein assembly takes place at about 20 amino acids per second. A protein of 10^6 bases will require about 13.9 hours for assembly.

Although these processing times for the nucleotides and proteins are impressive, we wouldn't be here if they were done sequentially—the processing time would be too great. Many different biopolymers are assembled in parallel. Bacteria can reproduce the entire organism in about 20 minutes.

During the reproduction of a lifeform, errors resulting from mutation might occur. Higher lifeforms that use genetic crossover (sex) in the reproduction swap segments of genetic material. In our artificial genetic algorithms, we will be concerned with these two high-level operations: crossover and mutation. We will not concern ourselves with the mechanics of molecular reproduction.

Genetic algorithms in artificial life

Artificial genetic algorithms pale by comparison with biological algorithms. The artificial genetic algorithms, hereinafter called genetic algorithms, are very abstracted from the biological version. The biological genetic algorithms shown earlier result in evolutionary changes in the phenotype. The net result is that the genotype is said to search a *complex fitness landscape*. The classical genetic algorithm of artificial life was developed by Holland (1975) and later by Goldberg (1989) and others (Davis, 1991, Below and Booker, 1991).

Genetic algorithms use random choice as a tool to search a complex fitness landscape. This is not a total random search, but rather a guided random search (gradient search) sometimes called *genetic hill climbing* (searching for the highest fitness peak). The fitness function is the guide through the landscape, so the landscape is called a *fitness landscape* (Eldridge, 1989). Like neural networks and cellular automata, genetic algorithms result in emergent behavior based on local dynamics (Forrest, 1991).

The objective of the genetic algorithm is to optimize a fitness function f(x). The parameter x is a string of bits called the *chromosome*. We want to find a bit string or chromosome with the best fitness value. This fitness value, often called the *phenotype*, is found by applying the fitness function to the string of bits.

The three primary steps in the artificial genetic algorithms are reproduction, crossover, and mutation. Another step, not usually called part of the genetic algorithm, is the evaluation of the chromosome by determining the fitness function, which is some measure of fitness of the organism. In the following text, I will explore a toy problem introduced by Goldberg (1989). This section will highlight the main steps of the genetic algorithm and explain why it is a successful search strategy.

We will select a simple fitness function: $f(x) = x^2$. The objective is to maximize this function for a given x. The constraint in the chromosome length is fixed. The chromosome is given as a bit string *(* * * * *)*, where * represents the "Don't care" state (i.e., wildcards) for binary bits [0,1]. To actually evaluate the fitness of the chromosome, the five-bit string vector is converted into decimal and squared, like so:

$$(1\,0\,0\,0\,0) = 16$$
$$f(16) = 256$$

We start the algorithm by selecting bits at random for the population. For a population of four organisms, we might get these strings:

#1 (0 1 1 0 1)
#2 (1 1 0 0 0)
#3 (0 1 0 0 0)
#4 (1 0 0 1 1)

These have the fitness values of 169, 576, 64, and 361 (respectively). The average fitness is 293. Therefore, organism #2 is the most fit. Organisms #1 and #4 are about equal and have medium fitness. Organism #3 is the most unfit. It should be clear that, by Darwinian selection, the unfit organism (#3) will become extinct and that the others will reproduce.

In selecting the mating, we first find fitness fractions of each organism in the entire population. The total fitness of the population is given by

$$\text{sum } f(x) = 1170$$

We now generate a biased roulette wheel with the fractions

169/1170, 576/1170, 64/1170, 361/1170

Selection of random numbers in the range [0,1170] are assigned to the organism number by choosing the organism with a fitness value greater or equal to our random number.

Random number	**Organism**
169	#1
359	#2
868	#2
274	#4

From this table, we can see that organisms #1 and #4 will each have one child, while #2 will have two children and #3 will have none.

So we have the following situation:

Initial population

#1	(0 1 1 0 1)
#2	(1 1 0 0 0)
#3	(0 1 0 0 0)
#4	(1 0 0 1 1)

Mating population

#1	(0 1 1 0 1)
#2	(1 1 0 0 0)
#3	(1 1 0 0 0)
#4	(1 0 0 1 1)

The crossover operator results in segments of chromosomes exchanging places. It is the equivalent of sex in the biological world. Two chromosomes are selected for mating. Part of each chromosome will swap places in order to create new chromosomes made up from parts of each of the two parents. The crossover sites for the offspring are selected at random with two mates also selected at random. The crossover sites selected at random for our chromosomes are

(0 1 1 0 ^ 1)
(1 1 0 ^ 0 0)
(1 1 0 ^ 0 0)
(1 0 ^ 0 1 1)

and the mates are organisms #4, #3, #1, #2. The crossover operation consists of swapping parts of the chromosome with the mate at the selected site. For our problem, we have:

(0 1 1 0 ^ 1) × (1 0 0 1 1) → (0 1 1 0 1)
(1 1 0 ^ 0 0) × (0 1 0 0 0) → (1 1 0 0 0)
(1 1 0 ^ 0 0) × (0 1 1 0 1) → (1 1 0 0 1)
(1 0 ^ 0 1 1) × (1 1 0 0 0) → (1 0 0 0 0)

This new population has fitness values 169, 576, 625, and 256. Some vectors (i.e., chromosomes) are more fit than others. In this problem, the vector (1 0 0 0 0) is more fit than, say, (0 0 0 0 1). For this example, it is clear that any vector (1 * * * *) is more fit than (0 * * * *), where * represents a "Don't care" state. These building blocks of genes are called *similarity templates* or *schema*.

For binary vectors, we have essentially extended the number of symbols in the vector by allowing the symbol * to represent the "Don't care" state. For a vector of length l, there are 3^l schemata or similarity templates. If we let S represent a schema within a population G(t) at time t, then there are M(G,t) examples of the schemata at time t. The average fitness is f(S). Thus, the growth is given by the difference equation

$$M(S,t+1) = M(S,t)\frac{f(S)}{\bar{f}}$$

where

$$\bar{f} = \frac{1}{n}\sum f_i$$

is the population average fitness.

This difference equation can be expanded to include schema that remain above average by a constant c:

$$M(S,t+1) = M(S,t)\left(\frac{\bar{f} + c\bar{f}}{\bar{f}}\right)$$

When crossover occurs, we expect some schema to remain intact with survival probability

$$p_s = 1 - \frac{\delta(S)}{l+1}$$

with δ(S) being the defining length of the schema. If we substitute this into the difference equation after assuming that crossover and reproduction are independent events and at random sites, then we get

$$M(S,t+1) \geq M(S,t)\frac{f(s)}{\bar{f}}\left(1 - p_c\,\frac{\delta(s)}{l-1}\right)$$

This difference equation considers both crossover and reproduction.

In our toy problem, we have introduced all the major steps of the genetic algorithm. After many generations, the population will begin to loose genetic diversity. All the members of a population will have a similar genotype and therefore a similar fitness value, which might be good or mediocre. When this occurs, the mutation rate can be increased. The mutation step may be performed after the crossover step and before the fitness values are computed. The mutation consists of random bit flipping for a small fraction of the total bits in the population.

The crossover operator just described implies that crossing at any random point is fine. In actual practice, for complex chromosomes, the gene location might be critical. Certain locations along the chromosome might be the only allowed crossover points.

We have focused on how a genetic algorithm will give rise to a population of one species or one type of phenotype. In nature the population often splits as some species dominate certain ecological niches (Eldredge, 1989). If we have a multi-peaked function landscape to optimize, we should like to develop a species for each of the ecological peaks or niches. Todd and Miller (1991) describe a genetic algorithm in which only the nearest neighbor, in the phenotype sense, can reproduce with each other. (Later in this chapter, I will describe a similar algorithm in a Pascal program.)

In nature, the payoff must often be shared with other members of a population. This acts as an apparent diluent of the individual fitness. If we have one shrimp in a flask of algae and water, this one shrimp will not need to share the algae with others. But if we have a population of shrimp, then each individual might consume less than what is needed for minimal survival. A mutant shrimp may become omnivorous and attack other shrimp. A new species will have thus evolved and have an advantage over a population of herbivores. Some dynamic equilibrium will evolve where each species now fills a niche in the small ecosystem. (I used this example because in Chapter 6 we will examine real closed aquatic ecosystems.)

If we have a complex function resembling the upper half of a damped oscillator and use a simple genetic algorithm to find the function maxima, we will find that the population might cluster near the top of one of the peaks. If we use a fitness function to consider the ideas of ecosystem niche fitting, then the fitness function for any individual must be diluted in the total population relative to the nearby population.

In summary, the first step in the genetic algorithm is random creation of bit strings with the appropriate representation. The fitness of these strings is then determined and mates are selected. Survival and mating success depends on global knowledge of the population

$$P(l) = \frac{f_i}{\sum_j f_i}$$

After mating and chromosome crossover, the mutation operator is applied and the new population is evaluated by the fitness function.

Genetic algorithms: Reviews

In this section, I will review some applications of genetic algorithms. The primary focus will be on neural network applications of genetic algorithms, primarily due to my own bias. There are many other applications (Goldberg, 1989; Davis, 1991; Langton et al. 1992; Forrest, 1991; Belew and Booker, 1991).

In Dewdney's book, The Armchair Universe (1988), there is a paper on the genetic algorithm's use to evolve artificial organisms called finite living blobs or *flibs*. The objective is to evolve organisms capable of predicting a finite state machine operation. The state transition table represents the organisms' chromosome. Dewdney also discusses a genetic algorithm solution of the traveling salesman problem and sequence prediction. The major difficulty with the traveling salesman problem is chromosome representation. This can be a very difficult problem and the results are not impressive.

Bramlette and Bouchard (1991) describe a genetic algorithm for parametric design of an aircraft. The chromosomes represent the loading, thrust-to-weight ratio, length of aft body, length of fore body, fuselage height, fuselage width, altitude outbound, and altitude inbound. These parameters are coded as real numbers in the chromosome.

Cox et al. (1991) describe a dynamic routing algorithm for telecommunications networks developed by a genetic algorithm. The network control problem is analogous to the traveling salesman problem. Calls arrive at a switching station and are to be routed to appropriate circuits. Each call has six parameters: source, destination, start time, duration request, bandwidth, and priority class, all of which would represent the genes within the chromosome. The fitness function is to find the shortest path from the origin to the destination node. For small networks, the genetic algorithm performs reasonably well; but for large networks, the computation time becomes excessive for real-time switching.

Lucasin et al. (1991) use a genetic algorithm for confirmational analysis of DNA. Large molecules such as DNA and proteins fold into complex three-dimensional shapes. These structures are at an energy minimum, and computation of these minimal structures can be very difficult. The parameters coded into the chromosome are bond angles, torsion angles, and atomic arrangement. These chromosomes are optimized by a fitness function to find the energy minimum of the molecular conformation. The chromosome fitness can be evaluated by nuclear magnetic resonance data for the given molecular structure.

Whitley et al. (1991) used a genetic algorithm technique for the traveling salesman problem. The possible edges for a graphic representation became the genes for the chromosome. They report good optimization results for 30 and 105 city tours. Muhlenbein et al. (1991) also worked on the same problem; they solved it for a 442-city tour. The solution is similar to the greedy algorithm solution. They point out that the evolution process can be modeled by two approaches; the differential/difference equations method—similar to Fontana et al. (1989)—and the DeJong (1980) and Booker, Goldberg, and Holland (1989) with a genetic algorithm approach.

Rizki et al. (1991) describe a genetic algorithm used to assist in a design process. The design space for a system is coded as a bit string for the chromosome. By applying the genetic algorithm to the bit string, the authors designed a pattern-recognition system. One point stressed is the primary difference between biological and computer genetic algorithms. In biological algorithms, new functions can emerge during the Darwinian evolution; in artificial genetic algorithms, the set of structures is pre-defined and no new structure can appear. (This assumption is not true, as we will see later in this chapter.)

Louis and Rawlins (1991) also used a genetic algorithm for design. In their case, they were designing an n-bit adder and an n-bit parity checker. These researchers let the artificial evolution develop the desired structure or circuit.

Smith (1991) used a genetic algorithm to interactively develop Dawkins-type biomorphs. His biomorphs were encoded as two chromosomes each. The first chromosome consisted of the Fourier coefficients for the cos series, and the second chromosome was comprised of the Fourier coefficients for the sin series. The biomorphs resembled Lissajous figures.

Other researchers are also using genetic algorithms for developing biomorphs. Shonkwiler et al. (1991) has developed iterated function systems and fractal trees, and Wyard (1991) and Wilson (1987) have developed L-system-like structures using genetic algorithms.

Schaffer et al. (1990) is one of the research groups using the genetic algorithm to optimize neural network architectures. One of the most difficult tasks in developing a neural network is finding the optimum tradeoff with the number of nodes, number of hidden layers, the best starting weights, and the learning rate. The architecture can be coded into a chromosome bit string as follows:

[0,1] $\eta \; \varepsilon \; \{0.5, 0.25, 0.125, 0.0625\}$ (learning rate)
[0,1] $\alpha \; \varepsilon \; \{0.9, 0.8, 0.7, 0.6\}$ (acceleration)
[0,1} $\omega_o \; \varepsilon \; \{\pm1.0, \pm0.5, \pm0.25, \pm0.125\}$
1st layer {1 bit present/absent, 4 bits for # of units}
2nd layer {1 bit present/absent, 4 bits for # of units}

The chromosome is converted into a network phenotype. The networks are evaluated by some fitness after training by the back-propagation algorithm. Training stops after a specific number of cycles through the database.

Whitley et al. (1990) experimented with the architecture evolution ideas similar to Schaffer et al. and extended the chromosomes to real number representations, rather than the binary representations that Goldberg (1989) claims are more robust. Whitley et al. also describe genetic algorithms for evolution of neural networks in which the learning algorithm is the genetic algorithm. This approach is more efficient, as I will show later in this chapter. However, Ackley and Littman (1992) demonstrate the Baldwin effect and show that better networks can evolve by using the Schaffer et al. approach.

Koza and Rice (1991) use a genetic algorithm to evolve networks of Boolean operators for one bit adders. This is similar to Louis and Rawlins (1991), who evolve *n*-bit adders. Dominic et al. (1991), Karr (1991), and Maricic (1991) develop systems for the control of an inverted pendulum. All these networks have in common that there is no learning process during the lifetime of the organism. The only learning that takes place is through evolution by crossover and mutation. In short, the organisms brains are hard-wired in their chromosomes.

In the networks explored in the next section, I have followed the work of De Garis (1991 a, b). He, like those mentioned in the previous paragraph, has developed neural networks with the genetic algorithm and no learning algorithm. In this case, the network is a fully connected network, including self-connections, and is coded as a bit string for the chromosome. Each connection is given as 7-bits for weight and one bit for sign. The chromosomes are evaluated for fitness, mates are subjected to crossover, and mutation occurs on the children. The fitness function is simply a feedforward operator evaluating the neural network. There is no learning except through the "genetic learning".

The tasks that De Garis's neural networks were supposed to perform were (in essence) feedforward controllers for robotics. One of the early experiments was a simulation of a two-legged walking robot. The inputs to the networks were the desired angles, and the outputs were the acceleration for the angles of the legs with respect to the hip joint and the knee angles.

A more advanced experiment was the evolution of an artificial insect. The inputs to the network were parameters representing antenna signals. These were connected to subnets, which were in turn connected to sensor fusion networks. The outputs were parameters such as direction and speed of walking, eating, and mating.

Genetic algorithms for neural networks

In this section, we will examine two genetic algorithm programs for neural network development. The first program, degaris1.pas, is similar to the work of de Garis (1991). In this particular case, we will not be evolving networks to act as controllers for robots but rather to just solve a simple pattern-recognition problem. The genetic algorithm will determine the synaptic connections for the networks. There is no learning in the network.

The second program, cagann1.pas, is set up on a cellular automata grid, and only nearest neighbors interact. In typical genetic algorithms, survival and mating success of each individual involves global knowledge of the total population,

$$P(i) = \frac{f_i}{\sum_j f_j}$$

In the approach we will take here, the local mating scheme requires only local information. It is therefore more biologically realistic.

The degaris1.pas genetic algorithm program is designed to evolve neural network organisms. The neural network organisms are Hopfield recurrent networks designed to act as content addressable memories. The evolution algorithm should evolve networks that improve on the recognition task. But, as we will see, the organisms tend to cluster about a mediocre level, which is due to the fact that mates are chosen based on the same hamming code. More fit individuals do not necessarily have more children. The number of children per hamming code (here referred to as species code) is determined by the number of parents of the same code. Thus, the mediocre end result is easily explained in hindsight.

The organisms exist as records that contain the organism chromosome array, fitness value, fitness fraction, species code and mate flag. The population of organisms consists of an array of organism records. Two other arrays declared are the test_structure array (an array of 8 binary values for use as the test input to the neural networks) and the synaps_matrix (an 8×array of binary integers used as the synapse matrix for the neural network).

There are two populations of organisms: one is the population being studied, and the other is the new population. Later in the program, after the gene crossover has occurred, the old population is set equal to the new population. Most of the other variables are declared with meaningful names or comments. Some of the variables are used in histogram studies of the population after the end of the run for that generation.

The program begins with assignment of the test vector and the variable mutfreq set equal to 20. This variable determines the number of bits per population to be mutated. In a population of 100 individuals, each with a 64-bit chromosome, the mutation fraction will be 0.003125. The next step is the initialization of the population by setting random bits in the chromosome.

Then, for a generation count of 250, a large while-loop begins. The first step is to increment the generation counter and zero the fitness sums used in the histograms at the end of each generation. A flag is then set (the mate flag) on each organism. This flag indicates who the mate will be at the time of reproduction.

Now that the organisms have been created, the next step is to fold the chromosome into the neural matrix to produce the phenotype. Each gene codes for one synapse. The organism is then exposed to the environment by a vector matrix multiplication. The output value of this dot product is compared with the input vector to determine the Hamming distance (which is the fitness value for this organism.)

After the fitness of each of the 100 organisms in the population has been determined, the fitness distribution (histogram) is determined and printed to the screen. If each Hamming code is thought of as a separate species, then we want only members of the same species to mate with each other. Organisms of code 1 will mate with others of code 1, organisms of code 2 will mate with others of code 2, etc.

The next section of the program is used to determine the species codes and the mate for each organism. Only those of Hamming code 8 (i.e., the worst) will reproduce with an improved species (e.g., 7). This will allow some degree of genetic diversity. The organisms are then reproduced into the new-pop array by a crossover operation where the crossover site is selected at random. The population array pop-member is then set equal to the new-pop array, and the process is repeated by reaching the end of the while loop.

The population at generation 0 has a distribution similar to Fig. 5-1. This is almost a perfect random distribution that should be centered at a fitness value of 4. The fitness represents the Hamming distance.

Figure 5-2 through Fig. 5-5 show distributions at generations 5, 25, 50, and 100 respectively. The most notable feature is in Fig. 5-5. This skewed distribution is similar to biological species distributions and will be examined later in this chapter. The population tends to cluster about a slightly better than mediocre level,

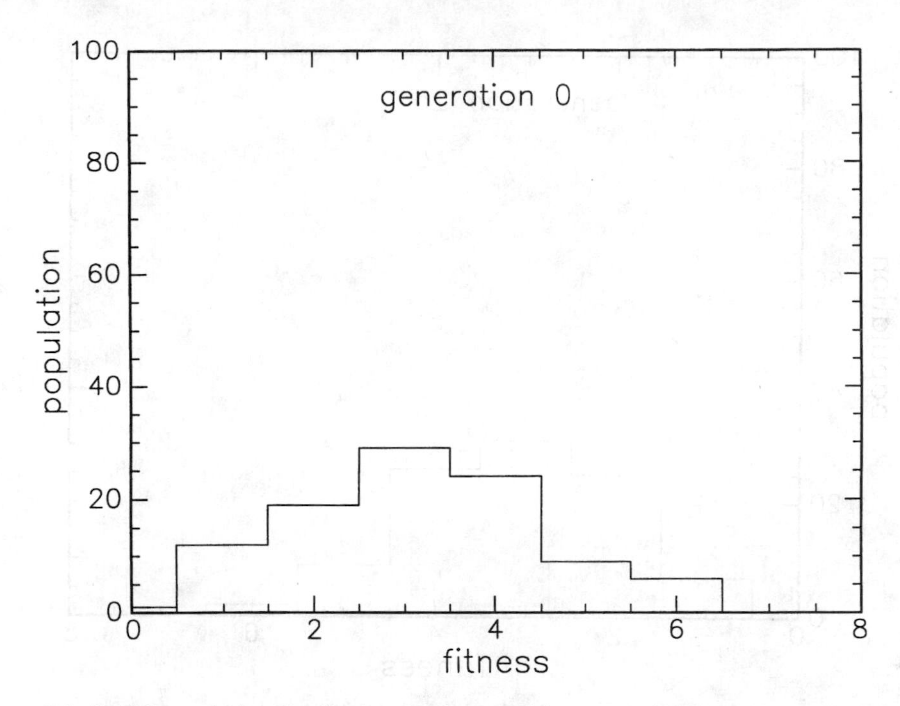

5-1 Initial population distribution for the degarisl.pas experiment.

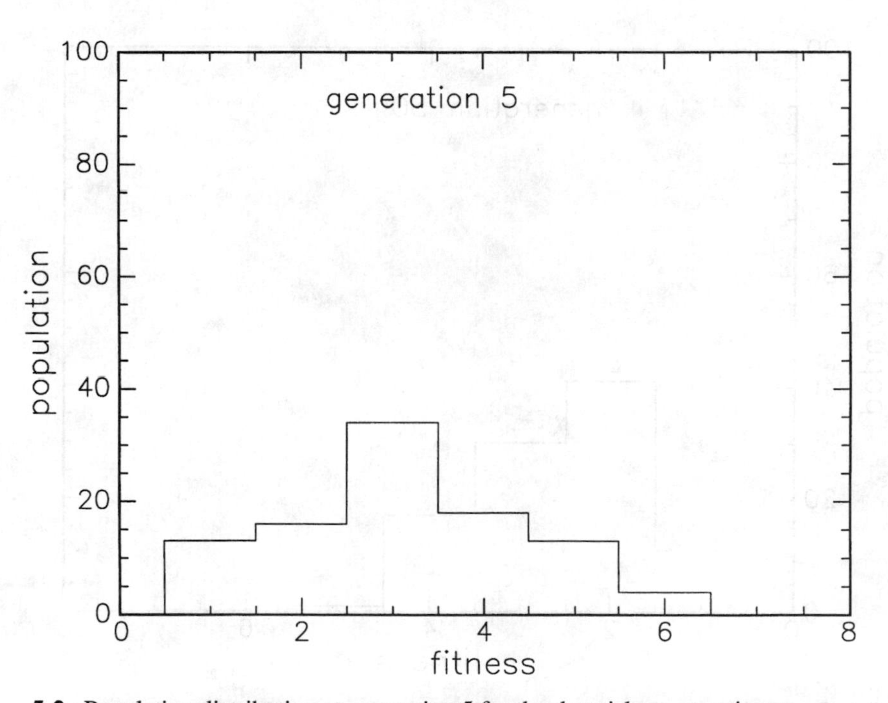

5-2 Population distribution at generation 5 for the degarisl.pas experiment.

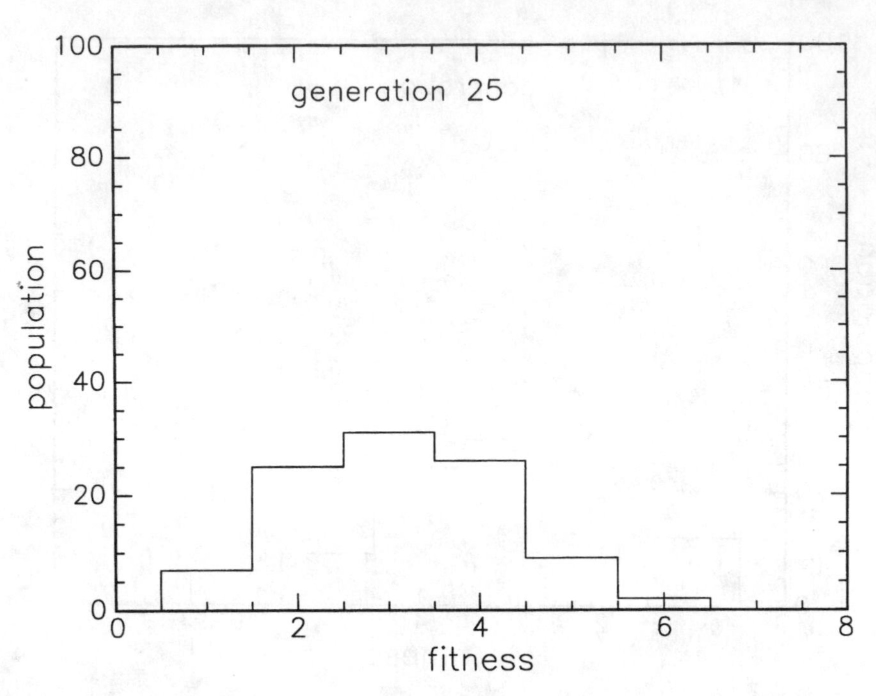

5-3 Population distribution at generation 25 for the degarisl.pas experiment.

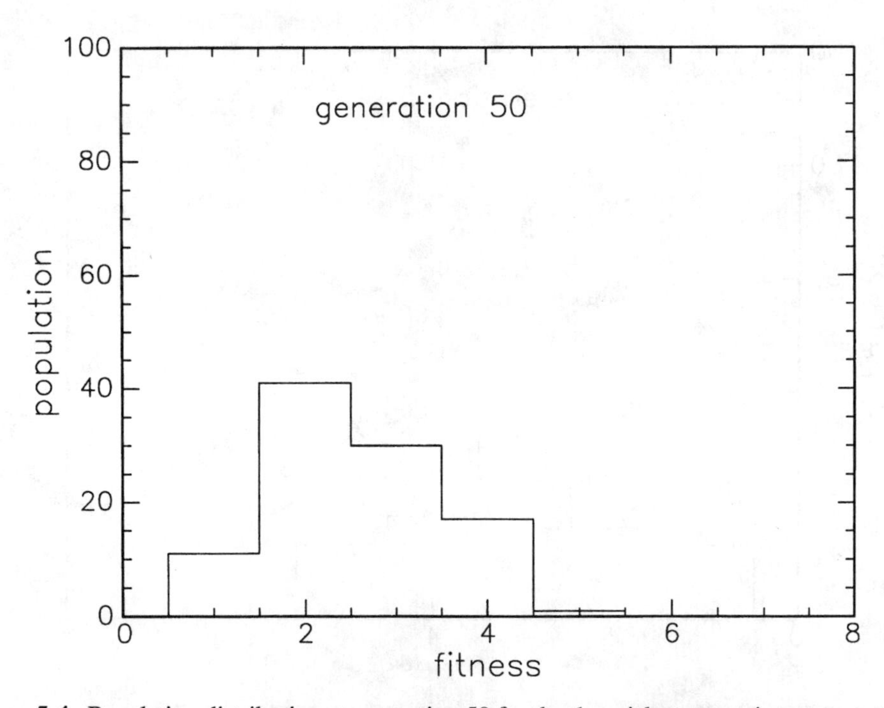

5-4 Population distribution at generation 50 for the degarisl.pas experiment.

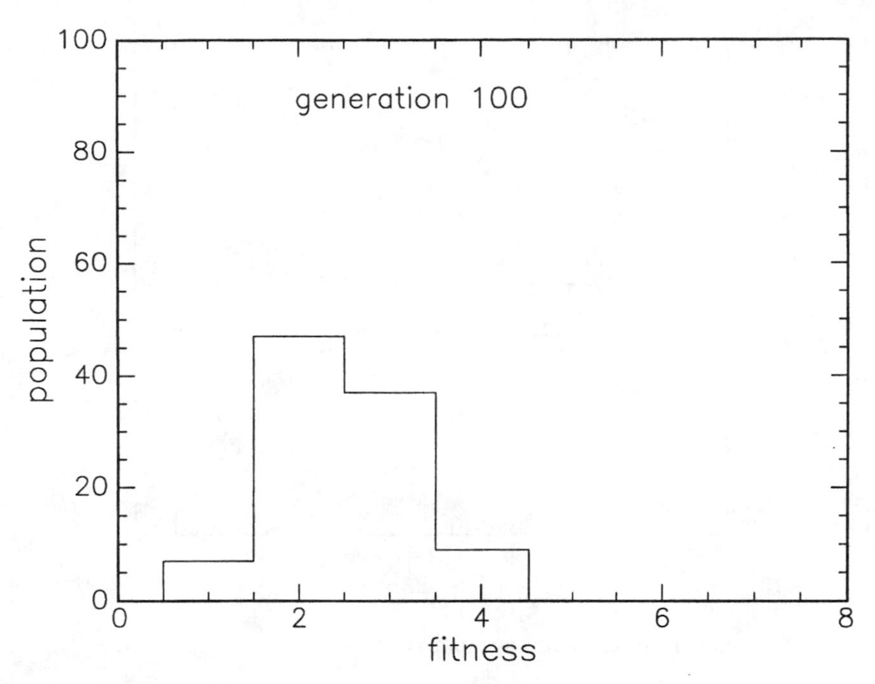

5-5 Population distribution at generation 100 for the degarisl.pas experiment.

something that is caused by the mates being chosen based on the same Hamming distance and that is analogous to sexual dynamics being species-confined, just like in nature. Another cause for the mediocre level achieved by the most fit species is the crossover operator itself. Recall, the chromosome is a folded synapse matrix, and this synapse matrix is a binary Hopfield type. Swapping large chunks of the matrix will have an adverse effect. In essence, the crossover acts this way on the synaptic matrices.

The second program is more complex in the mate selection and also more biologically plausible. The mate selection is based on a nearest neighborhood, and the degree of fitness of the organisms will determine the number of children. Figure 5-6 shows this children function. The crosses are the actual values, and the curved line is a second-order polynomial fit of the function.

The program is designed for genetic evolution of Hopfield recurrent neural network organisms. The task that each organism must master is a content-addressable operation. The entire population is considered to be a single species, with the most fit individual having a lower Hamming code. The most fit individual will have more children than the least fit. The function to determine the number of children is based on a second-order equation. The most fit will have more and the least fit none, while the mediocre individuals will have a small number of children. The population is set up on a cellular automata grid with the organisms reproducing with nearest neighbors. Both the crossover site and the neighbors are selected at random.

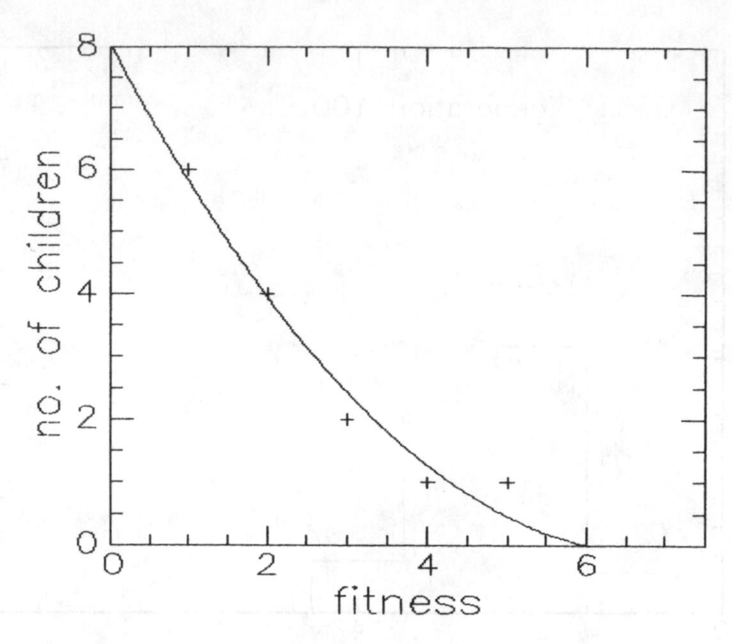

5-6 Fitness relation for the cagannl.pas experiment.

The cellular automata analogy here is good. The state of the automata is determined by the fitness, which is determined by the Hamming distance from a vector matrix product of a Hopfield neural network and some test vector. The new state at time t is always the same as at time $t - 1$. However, the neighbors state at time t is changed, determined by the genetic crossover operation. When all the members of the population have been changed (a synchronous parallel CA) for each generation, then the pop-array is set equal to the new-pop array and the process of living is repeated.

The organisms are, again, records that contain chromosome arrays, fitness values, fitness fractions, and a flag value. The organisms are in an organism array of 15x15 (the cellular automata array), with the left edge joined to the right edge and the top joined to the bottom. The array variables now have new names. Mother is the primary population array in the cellular automata configuration. Daughter is the new array created from the crossover operation. All the other variables are as in the degaris1.pas program or have clear, descriptive names and/or comments in the source code.

After the mother population and the fitness counters for the distribution have been initialized, the fitness of each individual is determined by phenotype construction, exposing the organism to the environment of the vector-matrix multiplication and measuring the fitness of the organism by the Hamming distance. After the fitness of all members of the population is determined, the distribution is computed and printed to the screen.

The directions of the compass are assigned integer values for random selection of the mates. Mate selection consists of first determining the number of neighbor cells that the center cell will mate with; this is equivalent to the number of children. The number of children is directly related to the fitness of the center cell by a second-order function. Random numbers are selected to determine the neighbor to mate with. After the mating and genetic crossover, the center cell is left unchanged but the neighbor cells are changed. The disadvantage in this approach (though not a big one) is that the last mother cell to access a particular daughter will determine the daughter cell state.

The distribution for the initial population is shown in Fig. 5-7. The shape of the distribution is similar to Fig. 5-1, as you would expect. However, immediately at generation 1 (Fig. 5-8), there is a noticeable improvement. Each successive generation (Fig. 5-9 through Fig. 5-22) shows increasing improvement. The population settles to a single species at a Hamming distance/fitness of 1 (see Fig. 5-22), due to the earlier mentioned effect that swapping pieces of Hopfield networks will result in one-bit of error (Hamming distance 1) on the average. Figure 5-23 presents the evolutionary landscape for this experiment, showing a smooth increase in fitness.

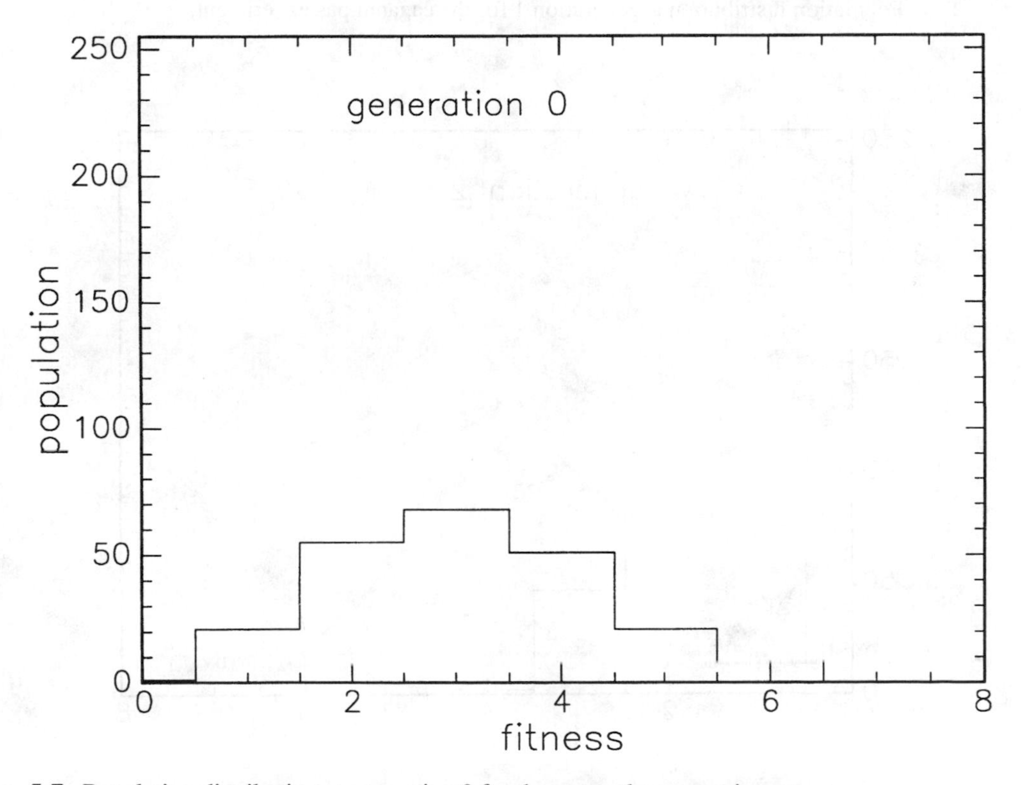

5-7 Population distribution at generation 0 for the cagannl.pas experiment.

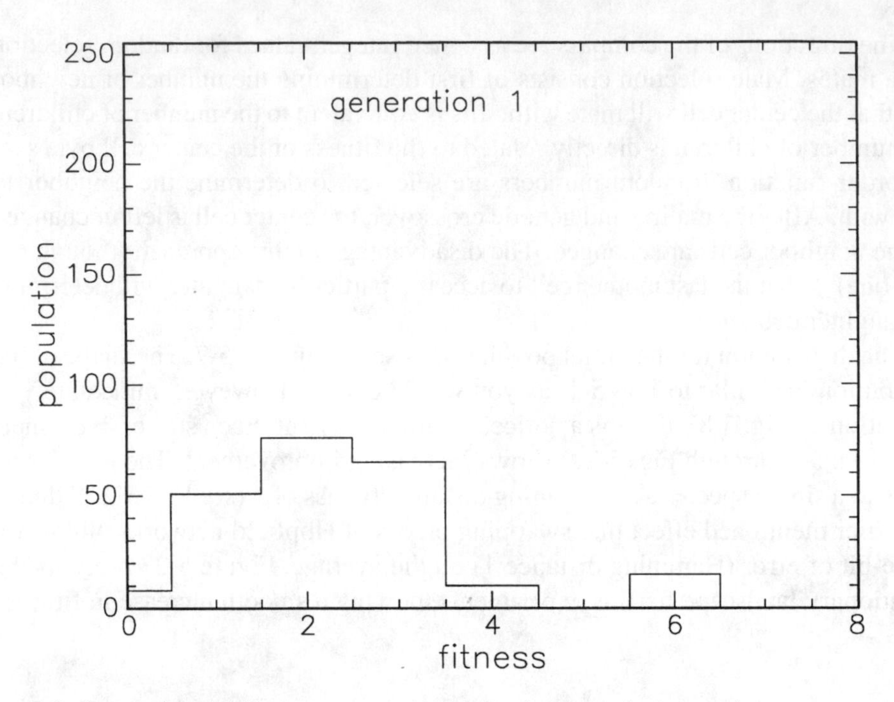

5-8 Population distribution at generation 1 for the cagannl.pas experiment.

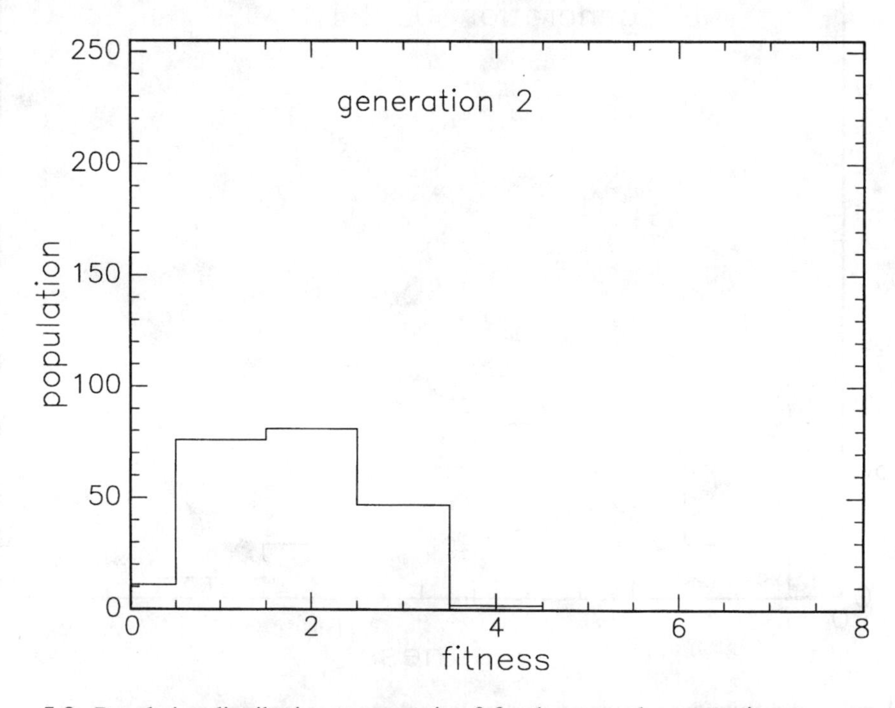

5-9 Population distribution at generation 2 for the cagannl.pas experiment.

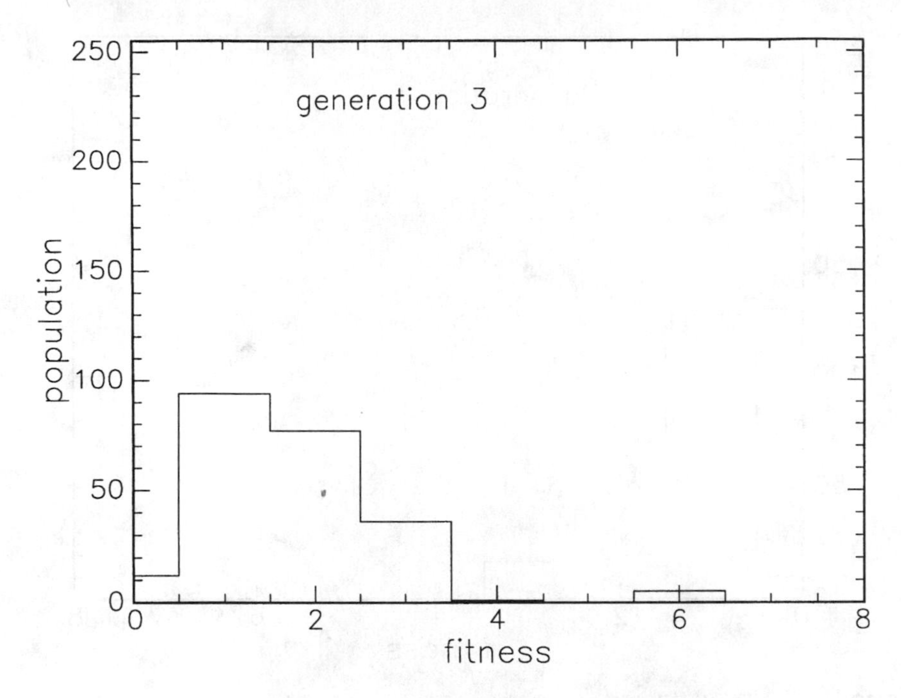

5-10 Population distribution at generation 3 for the cagannl.pas experiment.

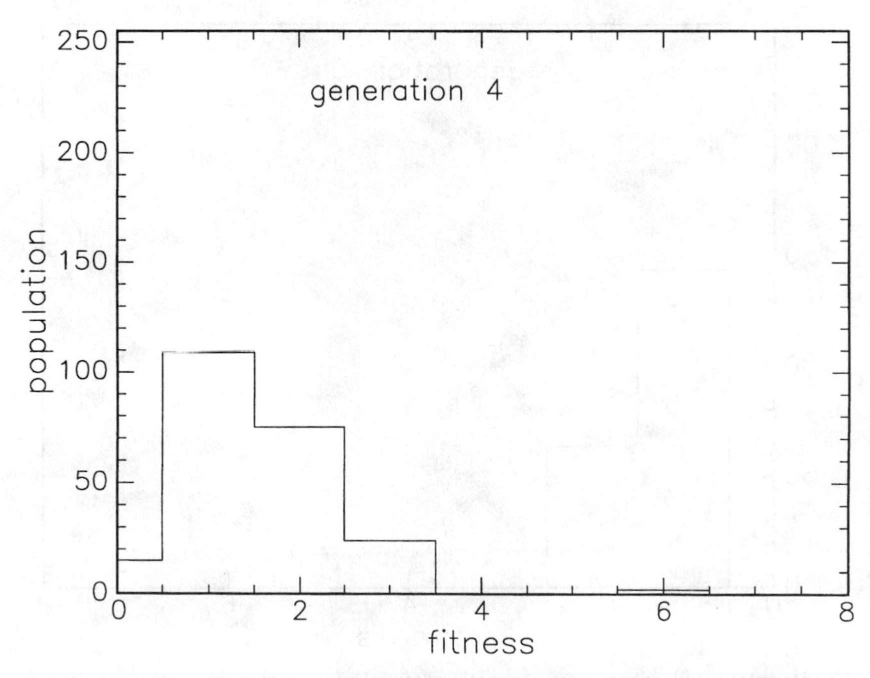

5-11 Population distribution at generation 4 for the cagannl.pas experiment.

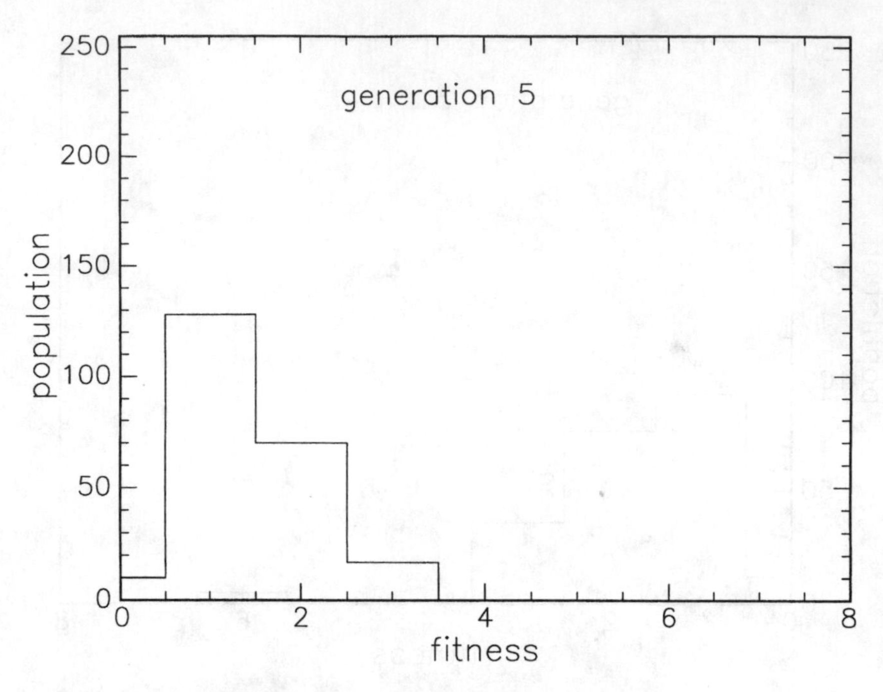

5-12 Population distribution at generation 5 for the cagannl.pas experiment.

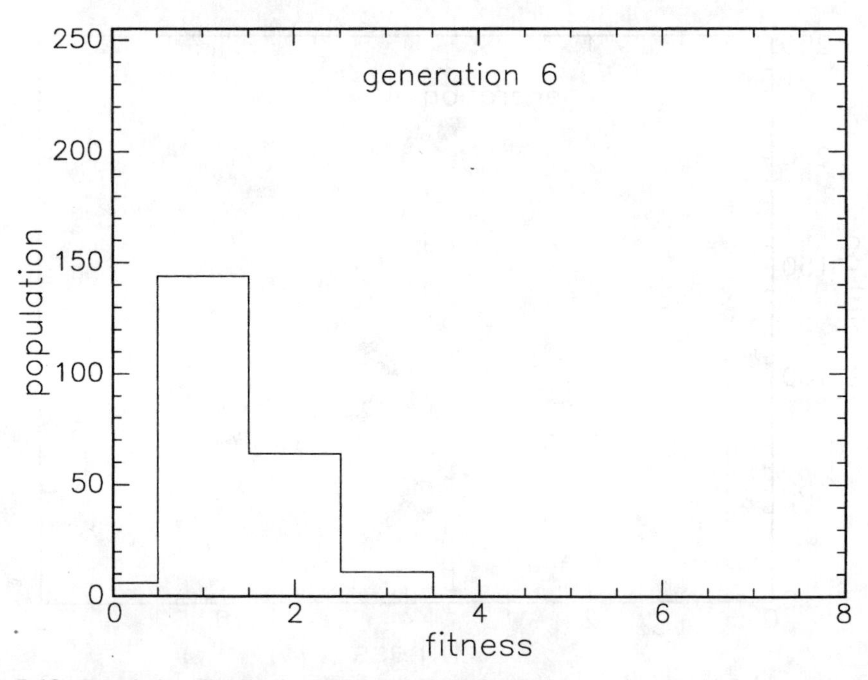

5-13 Population distribution at generation 6 for the cagannl.pas experiment.

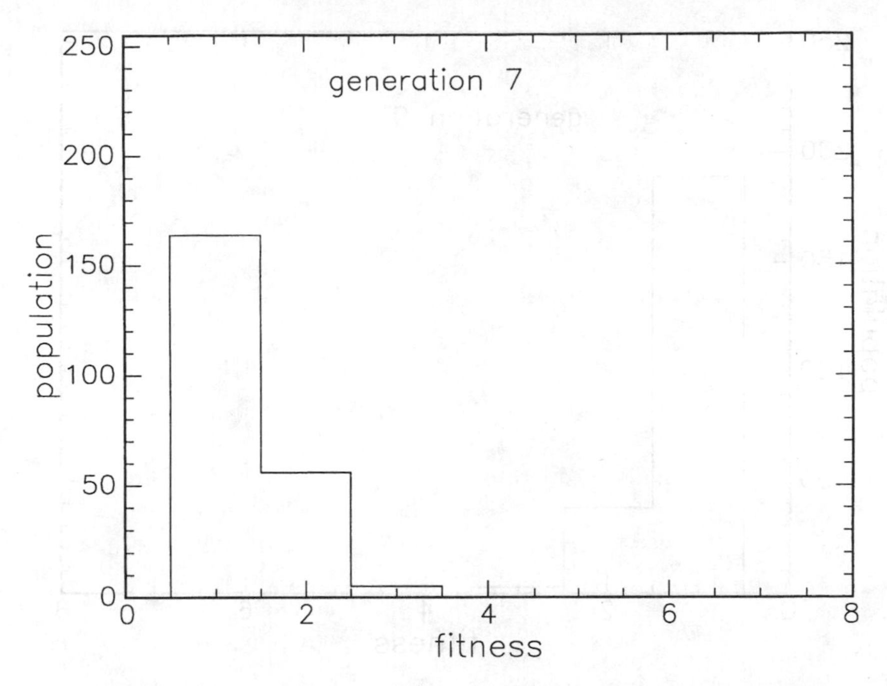

5-14 Population distribution at generation 7 for the cagannl.pas experiment.

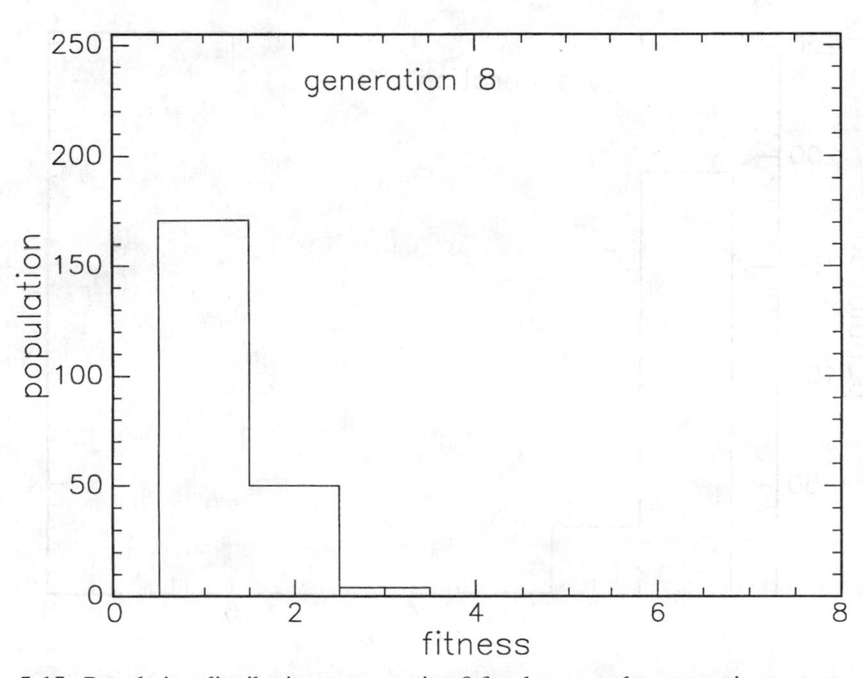

5-15 Population distribution at generation 8 for the cagannl.pas experiment.

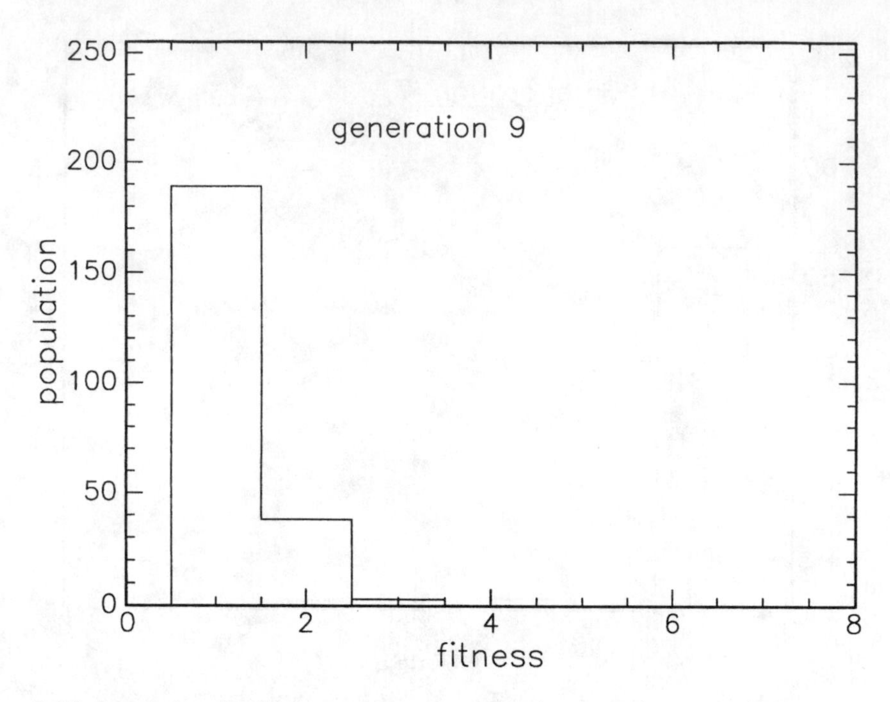

5-16 Population distribution at generation 9 for the cagannl.pas experiment.

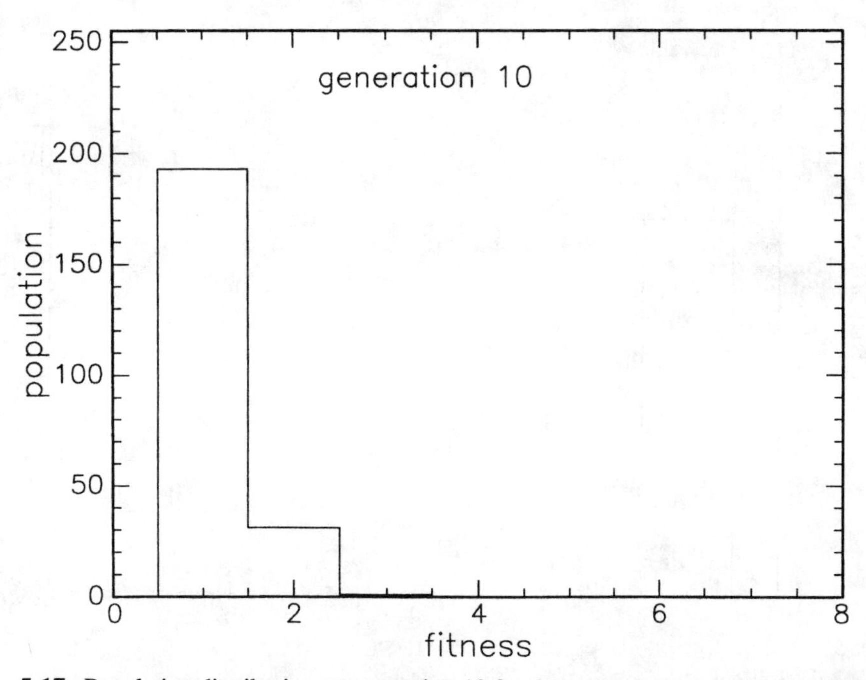

5-17 Population distribution at generation 10 for the cagannl.pas experiment.

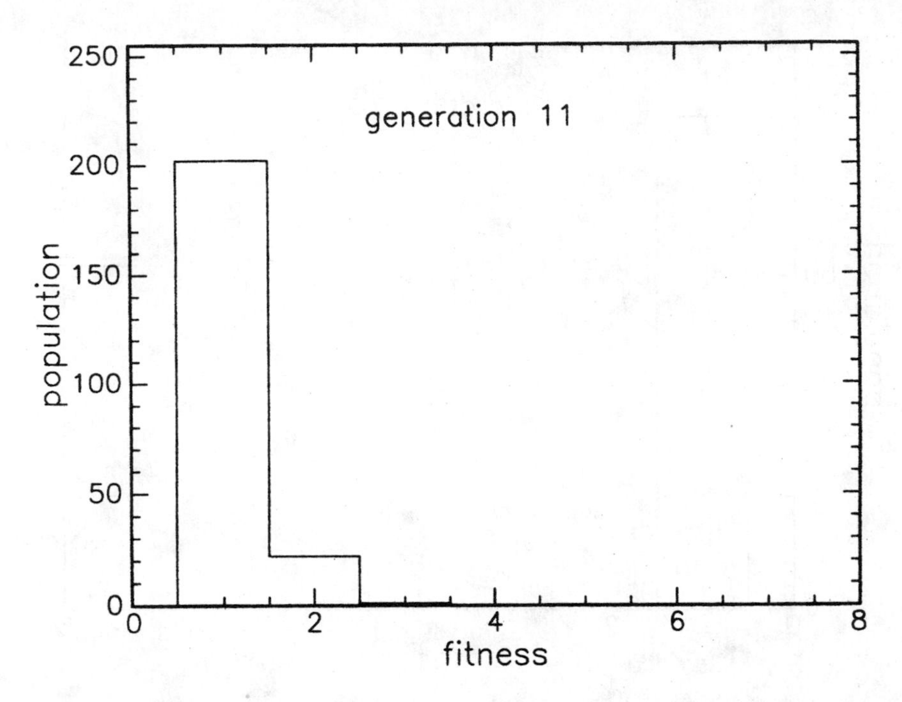

5-18 Population distribution at generation 11 for the cagannl.pas experiment.

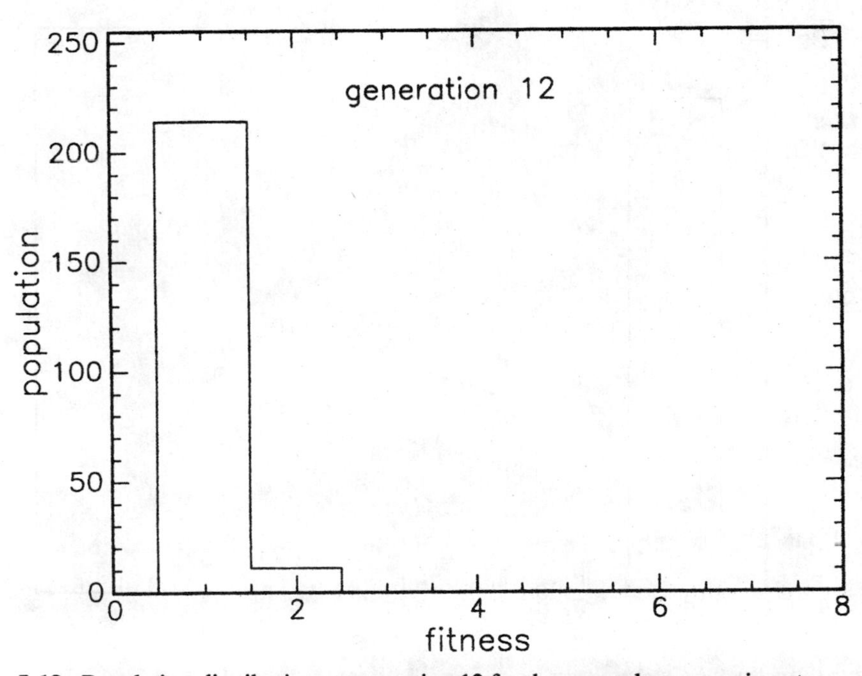

5-19 Population distribution at generation 12 for the cagannl.pas experiment.

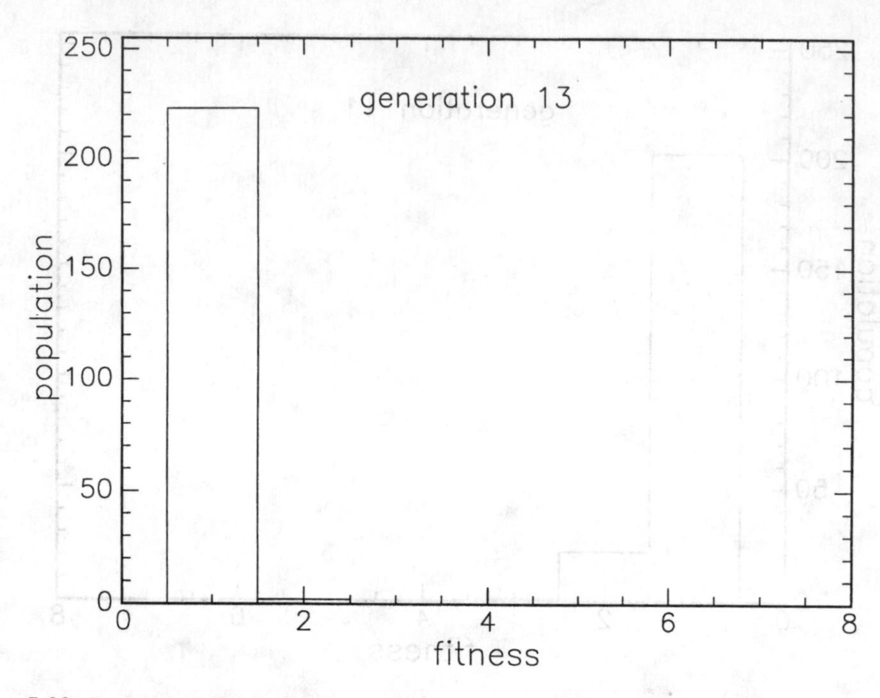

5-20 Population distribution at generation 13 for the cagannl.pas experiment.

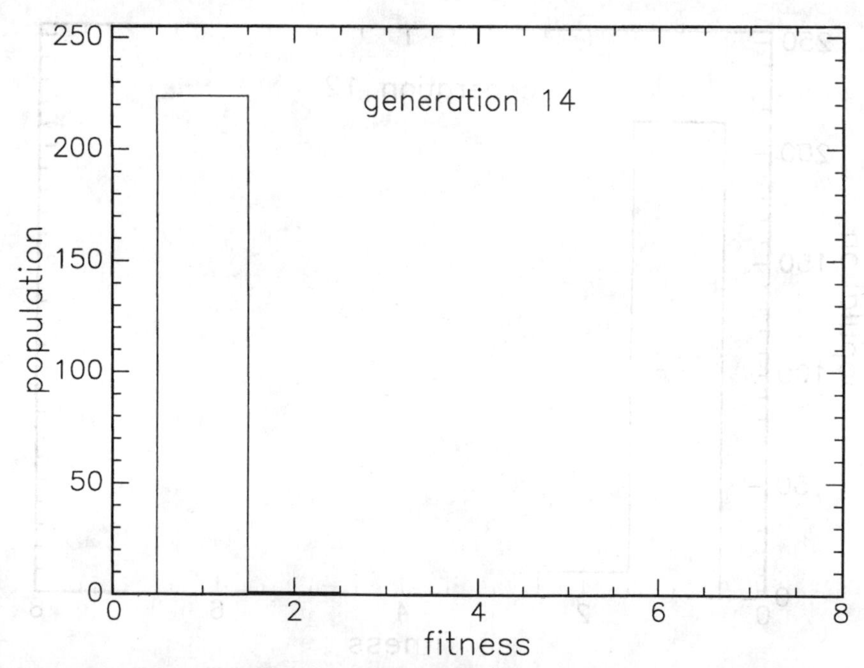

5-21 Population distribution at generation 14 for the cagannl.pas experiment.

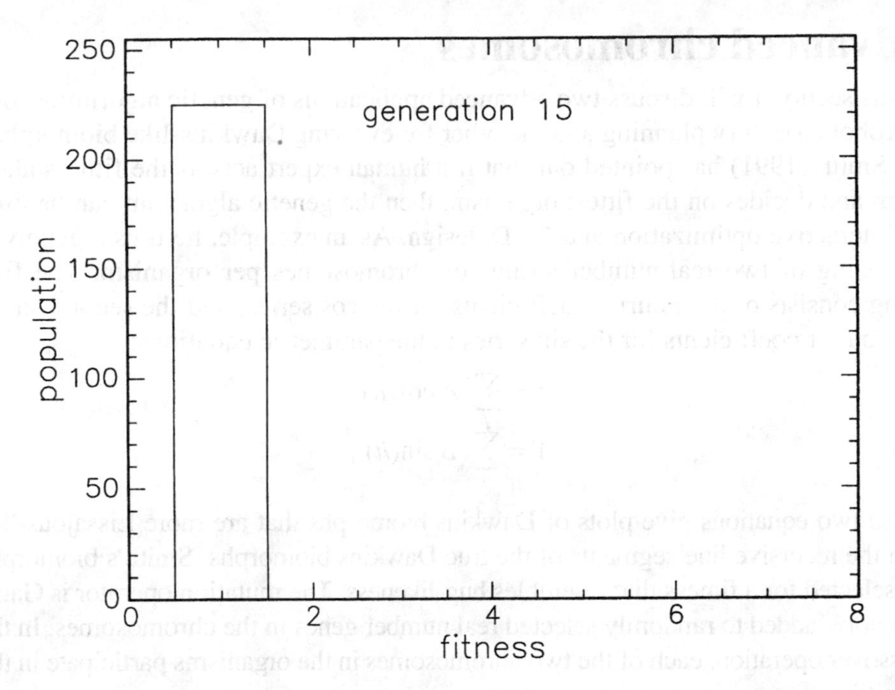

5-22 Population distribution at generation 15 for the cagannl.pas experiment.

Evolutionary Landscape

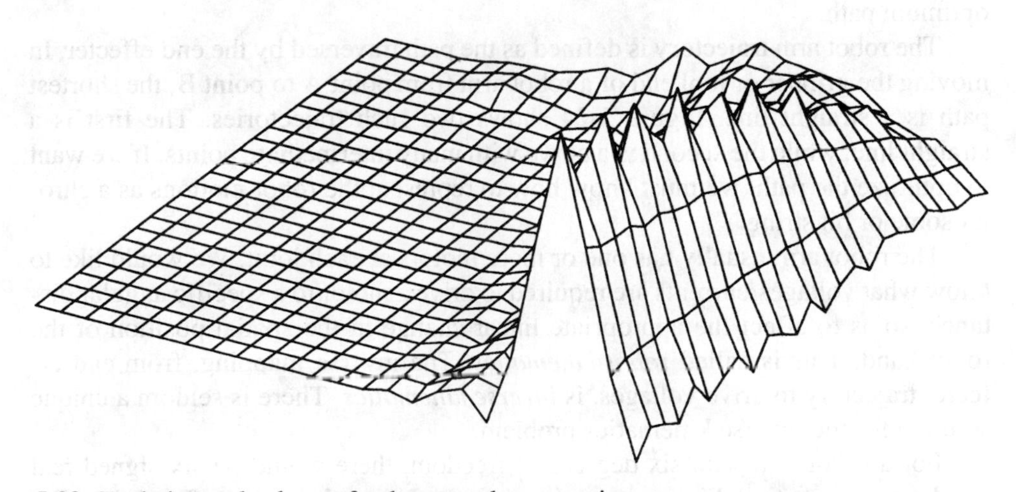

5-23 Evolutionary landscape for the cagannl.pas experiment.

Advanced chromosomes

In this section I will discuss two advanced applications of genetic algorithms: one for robot trajectory planning and the other for evolving Dawkins-like biomorphs.

Smith (1991) has pointed out that if a human expert acts as the fitness algorithm and decides on the fittest organism, then the genetic algorithm can be used for interactive optimization and CAD design. As an example, he uses a genotype consisting of two real-number strings or chromosomes per organism. The first string consists of the Fourier coefficients for the cos series, and the second string the Fourier coefficients for the sin series in the parametric equations:

$$X = \sum_i A_i \cos(it)$$

$$Y = \sum_i B_i \sin(it)$$

These two equations give plots of Dawkins biomorphs that are more Lissajous-like than the recursive line segments of the true Dawkins biomorphs. Smith's biomorphs are selected for a fitness that resembles bug-likeness. The mutation operator is Gaussian noise added to randomly selected real number genes in the chromosomes. In the crossover operation, each of the two chromosomes in the organisms participate in the operation.

Davidor (1991) has a paper in the *Genetic Algorithm Handbook*, edited by Davis (1991). This paper is a summary of his research on the genetic algorithm for robotic trajectory planning. His Ph.D. dissertation has been published (Davidor, 1991) by World Scientific Publishers. The objective of his study was to use genetic algorithms to generate feedforward controllers for control of a robot arm. In designing the trajectory or path for a robot arm to traverse, it is necessary to consider any obstacles in the path. Furthermore, the shortest path is most likely to be the optimum path; therefore, it is reasonable to use a genetic algorithm to "discover" this optimum path.

The robot arm trajectory is defined as the path traversed by the end effecter. In moving the gripper or tool end of a robot arm from point A to point B, the shortest path is a straight line. Figure 5-24 shows two such trajectories. The first is a straight line, while the second is a curve with many intermediate points. If we want to optimize the path, we must know how to represent the robot motions as a chromosome or bit string.

The robot arm usually has one or more motors at each joint. We would like to know what voltages/currents are required to drive a motor to a specific angular distance, so as to affect the appropriate linear change in the spatial position of the robot hand. This is called *robot kinematics*. The inverse mapping, from end effecter trajectory to drive voltages, is *inverse kinematics*. There is seldom a unique solution for the inverse kinematics problem.

For a robot arm with six degrees of freedom, there would be six signed real numbers representing drive voltages/currents to change the arm configuration. For

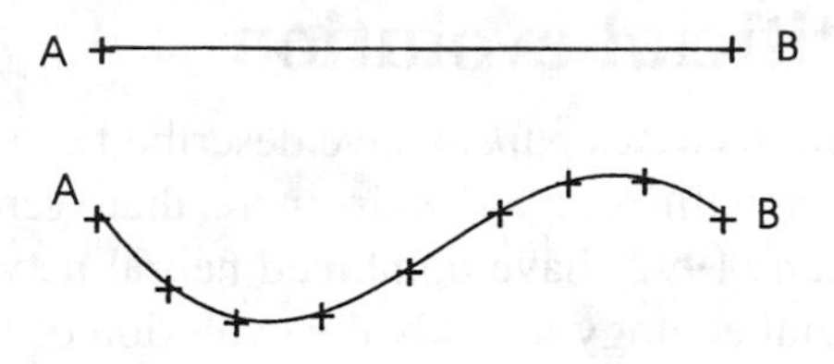

5-24 Examples of two robot trajectories. The second trajectory requires many vector changes.

the end effecter to traverse a trajectory, it might require the application of many such voltage vectors, as shown in Fig. 5-24. If we assume that l such vectors are required then, there would be 6xl voltage signals. The two trajectories shown in Fig. 5-24 would require a different number of voltage vectors.

The voltage vectors can represent alleles or segments of chromosomes. The l voltage vectors concatenated together, in the appropriate time sequence, represents the chromosomes. It is obvious that the voltage vectors must not be broken at crossover time and that crossover can only occur between the vectors or alleles. Furthermore, as can be seen from Fig. 5-24, it will be necessary to have variable length chromosomes for the adaptive path generation.

The selection of crossover sites could be a problem with chromosomes representing a time sequence of voltages. Davidor developed the solution that crossover is allowed only at physically meaningful sites. Figure 5-25 shows two trajectories that represent the chromosomes. If the end effecter is traveling from left to right along one of these paths, then only the circled regions should correspond to crossover sites. If crossover occurred at some random position, the end effecter could end up following a very chaotic trajectory.

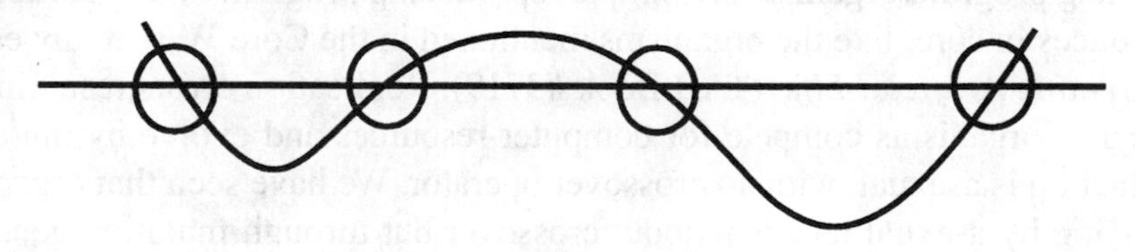

5-25 Examples of two robot trajectories. The circled regions represent places where crossover would be allowed between the chromosomes describing the two trajectories.

The mutation operator typically consists of adding Gaussian noise to a real number in the chromosome vectors. Davidor chose to delete or clone one voltage allele in the chromosome as the mutation operator. The effect is the same as mutation with Gaussian noise. Davidor (1991) should be consulted for the learning curves by this approach; the results are very effective.

Artificial evolution

Several research papers have described evolution of populations of organisms that interact with each other in more than genetic ways. For example, Ackley and Littman (1992) have combined neural network learning, genetic algorithms, and artificial ecology to study the evolution of intelligent agents. The agents have two neural networks: one maps the organism's sensory input to a scalar value representing fitness, while the other network maps sensory input to behavioral output. Both networks are determined by the chromosome pattern, but the second network has the capability of learning by back-propagation.

The agents operate in an environment consisting of walls, plants, and carnivores. The agents eat the plants, and the carnivores eat the agents. The agents must adopt/learn to stay alive by not starving (an energy level counter) and by not becoming dinner for a carnivore. The world is a 100x100 cellular array populated with the agents, plants, carnivorous, and walls. The artificial physics is set at the start and remains invariant through the lifetime of the world.

Ackley and Littman found oscillations in the population just like the Lotka-Volterra logistic equation discussed in #3719. This, however, was an emergent property because there were no different equations determining the dynamics. They also found that agents with learning and genetic algorithms performed better than agents using only one type. The agents were, in effect, using the learning to stay alive until the genetics hardwired it for them (an interesting point to note for man's future.)

Another interesting evolution simulation is by Ray (1991, 1992a, b) with his Tierra Simulator, which can be ordered from Media Magic (800-882-8284). Most genetic algorithms have a goal to optimize a fitness function. Ray's evolving organisms have an open-ended evolution, and many species can evolve and coexist. The starting program/organism is a simple reproducing program of 80 instructions. It reproduces in core, like the organisms mentioned in the Core Wars in my earlier book *Creating Artificial Life* (TAB Book #3719). Replication errors result in mutations. The organisms compete for computer resources and evolve by mutation. Reproduction is asexual, with no crossover operator. We have seen that organisms reproducing by asexual means without crossover but through mutation generally do not evolve as fast as organisms that reproduce sexually, due to the reduction of genetic diversity.

With a typical genetic algorithm, the evolution is toward fitness. With an open-ended evolution, the unfit individuals die young and only the strong survive. The weak do not inherit anything except an early grave. This is much like real biological evolution and is the approach Ray takes with his organisms. Counters are used to record the energy level and life span of each organism. CPU time is the energy resource, and memory is the material resource used by the organisms in order to stay live.

The chromosome of the initial organism consists of 80 operations. One of these operations is self-examination, to count the number of operations or chromosome length. It then locates a block of core of the appropriate length and copies itself into this block as a means of reproduction. Forty-eight of the 80 instructions are null operators or no-ops, which are used as templates to locate the beginnings and endings of code, as well as specific code segments. These no-op segments can be mutated by bit flipping and/or errors in counting of the genome length. The counting error can result in new daughter organisms with different code lengths. Thus new species can evolve, not just creating mutations of the same species. When the resources begin to run out, another program (called the Reaper) begins to remove long-lived individuals. A program called the Gene Bank keeps records of the number of organisms of each genome length.

In one experiment, Ray, let the program run for almost 3 billion instruction cycles. He presents the gene bank results as a table in his paper. I have converted that into a log-log distribution shown in Fig. 5-26.

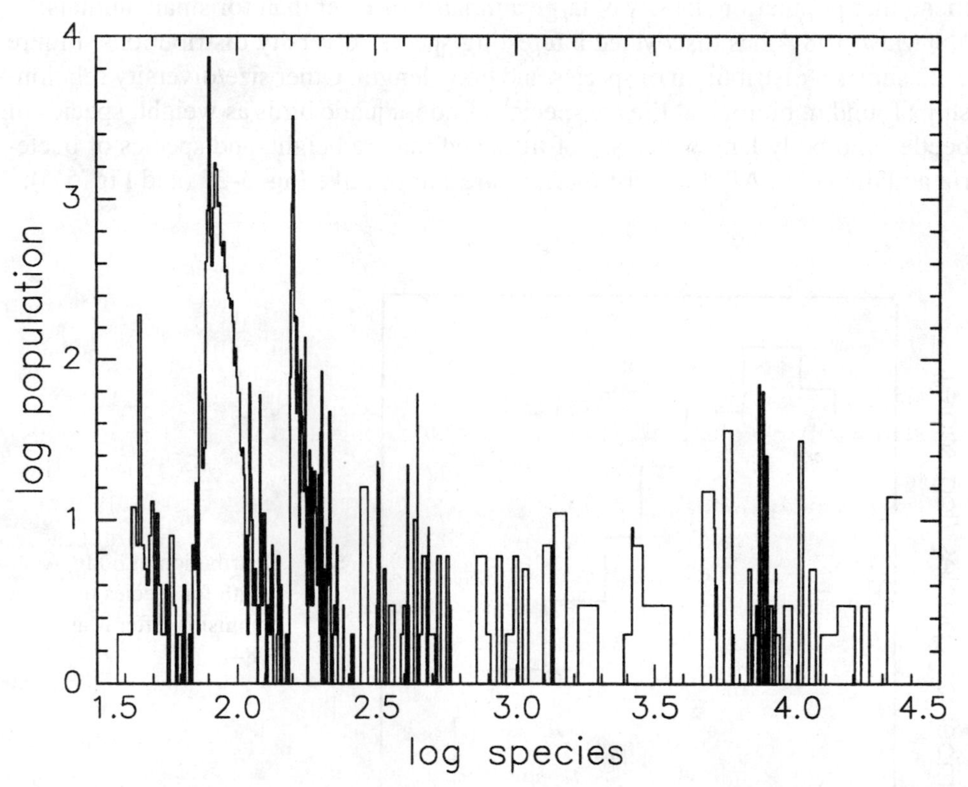

5-26 Log-Log plot of Ray's (1992) data.

Recall that the original organism used to inoculate the soup had a genome length of 80. The bulk of the artmass (artificial biomass, analog of biomass) is organisms with genomes of 80 (log 80 = 1.9) instructions. There is another large cluster at a length of 160 (log 160 = 2.2). The most surprising feature of this plot is the species diversity. But, before making further comments on this, I'd like to look at some work of May (1984).

Bonner (1988) discusses the evolution of complexity in biology. A simple observation is, that over the course of an evolutionary time scale, organisms (plants and animals) have become more complex in both the neural networks of the organisms and in the large physical size. In every case, the larger organisms are more complex. The main reason for increase in complexity is that there are always more ecological niches at the top; large organisms can always dominate smaller organisms. Thus, there are always species niches at the large (or more complex) end of the spectrum.

Evolution has not been (and isn't) a one-dimensional increase in complexity. Eldridge (1989) points out that it is a parallel convergent phenomenon. Even though evolution has resulted in greater complexity over the years, this does not mean that population density of large animals is greater than for small animals.

May (1984) has discovered interesting species diversity distributions. Figure 5-27 shows a distribution of species and body length. Other size/diversity relationships found in biological life are species of non-aquatic birds as weight, species of beetles and body length, species of trees and mature height, and species of bacteria and diameter. All these distributions are skewed like Fig. 5-27 (and Fig. 5-5).

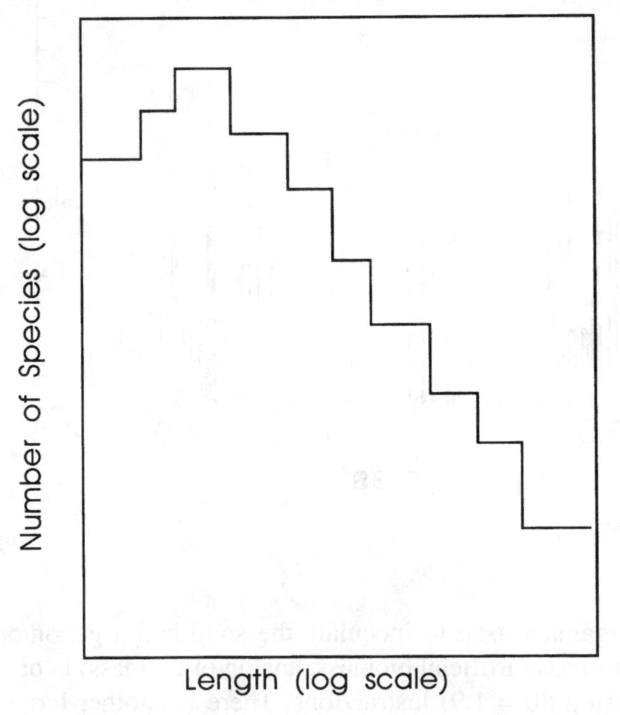

5-27 Distribution of body length for species of organisms (after May, 1984).

A notable feature of the Ray experiments is that his distributions are also skewed the same way. Figure 5-28 is a scaled linear distribution. Notice the distributions are exactly as those for the real biological diversity histograms of Fig. 5-27. Ray's artificial evolution programs have clearly modeled a real world ecological observation.

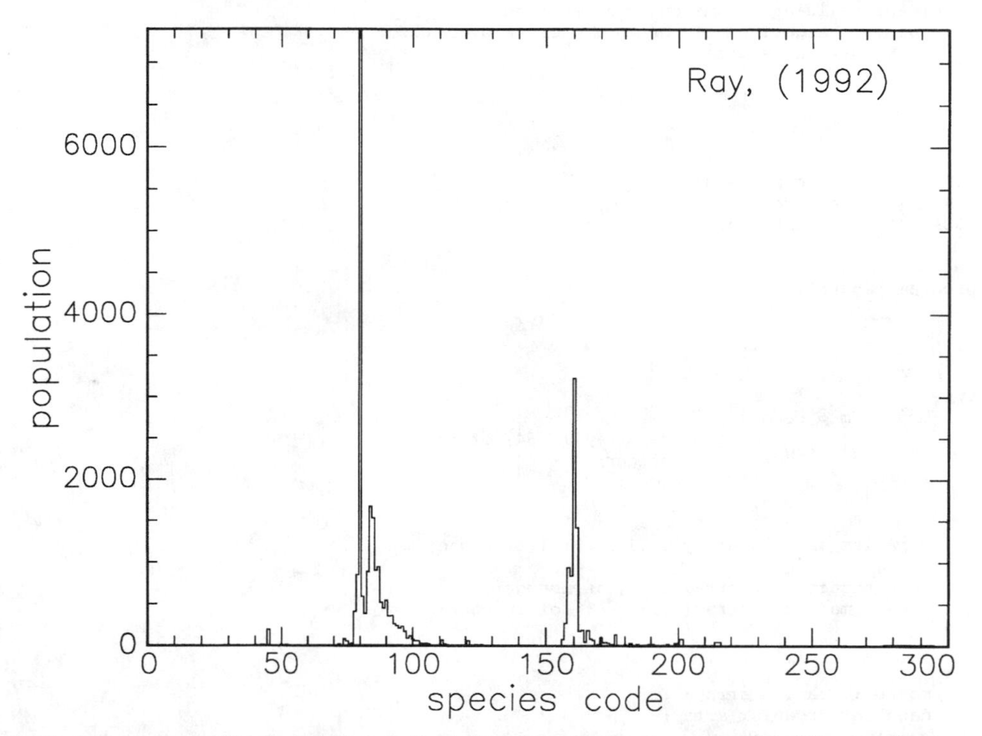

5-28 Linear plot of Ray's (1992) data. Scaled to show the maximum height of the distributions.

Summary

In this chapter, we have covered much ground. We started with an introduction to molecular biological genetic algorithms and a simple artificial genetic algorithm to optimize a toy problem. We then examined a biased sample of applications and neural network evolution by genetic algorithms. Two Pascal programs were examined, along with their results for development of Hopfield neural networks by a genetic algorithm.

Finally we concluded with two applications using complex chromosomes, biomorph evolution and robot control, and artificial evolution modeled by Ackley and Littman and by Ray. We discovered the surprising result that the artificial genetic algorithms applied to artificial evolution will produce diverse species with genome distributions resembling biological populations.

cagann1.pas

```
(* Program designed to evolve neural organisms.  The
   population is set up as two 2-d arrays of mother
   and daughter organisms.  These act as a cellular
   automata.  The state of the cell/orgainsm is
   determined by the fitness.  The fitness is a
   measure of the organisms neural network ability
   to solve a particular problem.  The new state
   of the cellular automaton is determined by
   a genetic algorithm of chromosome cross over and
   mutation with neighbors.

              Program date 021592                  *)



{$N+} (* numeric processing *)
{$R+} (* range processing *)
{$S+} (* stack checking *)


program cagann1;


(* data structures *)

type
    organism = record
                  chromosome:array[1..64] of integer;
                  fitval:integer;
                  flag:integer;
               end;

    organism_array = array[1..15,1..15] of organism;

    test_structure = array[1..8] of integer;
    synaps_matrix = array[1..8,1..8] of integer;


var
    mother:organism_array;
    daughter:organism_array;
    hamming:integer;
    generation:integer; (* genration counter *)


    site:integer; (* crossover point *)
    gene:real; (* used with random *)
    vector:test_structure;
    output:test_structure;
    neural_weights:synaps_matrix;

    fit0:integer; (* used in fitness histogram *)
    fit1:integer;
    fit2:integer;
    fit3:integer;
    fit4:integer;
    fit5:integer;
    fit6:integer;
    fit7:integer;
    fit8:integer;

    mutfreq:integer; (* number of mutations *)
    mutant1,mutant2:integer; (* used in mutations *)
    mutsite:integer; (* used in mutations *)
    value:integer; (* used in mutations *)

    x:integer; (* used in debugging *)
```

```
    imod,jmod:integer; (* used in mod operation *)
    enab1,enab2,enab3,enab4:integer; (* excluded neighbors *)

    i,j,k,m,n: integer; (* counters *)



begin (* main *)

      vector[1] := 0; (* the test vector *)
      vector[2] := 1;
      vector[3] := 0;
      vector[4] := 0;
      vector[5] := 0;
      vector[6] := 0;
      vector[7] := 0;
      vector[8] := 1;


      (* set the number to be mutated *)
      mutfreq := 20;

      (* initialize the population *)

      for i := 1 to 15 do (* each member of population *)
      begin
           for j := 1 to 15 do
           begin
                for k := 1 to 64 do (* for each gene *)
                begin

                     gene := random;
                     if gene <= 0.5 then
                        mother[i,j].chromosome[k] := 1
                     else
                        mother[i,j].chromosome[k] := 0;

                     {
                     (* for dubugging only *)
                     x := mother[i,j].chromosome[k];
                     write(x:5);
                     }

                end;

           end;

      end;


      (* main algorithm *)

      while(generation < 1000 ) do
      begin
           writeln;
           writeln(generation);
           generation := generation + 1;

           fit0 := 0; (* used in histograms *)
                fit1 := 0;
           fit2 := 0;
                fit3 := 0;
                fit4 := 0;
                fit5 := 0;
                fit6 := 0;
                fit7 := 0;
```

```
    fit8 := 0;


  (* compute fitness (cell state) values *)

  for i := 1 to 15 do (* each member of population *)
  begin
      for j := 1 to 15 do
      begin

          (* fold the chromosome into neural matrix *)
          for k := 1 to 8 do
          begin
            for m := 1 to 8 do
            begin
              neural_weights[k,m] :=
                     mother[i,j].chromosome[(k-1)*8+m];

            end;
          end;


          (* compute the vector matrix product *)
          for k := 1 to 8 do
          begin
              output[k] := 0;
              for m := 1 to 8 do
              begin
                   output[k] := output[k] +
                          neural_weights[k,m]*vector[m];
              end;


              (* threshold the output *)
              if output[k] > 1 then output[k] := 1;

              {
              (* for debugging only *)
              write(output[k]);
              }

          end; (* end vector product *)

          {
          (* for debugging only *)
          writeln;
          }



          (* compute the hamming distance *)
          hamming := 0;
          for k := 1 to 8 do
          begin
               if vector[k] = output[k] then
                  hamming := hamming + 1;
          end;

          mother[i,j].fitval := hamming;

          (* compute the distribution *)
          if mother[i,j].fitval = 0 then
             fit0 := fit0 + 1;
          if mother[i,j].fitval = 1 then
             fit1 := fit1 + 1;
          if mother[i,j].fitval = 2 then
             fit2 := fit2 + 1;
          if mother[i,j].fitval = 3 then
```

```
                fit3 := fit3 + 1;
             if mother[i,j].fitval = 4 then
                fit4 := fit4 + 1;
             if mother[i,j].fitval = 5 then
                fit5 := fit5 + 1;
             if mother[i,j].fitval = 6 then
                fit6 := fit6 + 1;
             if mother[i,j].fitval = 7 then
                fit7 := fit7 + 1;
             if mother[i,j].fitval = 8 then
                fit8 := fit8 + 1;
         end; (* end j loop *)

    end;  (* end i loop *)


    (* print the distribution *)
    writeln('0  ',fit0);
    writeln('1  ',fit1);
    writeln('2  ',fit2);
    writeln('3  ',fit3);
    writeln('4  ',fit4);
    writeln('5  ',fit5);
    writeln('6  ',fit6);
    writeln('7  ',fit7);
    writeln('8  ',fit8);


    (* neighbor order
       ne = 0,
       n  = 1,
       nw = 2,
       e  = 3,
       w  = 4,
       se = 5,
       s  = 6,
       sw = 7 and

       n  = 0,
       e  = 1,
       w  = 2,
       s  = 3
       used in computing excluded neighbors for mating *)

    (* select mate and reproduce *)
    for i := 1 to 15 do
    begin
         for j := 1 to 15 do
         begin

              if mother[i,j].fitval = 0 then
              begin
                 (* mate with eight neighbors *)
                 (* N.E. neighbor *)
                 imod := i mod 15 - 1;
                 if imod <= 0 then imod := 1;
                 jmod := j mod 15 - 1;
                 if jmod <= 0 then jmod := 1;
                 site := round(random*63) + 1;
                 for k := 1 to site do
                 begin
                      daughter[imod,jmod].chromosome[k]
:=
                           mother[i,j].chromosome[k];
                 end;
                 for k := site to 64 do
                 begin
                      daughter[imod,jmod].chromosome[k]
```

```
:=
          mother[i,j].chromosome[k];
end;

(* N. neighbor *)
imod := i mod 15 - 1;
if imod <= 0 then imod := 1;
jmod := j mod 15;
if jmod <= 0 then jmod := 1;
site := round(random*63) + 1;
for k := 1 to site do
begin
     daughter[imod,jmod].chromosome[k] :=
          mother[i,j].chromosome[k];
end;
for k := site to 64 do
begin
     daughter[imod,jmod].chromosome[k] :=
          mother[i,j].chromosome[k];
end;

(* N.W. neighbor *)
imod := i mod 15 - 1;
if imod <= 0 then imod := 1;
jmod := j mod 15 + 1;
if jmod <= 0 then jmod := 1;
site := round(random*63) + 1;
for k := 1 to site do
begin
     daughter[imod,jmod].chromosome[k] :=
          mother[i,j].chromosome[k];
end;
for k := site to 64 do
begin
     daughter[imod,jmod].chromosome[k] :=
          mother[i,j].chromosome[k];
end;

(* E. neighbor *)
imod := i mod 15;
if imod <= 0 then imod := 1;
jmod := j mod 15 - 1;
if jmod <= 0 then jmod := 1;
site := round(random*63) + 1;
for k := 1 to site do
begin
     daughter[imod,jmod].chromosome[k] :=
          mother[i,j].chromosome[k];
end;
for k := site to 64 do
begin
     daughter[imod,jmod].chromosome[k] :=
          mother[i,j].chromosome[k];
end;

(* W. neighbor *)
imod := i mod 15;
if imod <= 0 then imod := 1;
jmod := j mod + 1;
if jmod <= 0 then jmod := 1;
site := round(random*63) + 1;
for k := 1 to site do
begin
     daughter[imod,jmod].chromosome[k] :=
          mother[i,j].chromosome[k];
```

```
                        end;
                        for k := site to 64 do
                        begin
                            daughter[imod,jmod].chromosome[k] :=
                                mother[i,j].chromosome[k];
                        end;

                        (* S.E. neighbor *)
                        imod := i mod 15 + 1;
                        if imod <= 0 then imod := 1;
                        jmod := j mod 15 - 1;
                        if jmod <= 0 then jmod := 1;
                        site := round(random*63) + 1;
                        for k := 1 to site do
                        begin
                            daughter[imod,jmod].chromosome[k] :=
                                mother[i,j].chromosome[k];
                        end;
                        for k := site to 64 do
                        begin
                            daughter[imod,jmod].chromosome[k] :=
                                mother[i,j].chromosome[k];
                        end;

                        (* S. neighbor *)
                        imod := i mod 15 + 1;
                        if imod <= 0 then imod := 1;
                        jmod := j mod 15;
                        if jmod <= 0 then jmod := 1;
                        site := round(random*63) + 1;
                        for k := 1 to site do
                        begin
                            daughter[imod,jmod].chromosome[k] :=
                                mother[i,j].chromosome[k];
                        end;
                        for k := site to 64 do
                        begin
                            daughter[imod,jmod].chromosome[k] :=
                                mother[i,j].chromosome[k];
                        end;


                        (* S.W. neighbor *)
                        imod := i mod 15 + 1;
                        if imod <= 0 then imod := 1;
                        jmod := j mod 15 + 1;
                        if jmod <= 0 then jmod := 1;
                        site := round(random*63) + 1;
                        for k := 1 to site do
                        begin
                            daughter[imod,jmod].chromosome[k] :=
                                mother[i,j].chromosome[k];
                        end;
                        for k := site to 64 do
                        begin
                            daughter[imod,jmod].chromosome[k] :=
                                mother[i,j].chromosome[k];
                        end;
                    end;


                    if mother[i,j].fitval = 1 then
                    begin
                       (* mate with six neighbors *)
                       enab1 := round(random*7);
                       while enab2 <> enab1 do
```

```
begin
     enab2 := round(random*7);
end;

(* N.E. neighbor *)
if (enab1 <> 0) or (enab2 <> 0) then
begin
     imod := i mod 15 - 1;
     if imod <= 0 then imod := 1;
     jmod := j mod 15 - 1;
     if jmod <= 0 then jmod := 1;
     site := round(random*63) + 1;
     for k := 1 to site do
     begin
          daughter[imod,jmod].chromosome[k] :=
             mother[i,j].chromosome[k];
     end;
     for k := site to 64 do
     begin
          daughter[imod,jmod].chromosome[k] :=
            mother[i,j].chromosome[k];
     end;
end;

(* N. neighbor *)
if (enab1 <> 1) or (enab2 <> 1) then
begin
     imod := i mod 15 - 1;
     if imod <= 0 then imod := 1;
     jmod := j mod 15;
     if jmod <= 0 then jmod := 1;
     site := round(random*63) + 1;
     for k := 1 to site do
     begin
          daughter[imod,jmod].chromosome[k] :=
             mother[i,j].chromosome[k];
     end;
     for k := site to 64 do
     begin
          daughter[imod,jmod].chromosome[k] :=
             mother[i,j].chromosome[k];
     end;
end;

(* N.W. neighbor *)
if (enab1 <> 2) or (enab2 <> 2) then
begin
     imod := i mod 15 - 1;
     if imod <= 0 then imod := 1;
     jmod := j mod 15 + 1;
     if jmod <= 0 then jmod := 1;
     site := round(random*63) + 1;
     for k := 1 to site do
     begin
          daughter[imod,jmod].chromosome[k] :=
             mother[i,j].chromosome[k];
     end;
     for k := site to 64 do
     begin
          daughter[imod,jmod].chromosome[k] :=
              mother[i,j].chromosome[k];
     end;
end;

(* E. neighbor *)
if (enab1 <> 3) or (enab2 <> 3) then
begin
     imod := i mod 15;
     if imod <= 0 then imod := 1;
```

```
        jmod := j mod 15 - 1;
        if jmod <= 0 then jmod := 1;
        site := round(random*63) + 1;
        for k := 1 to site do
        begin
             daughter[imod,jmod].chromosome[k] :=
                mother[i,j].chromosome[k];
        end;
        for k := site to 64 do
        begin
             daughter[imod,jmod].chromosome[k] :=
                mother[i,j].chromosome[k];
        end;
   end;

   (* W. neighbor *)
   if (enab1 <> 4) or (enab2 <> 4) then
   begin
        imod := i mod 15;
        if imod <= 0 then imod := 1;
        jmod := j mod + 1;
        if jmod <= 0 then jmod := 1;
        site := round(random*63) + 1;
        for k := 1 to site do
        begin
             daughter[imod,jmod].chromosome[k] :=
                mother[i,j].chromosome[k];
        end;
        for k := site to 64 do
        begin
             daughter[imod,jmod].chromosome[k] :=
                mother[i,j].chromosome[k];
        end;
   end;

   (* S.E. neighbor *)
   if (enab1 <> 5) or (enab2 <> 5) then
   begin
        imod := i mod 15 + 1;
        if imod <= 0 then imod := 1;
        jmod := j mod 15 - 1;
        if jmod <= 0 then jmod := 1;
        site := round(random*63) + 1;
        for k := 1 to site do
        begin
             daughter[imod,jmod].chromosome[k] :=
                mother[i,j].chromosome[k];
        end;
        for k := site to 64 do
        begin
             daughter[imod,jmod].chromosome[k] :=
                mother[i,j].chromosome[k];
        end;
   end;

   (* S. neighbor *)
   if (enab1 <> 6) or (enab2 <> 6) then
   begin
        imod := i mod 15 + 1;
        if imod <= 0 then imod := 1;
        jmod := j mod 15;
        if jmod <= 0 then jmod := 1;
        site := round(random*63) + 1;
        for k := 1 to site do
        begin
             daughter[imod,jmod].chromosome[k] :=
                mother[i,j].chromosome[k];
        end;
        for k := site to 64 do
```

```
                    begin
                        daughter[imod,jmod].chromosome[k] :=
                           mother[i,j].chromosome[k];
                    end;
                end;
                (* S.W. neighbor *)
                if (enab1 <> 7) or (enab2 <> 7) then
                begin
                    imod := i mod 15 + 1;
                    if imod <= 0 then imod := 1;
                    jmod := j mod 15 + 1;
                    if jmod <= 0 then jmod := 1;
                    site := round(random*63) + 1;
                    for k := 1 to site do
                    begin
                        daughter[imod,jmod].chromosome[k] :=
                           mother[i,j].chromosome[k];
                    end;
                    for k := site to 64 do
                    begin
                        daughter[imod,jmod].chromosome[k] :=
                           mother[i,j].chromosome[k];
                    end;
                end;
            end;

            if mother[i,j].fitval = 2 then
            begin
                (* mate with four neighbors *)
                (* N. neighbor *)
                    imod := i mod 15 - 1;
                    if imod <= 0 then imod := 1;
                    jmod := j mod 15;
                    if jmod <= 0 then jmod := 1;
                    site := round(random*63) + 1;
                    for k := 1 to site do
                    begin
                        daughter[imod,jmod].chromosome[k] :=
                           mother[i,j].chromosome[k];
                    end;
                    for k := site to 64 do
                    begin
                        daughter[imod,jmod].chromosome[k] :=
                           mother[i,j].chromosome[k];
                    end;

                (* E. neighbor *)
                    imod := i mod 15;
                    if imod <= 0 then imod := 1;
                    jmod := j mod 15 - 1;
                    if jmod <= 0 then jmod := 1;
                    site := round(random*63) + 1;
                    for k := 1 to site do
                    begin
                        daughter[imod,jmod].chromosome[k] :=
                           mother[i,j].chromosome[k];
                    end;
                    for k := site to 64 do
                    begin
                        daughter[imod,jmod].chromosome[k] :=
                            mother[i,j].chromosome[k];
                    end;

                (* W. neighbor *)
                    imod := i mod 15;
                    if imod <= 0 then imod := 1;
```

```
        jmod := j mod + 1;
        if jmod <= 0 then jmod := 1;
        site := round(random*63) + 1;
        for k := 1 to site do
        begin
             daughter[imod,jmod].chromosome[k] :=
                mother[i,j].chromosome[k];
        end;
        for k := site to 64 do
        begin
             daughter[imod,jmod].chromosome[k] :=
                mother[i,j].chromosome[k];
        end;

   (* S. neighbor *)
        imod := i mod 15 + 1;
        if imod <= 0 then imod := 1;
        jmod := j mod 15;
        if jmod <= 0 then jmod := 1;
        site := round(random*63) + 1;
        for k := 1 to site do
        begin
             daughter[imod,jmod].chromosome[k] :=
                mother[i,j].chromosome[k];
        end;
        for k := site to 64 do
        begin
             daughter[imod,jmod].chromosome[k] :=
                mother[i,j].chromosome[k];
        end;
end;

if mother[i,j].fitval = 3 then
begin
   (* mate with two neighbors *)
   enab1 := round(random*3);
   while(enab2 <> enab1) do
   begin
        enab2 := round(random*3);
   end;
   while((enab3 <> enab2) or (enab3 <> enab1)) do
   begin
        enab3 := round(random*3);
   end;

   (* N. neighbor *)
   if (enab1 <> 0) or (enab2 <> 0) or (enab3 <> 0) then
   begin
        imod := i mod 15 - 1;
        if imod <= 0 then imod := 1;
        jmod := j mod 15;
        if jmod <= 0 then jmod :- 1;
        site := round(random*63) + 1;
        for k := 1 to site do
        begin
             daughter[imod,jmod].chromosome[k] :=
                mother[i,j].chromosome[k];
        end;
        for k := site to 64 do
        begin
             daughter[imod,jmod].chromosome[k] :=
                mother[i,j].chromosome[k];
        end;
   end;

   (* E. neighbor *)
   if (enab1 <> 1) or (enab2 <> 1) or (enab3 <> 1) then
   begin
        imod := i mod 15;
```

```
                if imod <= 0 then imod := 1;
                jmod := j mod 15 - 1;
                if jmod <= 0 then jmod := 1;
                site := round(random*63) + 1;
                for k := 1 to site do
                begin
                     daughter[imod,jmod].chromosome[k] :=
                        mother[i,j].chromosome[k];
                end;
                for k := site to 64 do
                begin
                     daughter[imod,jmod].chromosome[k] :=
                         mother[i,j].chromosome[k];
                end;
           end;

           (* W. neighbor *)
           if (enab1 <> 2) or (enab2 <> 2) or (enab3 <> 2) then
           begin
                imod := i mod 15;
                if imod <= 0 then imod := 1;
                jmod := j mod + 1;
                if jmod <= 0 then jmod := 1;
                site := round(random*63) + 1;
                for k := 1 to site do
                begin
                     daughter[imod,jmod].chromosome[k] :=
                        mother[i,j].chromosome[k];
                end;
                for k := site to 64 do
                begin
                     daughter[imod,jmod].chromosome[k] :=
                        mother[i,j].chromosome[k];
                end;
           end;

           (* S. neighbor *)
           if (enab1 <> 3) or (enab2 <> 3) or (enab3 <> 3) then
           begin
                imod := i mod 15 + 1;
                if imod <= 0 then imod := 1;
                jmod := j mod 15;
                if jmod <= 0 then jmod := 1;
                site := round(random*63) + 1;
                for k := 1 to site do
                begin
                     daughter[imod,jmod].chromosome[k] :=
                        mother[i,j].chromosome[k];
                end;
                for k := site to 64 do
                begin
                     daughter[imod,jmod].chromosome[k] :=
                        mother[i,j].chromosome[k];
                end;
           end;

     end;


     if mother[i,j].fitval = 4 then
     begin
           enab1 := round(random*3);
           (* mate with one neighbor *)

           (* N. neighbor *)
           if enab1 = 0 then
           begin
                imod := i mod 15 - 1;
```

```
        if imod <= 0 then imod := 1;
        jmod := j mod 15;
        if jmod <= 0 then jmod := 1;
        site := round(random*63) + 1;
        for k := 1 to site do
        begin
             daughter[imod,jmod].chromosome[k] :=
                mother[i,j].chromosome[k];
        end;
        for k := site to 64 do
        begin
             daughter[imod,jmod].chromosome[k] :=
                mother[i,j].chromosome[k];
        end;
   end;

   (* E. neighbor *)
   if enabl = 1 then
   begin
        imod := i mod 15;
        if imod <= 0 then imod := 1;
        jmod := j mod 15 - 1;
        if jmod <= 0 then jmod := 1;
        site := round(random*63) + 1;
        for k := 1 to site do
        begin
             daughter[imod,jmod].chromosome[k] :=
                mother[i,j].chromosome[k];
        end;
        for k := site to 64 do
        begin
             daughter[imod,jmod].chromosome[k] :=
                 mother[i,j].chromosome[k];
        end;
   end;

   (* W. neighbor *)
   if enabl = 2 then
   begin
        imod := i mod 15;
        if imod <= 0 then imod := 1;
        jmod := j mod + 1;
        if jmod <= 0 then jmod := 1;
        site := round(random*63) + 1;
        for k := 1 to site do
        begin
             daughter[imod,jmod].chromosome[k] :=
                mother[i,j].chromosome[k];
        end;
        for k := site to 64 do
        begin
             daughter[imod,jmod].chromosome[k] :-
                mother[i,j].chromosome[k];
        end;
   end;

   (* S. neighbor *)
   if enabl = 3 then
   begin
        imod := i mod 15 + 1;
        if imod <= 0 then imod := 1;
        jmod := j mod 15;
        if jmod <= 0 then jmod := 1;
        site := round(random*63) + 1;
        for k := 1 to site do
        begin
             daughter[imod,jmod].chromosome[k] :=
                mother[i,j].chromosome[k];
        end;
```

```
                    for k := site to 64 do
                    begin
                         daughter[imod,jmod].chromosome[k] :=
                            mother[i,j].chromosome[k];
                    end;
               end;

          end;


          if mother[i,j].fitval > 4 then
          begin
             (* do not mate *)
          end;

     end;
end;

(* mutate some genes *)
for i := 1 to mutfreq do
begin
     (* select an organism *)
     mutant1 := round(random*14) + 1;
     mutant2 := round(random*14) + 1;

     (* select mutation site *)
     mutsite := round(random*63) + 1;

     value := round(random*1);
     mother[mutant1,mutant2].chromosome[mutsite] := value;
end;

(* set daughters to mothers *)
for i := 1 to 15 do
begin
     for j := 1 to 15 do
     begin
          for k := 1 to 64 do
          begin
               mother[i,j].chromosome[k] :=
                 daughter[i,j].chromosome[k];
          end;
     end;
end;


end; (* end generation counter *)

end. (* end main *)
```

degaris1.pas

```
(* program designed to simulate the De Garis genetic algorithm
   for the evolution of Hopfield neural networks.  012592
   random selection with own species only i.e. same hamming code *)

{$N+} (* numeric processing *)
{$R+} (* range processing *)
{$S+} (* stack checking *)


program degaris1;


(* data structures *)

type
    organism = record
```

```
                    chromosome:array[1..64] of integer;
                    fitval:integer;
                    fitfract:real;
                              speciecode:integer;
                    mateflag:integer;
                 end;

    organism_array = array[1..100] of organism;

    test_structure = array[1..8] of integer;
    synaps_matrix = array[1..8,1..8] of integer;


var
    pop_member:organism_array;
    new_pop:organism_array;
    max_fitval:integer;
    hamming:integer;
    avg_hamming:real; (* population average -scaled- *)
    generation:integer; (* genration counter *)

    gene:real; (* used with random *)
    vector:test_structure;
    output:test_structure;
    neural_weights:synaps_matrix;
    specie_hamming:integer;
    site:integer;
    mutant:integer;
    mutsite:integer;
    mutfreq:integer;

    fit0:integer;
    fit1:integer;
    fit2:integer;
    fit3:integer;
    fit4:integer;
    fit5:integer;
    fit6:integer;
    fit7:integer;
    fit8:integer;

    i,j,k: integer; (* counters *)
    x:integer;


begin (* main *)

      vector[1] := 0;
      vector[2] := 1;
      vector[3] := 0;
      vector[4] := 0;
      vector[5] := 0;
      vector[6] := 0;
      vector[7] := 0;
      vector[8] := 1;


      (* set the number to be mutated *)
      mutfreq := 20;


      (* initialize the population *)

      for i := 1 to 100 do (* each member of population *)
      begin
           for j := 1 to 64 do (* each gene in chromosome
*)
```

```
          begin
               gene := random;
               if gene <= 0.5 then
                     pop_member[i].chromosome[j] := 1
               else
                     pop_member[i].chromosome[j] := 0;

          end;

     end;


     (* main algorithm *)

     while(generation < 250 ) do
     begin
          writeln;
          writeln(generation);
          generation := generation + 1;

          fit0 := 0;
               fit1 := 0;
          fit2 := 0;
               fit3 := 0;
               fit4 := 0;
               fit5 := 0;
               fit6 := 0;
               fit7 := 0;
               fit8 := 0;

          (* set the mate flags *)
          for i := 1 to 100 do
          begin
               pop_member[i].mateflag := -1;
          end;


          (* compute fitness values *)

          for i := 1 to 100 do (* each member of population *)
          begin


               (* fold the chromosome into neural matrix
                       i.e. produce the phenotype *)
               for j := 1 to 8 do
               begin
                    for k := 1 to 8 do
                    begin

                        neural_weights[j,k] :=
                              pop_member[i].chromosome[(j-1)*8+k];

                    end;

               end; (* end phenotype construction *)


               (* compute the vector matrix product
                     i.e. expose the organism to the environment *)
               for j := 1 to 8 do
               begin
                    output[j] := 0;
                    for k := 1 to 8 do
                    begin
                         output[j] := output[j] +
                                    neural_weights[j,k]*vector[k];
```

```
          end;
          (* threshold operation *)
          if output[j] > 1 then output[j] := 1;

     end; (* end vector product *)


     (* compute the hamming distance *)
     hamming := 0;
     for j := 1 to 8 do
     begin
          if vector[j] = output[j] then
             hamming := hamming + 1;
     end;

     pop_member[i].fitval := hamming;
     (* end hamming computation *)

     avg_hamming := (avg_hamming + hamming)/8.0;

end; (* next member of population (i loop) *)


(* find fitness fractions *)
for i := 1 to 100 do (* for each member of the population *)
begin

     pop_member[i].fitfract := pop_member[i].fitval/8.0;

end;


(* compute the distribution of species *)
for i := 1 to 100 do
begin
     if pop_member[i].fitval = 0 then
        fit0 := fit0 + 1;
     if pop_member[i].fitval = 1 then
        fit1 := fit1 + 1;
     if pop_member[i].fitval = 2 then
        fit2 := fit2 + 1;
     if pop_member[i].fitval = 3 then
        fit3 := fit3 + 1;
     if pop_member[i].fitval = 4 then
        fit4 := fit4 + 1;
     if pop_member[i].fitval = 5 then
        fit5 := fit5 + 1;
     if pop_member[i].fitval = 6 then
        fit6 := fit6 + 1;
     if pop_member[i].fitval = 7 then
        fit7 := fit7 + 1;
     if pop_member[i].fitval = 8 then
        fit8 := fit8 + 1;

end;


(* print the distribution *)
writeln('0  ',fit0);
writeln('1  ',fit1);
writeln('2  ',fit2);
writeln('3  ',fit3);
writeln('4  ',fit4);
writeln('5  ',fit5);
writeln('6  ',fit6);
writeln('7  ',fit7);
writeln('8  ',fit8);
```

```
(* set mate code within this species distribution *)
for i := 1 to 100 do
begin
     x := round(random*100);

     if (x > 0) and (x <= fit0)
     then pop_member[i].speciecode := 0;

               if (x > fit0) and (x <= fit0+fit1)
     then pop_member[i].speciecode := 1;

               if (x > fit0+fit1) and
                  (x <= fit0+fit1+fit2)
     then pop_member[i].speciecode := 2;

               if (x > fit0+fit1+fit2) and
                  (x <= fit0+fit1+fit2+fit3)
     then pop_member[i].speciecode := 3;

               if (x > fit0+fit1+fit2+fit3) and
                  (x <= fit0+fit1+fit2+fit3+fit4)
     then pop_member[i].speciecode := 4;

               if (x > fit0+fit1+fit2+fit3+fit4) and
                  (x <= fit0+fit1+fit2+fit3+fit4+fit5)
     then pop_member[i].speciecode := 5;

               if (x > fit0+fit1+fit2+fit3+fit4+fit5) and
                  (x <= fit0+fit1+fit2+fit3+fit4+fit5+fit6)
     then pop_member[i].speciecode := 6;

               if (x > fit0+fit1+fit2+fit3+fit4+fit5+fit6) and
                  (x <= fit0+fit1+fit2+fit3+fit4+fit5+fit6+fit7)
     then pop_member[i].speciecode := 7;

               if (x > fit0+fit1+fit2+fit3+fit4+fit5+fit6+fit7) and
         (x <= 100)
     then pop_member[i].speciecode := 7; (* mate exception *)


end;


(* select mate id *)
for i := 1 to 100 do
begin
     j := i;
     if j > 99 then j := 1;
     while pop_member[i].mateflag = -1 do
     begin
          if pop_member[i].speciecode =
             pop_member[j+1].speciecode then
             begin
                  pop_member[i].mateflag := j+1;
                  pop_member[j+1].mateflag := i;


             end
          else
              j := j+1;
              if j > 99 then j := 1;

     end;

end; (* end mate id selection *)

{
(* a test segment of code *)
```

```
for i := 1 to 100 do
begin
     k := pop_member[i].mateflag;
     writeln(i,k:6);
end;
}


(* set mate flag for the new population array *)
for i:= 1 to 100 do
begin
     new_pop[i].mateflag := -1;
end;


(* generate new organisms by crossover *)
for i := 1 to 100 do
begin
     while new_pop[i].mateflag = -1 do
     begin
          site := round(random*64);
          for j := 1 to site do
          begin
              new_pop[i].chromosome[j] :=
                pop_member[i].chromosome[j];
          end;

          k := pop_member[i].mateflag;

          for j := site+1 to 64 do
          begin
              new_pop[i].chromosome[j] :=
                 pop_member[k].chromosome[j];
               {
               write(i:5);
               }
          new_pop[i].mateflag := 100;

          end;
     end;


end;


{

(* test segment of code *)
for i := 1 to 64 do
begin

     j := new_pop[1].chromosome[i];
     k := pop_member[1].chromosome[i];
     writeln(i,j:5,k:5);
end;
}

(* set the old pop = new pop *)
for i := 1 to 100 do
begin
     for j := 1 to 64 do
     begin

          pop_member[i].chromosome[j] :=
            new_pop[i].chromosome[j];
     end;
end;
```

```
        (* mutate *)
        for i := 1 to mutfreq do
        begin
            mutant := round(random*100);
            if mutant = 0 then mutant := 1;

            mutsite := round(random*64);
            if mutsite = 0 then mutsite := 1;
            pop_member[mutant].chromosome[mutsite] := 1;
        end;


    end; (* main while loop *)

end. (* end main *)
```

6
CHAPTER

Ecosystem models and experiments

Ecosystems are complex systems of interacting autonomous agents. By complex systems, we mean systems in which there might be oscillations or limit cycles in the populations of the autonomous agents. These oscillations and limit cycles are an emergent propriety of the system. Ecosystems are by definition, complex. Application of physics, chemistry and biology are of limited utility in understanding the dynamics of ecosystems. The typical reductionist approach to system analysis is inadequate. A new philosophy of biology is needed in which the reductionist approach is replaced with a synthesis approach (cf. Rosen, 1992; Kampis, 1991; and the references sited in Kampis).

In this chapter, we will examine elementary population dynamics models in which two or more species are interacting. We'll approach it from a systems perspective. We will assume only that the populations of the agents under study are large and that the time frame is continuous. In short, we will use the calculus for our initial studies. The differential equations will be solved by the Newton method on a digital computer and therefore made discrete.

However, before our study of population dynamics, let's examine some issues concerning the energetics and thermodynamics of ecosystems. Then, unlike in the other chapters, we'll get our hands wet in actual experiments. Unless you did some prebiotic simulations in chem lab during #3719 or some biohacking in the lab while reading #3719, these will be the first wet experiments. We will examine materially-closed energetically-open micro ecosystems. The last part of the chapter will contain speculations on emergent behavior of ecosystems, which includes artificial computing networks and planetary scale systems.

Ecosystem energetics

All global ecosystems are closed, with respect to matter, open with respect to energy, and are either in or heading toward a global steady state. This type of system has no reservoir and no external disposal. The material source and sink must come about from the internal dynamics of the system. The by-products produced by one hierarchical level are utilized by another for production purposes, and the material pathways are mutually reinforcing. In a closed ecosystem, the autotrophs (plants) are the ultimate producers.

All self-organizing systems are organized in hierarchies (Odum, 1988) and thus maximize energy utilization. Production and consumption often occur in pulses. Growth in production is followed by growth in consumption. In the next section, we will see computer models of this phenomenon and observe steady state development and limit cycles.

Smil (1991), points out that biological organisms are not necessarily efficient in energy conversion but that they have maximized their power output (measured in the units of watts per gram). This measure is useful in describing autotrophs (plants), heterotrophs (animals), and inanimate machines. Our sun, a G-type star, releases 3.89×10^{26} J of energy. Given the size of our planet, 1370 W/m^2 of radiation reaches the surface. This so-called solar constant is changing as the sun ages, but the change is very slow with respect to human time frames.

During the billions of years since our planet came into existence, climate and life have coevolved (cf. Schneider and Londer, 1984) into a complex network of limit cycles, attractors, and strange attractors. The climate and all life on this planet exists as a result of the solar flux reaching the surface. The life is essentially a transmutation of the solar light.

Heterotroph life depends on the autotrophs. Plants and animals use ATP (adenosine triphosphate) for energy conversion. The amount of ATP generation in living organisms is phenomenal. A 60kg man consuming about 700g of carbohydrates would manufacture and utilize about 70,000g of ATP, which is about 36 ATP molecules per molecule of sugar consumed (Smil, 1991). The power output is thus enormous.

For example, the sun—with a mass of about 2×10^{33}g has a power intensity of about 200 nW/g. The daily metabolism of a child proceeds at a rate of about 3mW/g (15,000 times greater than the sun). And Azotobacter can achieve 100W/g (i.e. 500 million times greater than the sun) power output.

The carbon cycle is an example of the type of material/energy balances in ecosystems. Figure 6-1 is a block diagram of the carbon cycle. This cycle is driven by solar energy input. Autotrophs convert the carbon dioxide in the atmosphere and ocean to carbohydrates. The detritus feeders are the ultimate converters of these carbohydrates back into carbon dioxide, which goes back into the atmo-

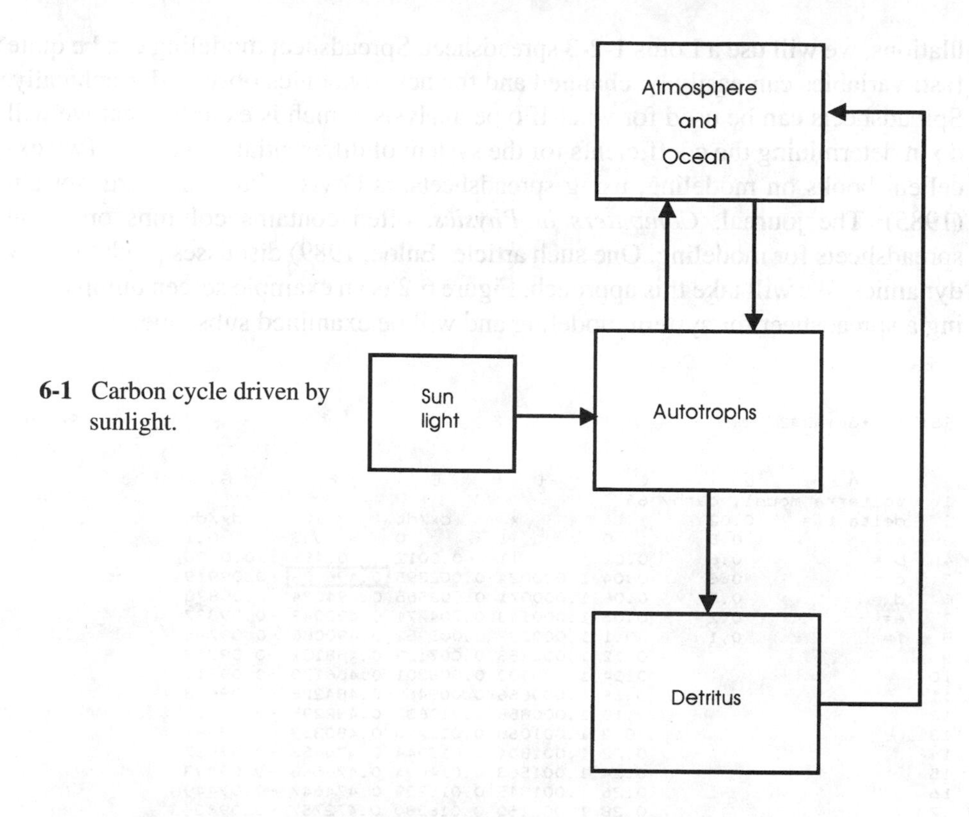

6-1 Carbon cycle driven by sunlight.

sphere and/or redissolves in the ocean. There are many other elemental cycles but the carbon cycle is the most important for carbon based life.

Population dynamics

In Chapter 1 back in *Creating Artificial Life* (TAB Book) #3719, we examined the logistic equation at length. This equation describes the dynamics (birth and death) of a single species interacting with the food supply. In that study, the ages of the population were integrated over all the ages to give an ordinary difference equation. We found that the degree of chaos varies continuously as the birth rate parameter changed. Although our examination of the logistic equation concentrated on the mathematics and dynamics, other authors have discussed the population aspects of the same equation (cf. May, 1976; and Oster and Ipaktchi, 1978). In this section, we will focus on the dynamics of two or more species. There is extensive literature on the subject (cf. Maguire, 1975, 1978; May, 1975; Dubois, 1978; Bentsman and Hannon, 1988; Posch et al., 1990; and Rai et al., 1991). Some of these we will review, but most of our studies will be from Danby (1985). An excellent summary can be found in Murray (1989).

Before reviewing a few papers, I'd like to examine some simulations of the dynamics of simple ecosystems, represented by differential equations. For these simulations, we will use a Lotus 1-2-3 spreadsheet. Spreadsheet modeling can be quite fast; variables can easily be changed and the new dynamics observed graphically. Spreadsheets can be used for what-if type analysis, which is exactly what we will do in determining the coefficients for the system of differential equations. Two excellent books on modeling, using spreadsheets, is Orvis (1987) and Arganbright (1985). The journal, *Computers in Physics*, often contains columns on using spreadsheets for modeling. One such article (Enloe, 1989) discusses predator-prey dynamics. We will take this approach. Figure 6-2 is an example screen dump of using a spreadsheet for system modeling and will be examined subsequently.

F5: +F4+G4*B2 READY

	A	B	C	D	E	F	G	H
1	volterra model, danby 65							
2	delta t =	0.02	t	x	dx/dt	y	dy/dt	
3	a =	0.5	0	1	0	0.5	-0.1	
4	b =	0.6	0.02	1	0.0012	0.498	-0.0996	
5	c =	0.6	0.04	1.000024	0.002395	0.496008	-0.09919	
6	d =	0.5	0.06	1.000071	0.003585	0.494024	-0.09878	
7	e=	0.2	0.08	1.000143	0.004771	0.492048	-0.09837	
8	f=	0.1	0.1	1.000239	0.005952	0.490080	-0.09795	
9			0.12	1.000358	0.007129	0.488121	-0.09753	
10			0.14	1.000500	0.008301	0.486170	-0.09711	
11			0.16	1.000666	0.009469	0.484228	-0.09668	
12			0.18	1.000856	0.010632	0.482295	-0.09625	
13			0.2	1.001068	0.011790	0.480369	-0.09581	
14			0.22	1.001304	0.012944	0.478453	-0.09537	
15			0.24	1.001563	0.014094	0.476546	-0.09493	
16			0.26	1.001845	0.015239	0.474647	-0.09449	
17			0.28	1.002150	0.016380	0.472757	-0.09404	
18			0.3	1.002477	0.017517	0.470876	-0.09359	
19			0.32	1.002828	0.018649	0.469004	-0.09313	
20			0.34	1.003201	0.019777	0.467142	-0.09268	

23-May-92 03:50 PM

6-2 Screen dump of spreadsheet solution for system of differential equations.

The conventional Volterra-Lotka system for predator-prey dynamics is given by

$$\frac{dx}{dt} = ax - bxy$$

$$\frac{dy}{dt} = -cy + dxy$$

The first term in each equation is the population tendency to increase or decrease if left alone (similar to logistic equation of Chapter 1). The second term in each is the effect of interactions between the two populations. As demonstrated by Haberman (1977), Dubois (1978), and Feistel and Ebeling (1989), this system is not stable. If we extend the relations to including the actions of a third force (man), removing both species at a different rate, then the relations become

$$\frac{dx}{dt} = ax - bxy - ex$$

$$\frac{dy}{dt} = -cy + dxy - fy$$

This is the system represented in the spreadsheet segment shown in Fig. 6-2. In this system, x is the prey population and y is the predator population. This system is also unstable.

Several solutions for the system are given in Fig. 6-3 through Fig. 6-9. The instability can be clearly seen in these phase plane plots. The trajectory representing the dynamics is spiraling outward.

The first four of these (Fig. 6-3 through Fig. 6-6) are equivalent to the Volterra-Lotka system with e and f equal to 0. The next three systems include fishing or removal of one and/or both of the species. All of these systems were studied with the spreadsheet shown in Fig. 6-2. Let's examine this in more detail.

Cells B2-B8 contain the change in the time or the time increment (Δt). The other constant cells are the coefficients of the terms in the differential equations. The C column contains the time increments and was prepared by the data fill command. The D column is the x solution and is given by

$$x_t = x_{t-1} + \frac{dx}{dt} \Delta t$$

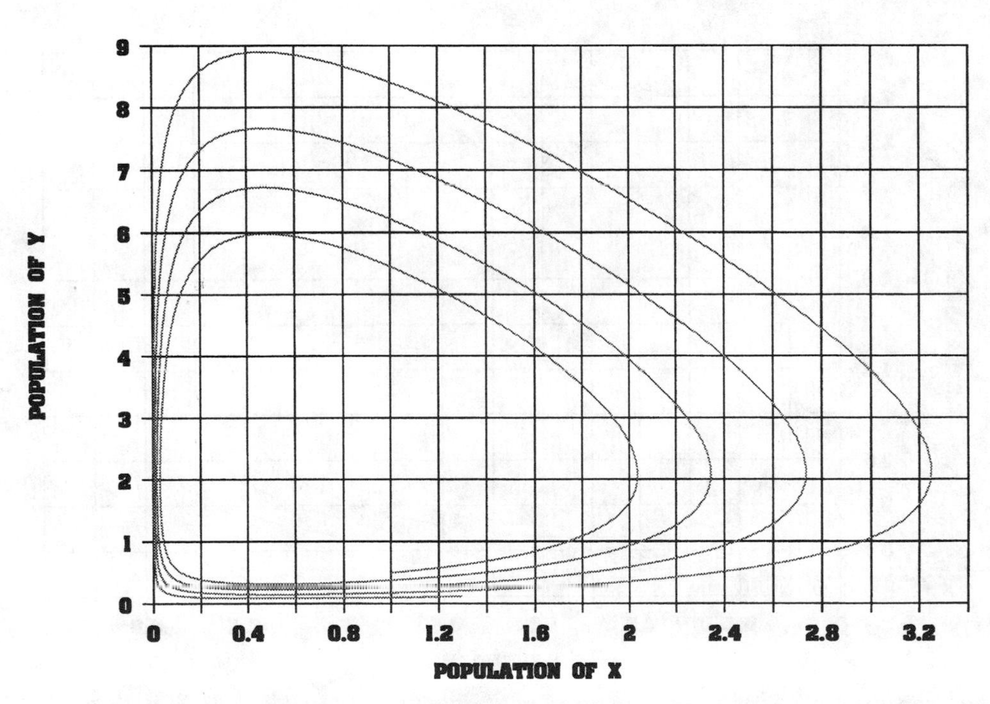

6-3 A solution to the Volterra-Lotka system. $a = 1$, $b = 0.5$, $c = 0.5$, $d = 1$, $e = 0$, $f = 0$.

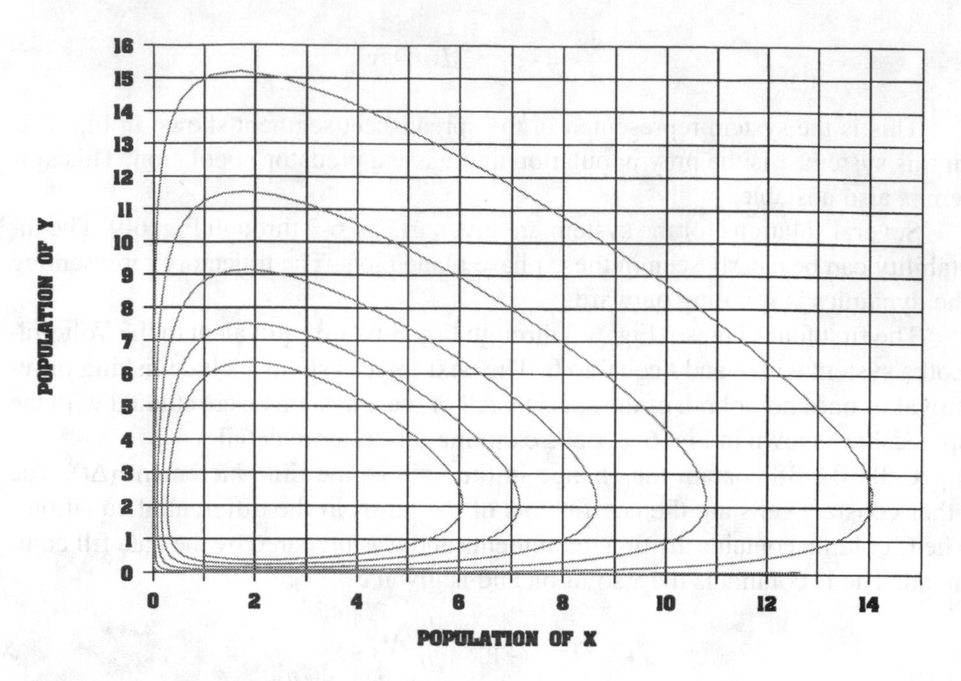

6-4 A solution to the Volterra-Lotka system. $a = 1$, $b = 0.5$, $c = 1$, $d = 0.5$, $e = 0$, $f = 0$.

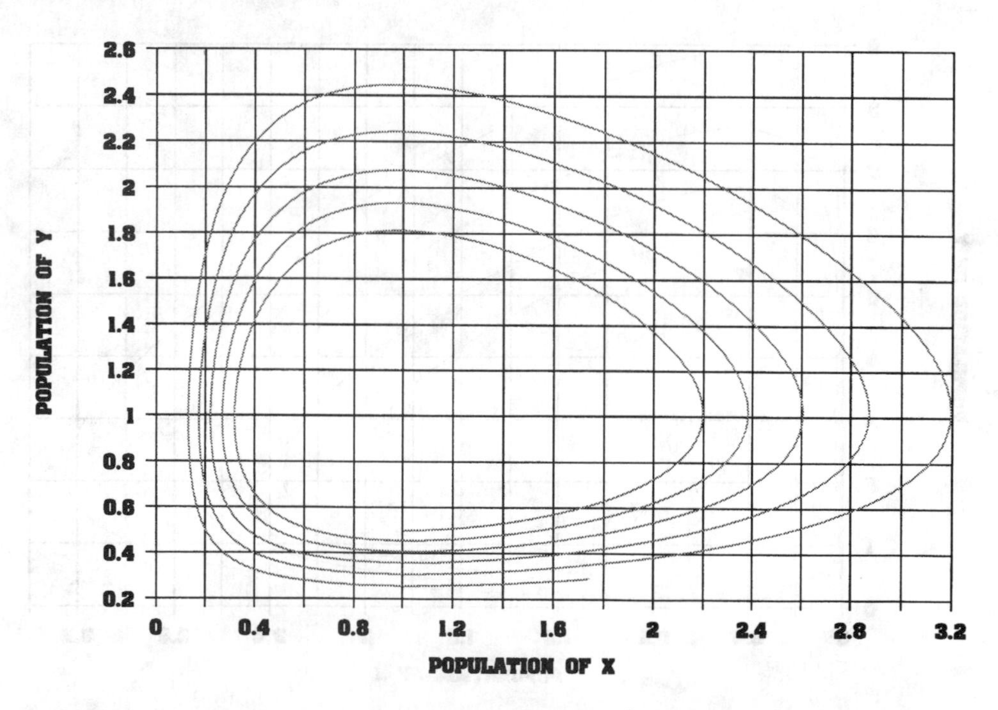

6-5 A solution to the Volterra-Lotka system. $a = 1$, $b = 1$, $c = 0.5$, $d = 0.5$, $e = 0$, $f = 0$.

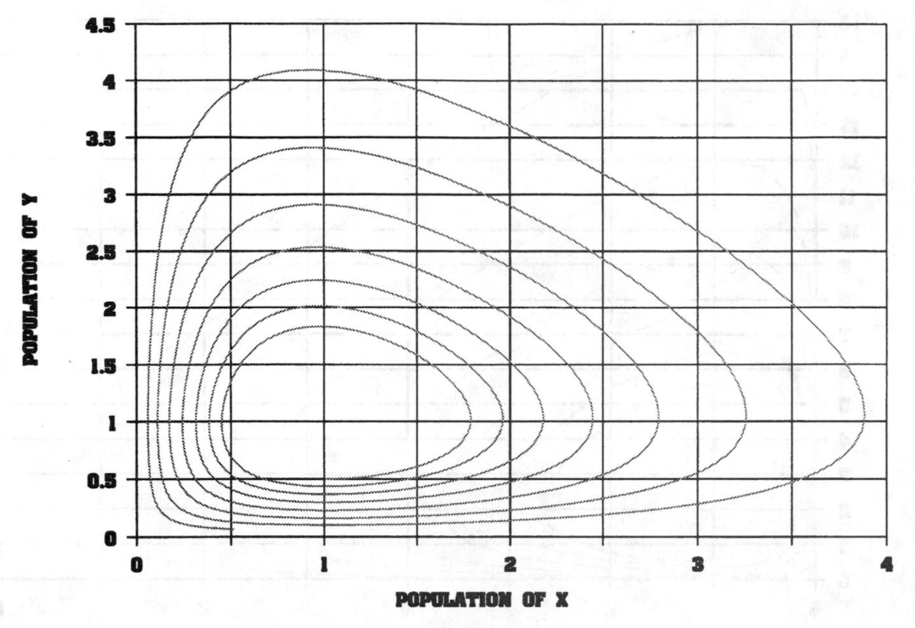

6-6 A solution to the Volterra-Lotka system. $a = 1$, $b = 1$, $c = 1$, $d = 1$, $e = 0$, $f = 0$.

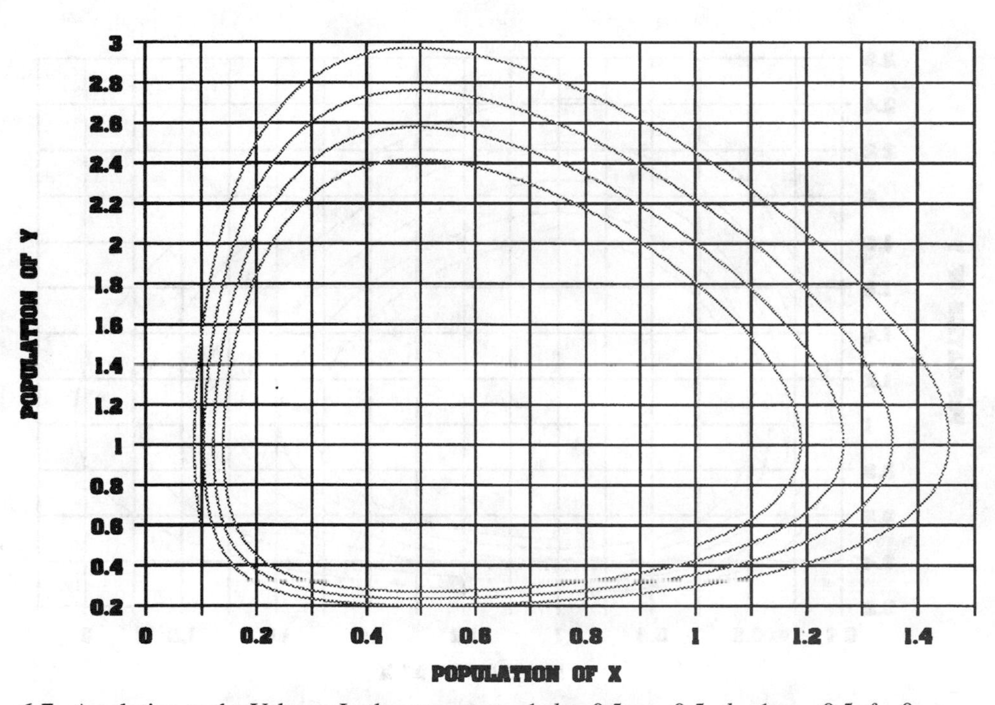

6-7 A solution to the Volterra-Lotka system. $a = 1$, $b = 0.5$, $c = 0.5$, $d = 1$, $e = 0.5$, $f = 0$.

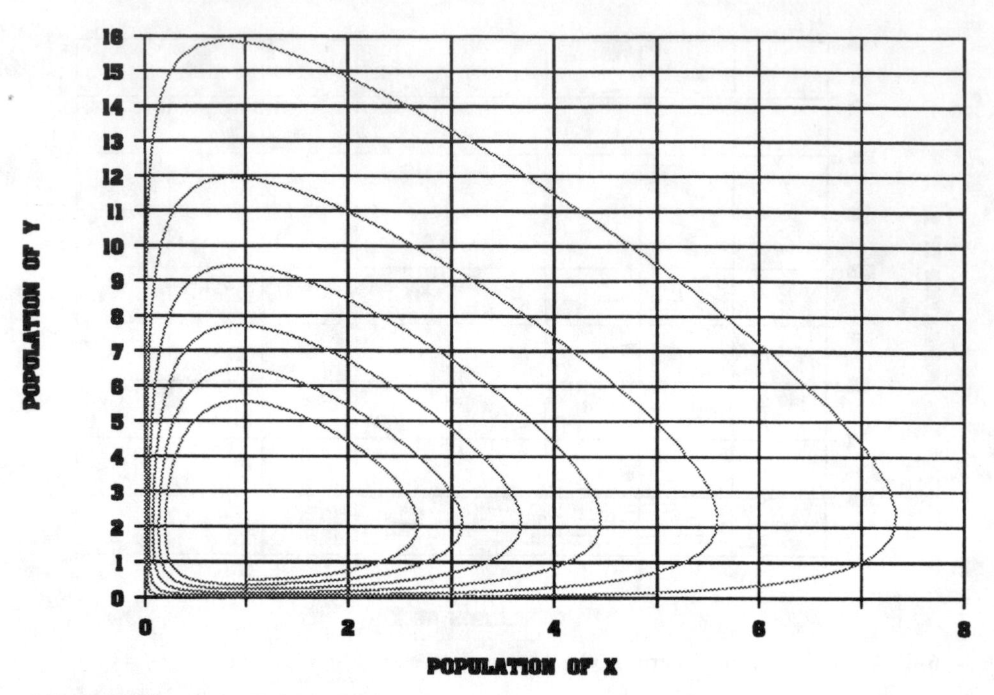

6-8 A solution to the Volterra-Lotka system. $a = 1$, $b = 0.5$, $c = 0.5$, $d = 1$, $e = 0$, $f = 0.5$.

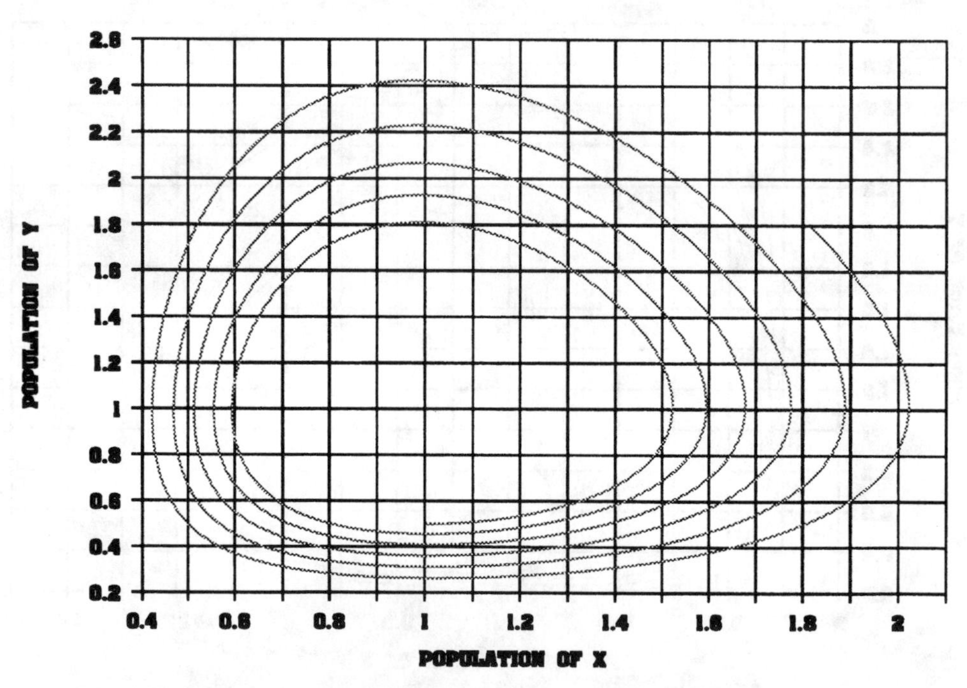

6-9 A solution to the Volterra-Lotka system. $a = 1$, $b = 0.5$, $c = 0.5$, $d = 1$, $e = 0.5$, $f = 0.5$.

The E and G columns are the differential equations given earlier for the system we are trying to solve. The F column is analogous to the D column. In this case, one of the cells is highlighted and the formula is shown in the upper left corner.

In my spreadsheet, I allowed the columns to reach to row 1000. All the other spreadsheets are analogous and won't be reviewed.

Danby (1985), gives the relations for another model ecosystem. In this case, we have a predator-prey system with internal competition. The system is given by

$$\frac{dx}{dt} = ax - bxy - ex^2$$

$$\frac{dy}{dt} = -cy + dxy - fy^2$$

Figure 6-10 through Fig. 6-12 are phase plane and time plots of this system with different values for the coefficients. All these systems spiral inward to an attractor point. This is clearly shown in Fig. 6-10 and Fig. 6-11. The time plot of Fig. 6-10b looks like two damped oscillators. Figure 6-11 also shows damped oscillations, but the time constant is greater than represented in this figure.

Figure 6-12 resembles a limit cycle but is not; I simply adjusted the damping time constant to give this appearance. In fact, this system of equations, representing predator-prey with internal competition, is always stable and attracted to a single point.

If we have three species interacting, the equations—given by Danby (1985)—are

$$\frac{dx}{dt} = ax - bxy - exz$$

$$\frac{dy}{dt} = -cy + dxy - fyz$$

$$\frac{dz}{dt} = -gz + hxz + iyz$$

Figure 6-13 is a time plot of the population of these species. The x species is the prey, and the y and z are the predators. In three dimensions, this phase plot would show a spiral toward an attractor point.

Danby (1985) reproduces May's (1973) model ecosystem. This system is attracted to a limit cycle. The system of differential equations is given by,

$$\frac{d\zeta}{dt} = \zeta - \left(\frac{bd}{a}\right)\zeta^2 - \left(\frac{c}{f}\right)\frac{\zeta\eta}{1+\zeta}$$

$$\frac{d\eta}{dt} = \left(\frac{e}{a}\right)\eta\left(1 - \frac{\eta}{\zeta}\right)$$

A limit cycle, shown in Fig. 6-14, occurs at

$$bd/a = 0.1$$

$$c/f = 1$$

$$e/a = 1/6$$

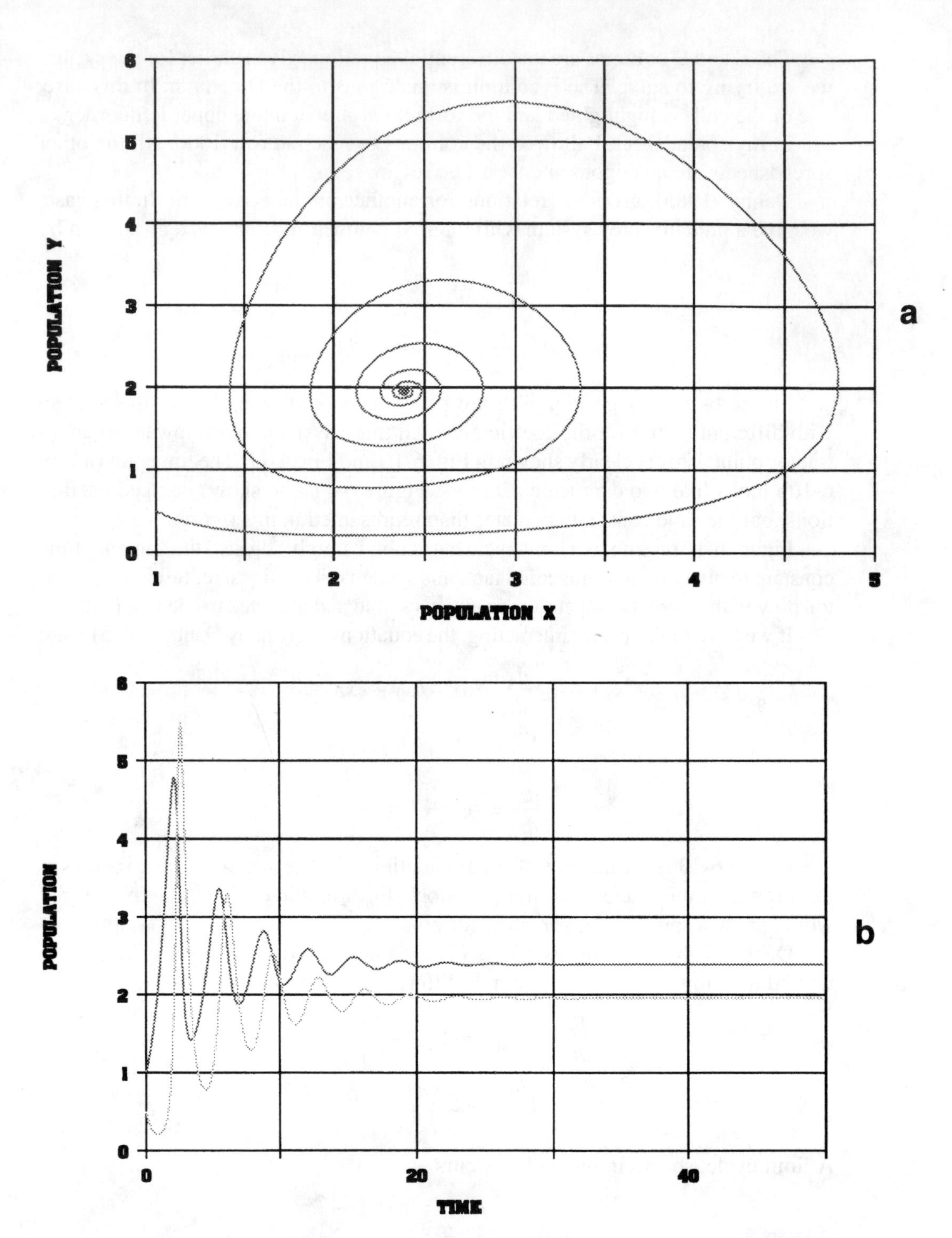

6-10a,b A solution to the Volterra-Lotka system. $a = 1$, $b = 0.5$, $c = 3$, $d = 1.5$, $e = 0.01$, $f = 0.3$.

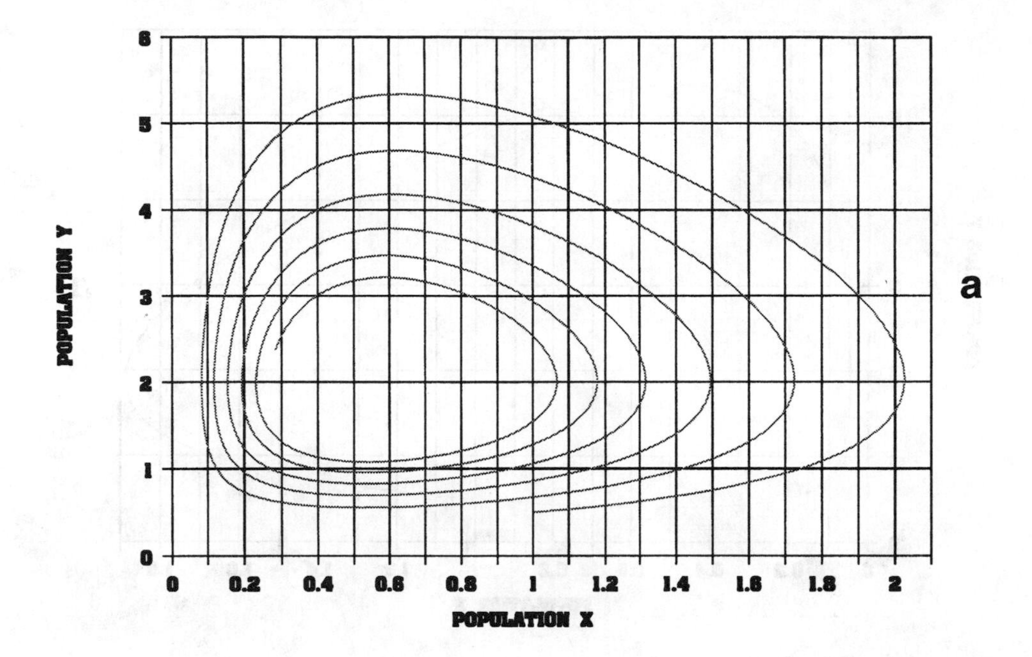

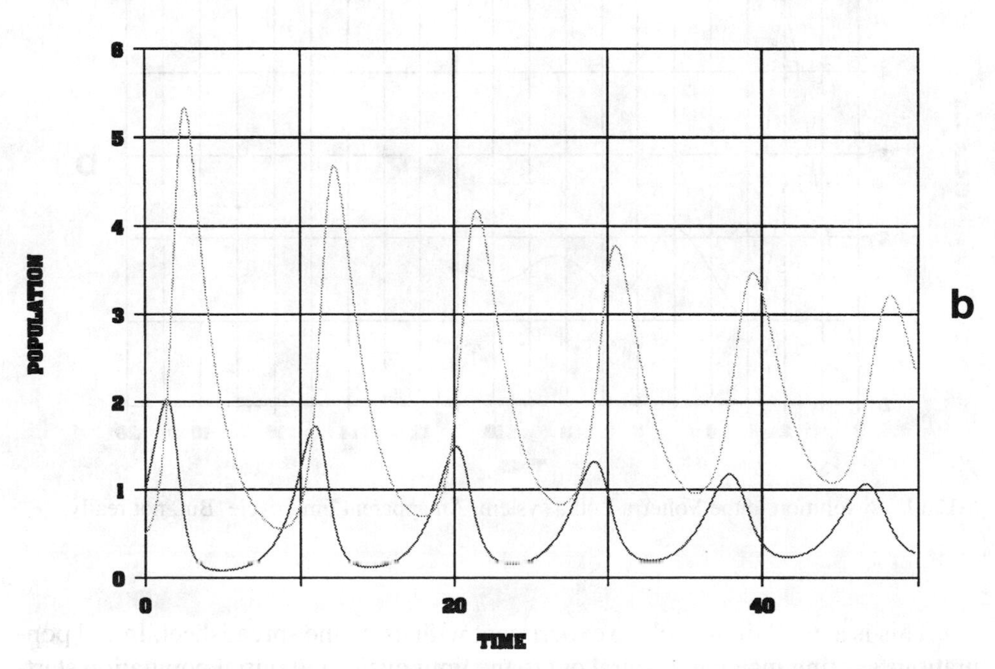

6-11a, b A solution to the Volterra-Lotka system. $a = 1$, $b = 0.5$, $c = 0.5$, $d = 1$, $e = 0.01$, $f = 0.3$.

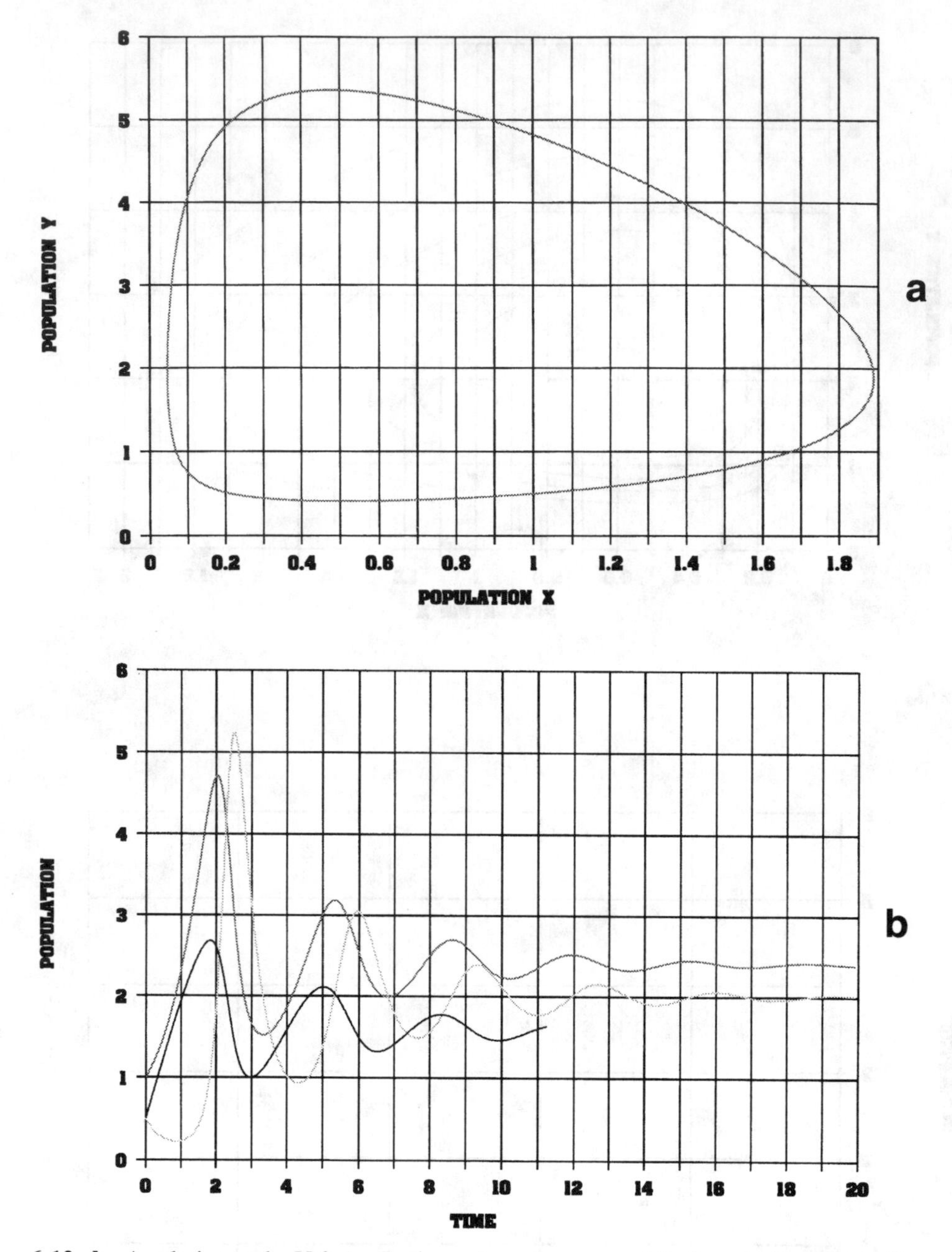

6-12a,b A solution to the Volterra-Lotka system. An apparent limit cycle. But, not really.

This is a good limit cycle to experiment with using the spreadsheet. Initial populations starting inside will spiral out to the limit cycle, and initial population starting outside will spiral into the limit cycle.

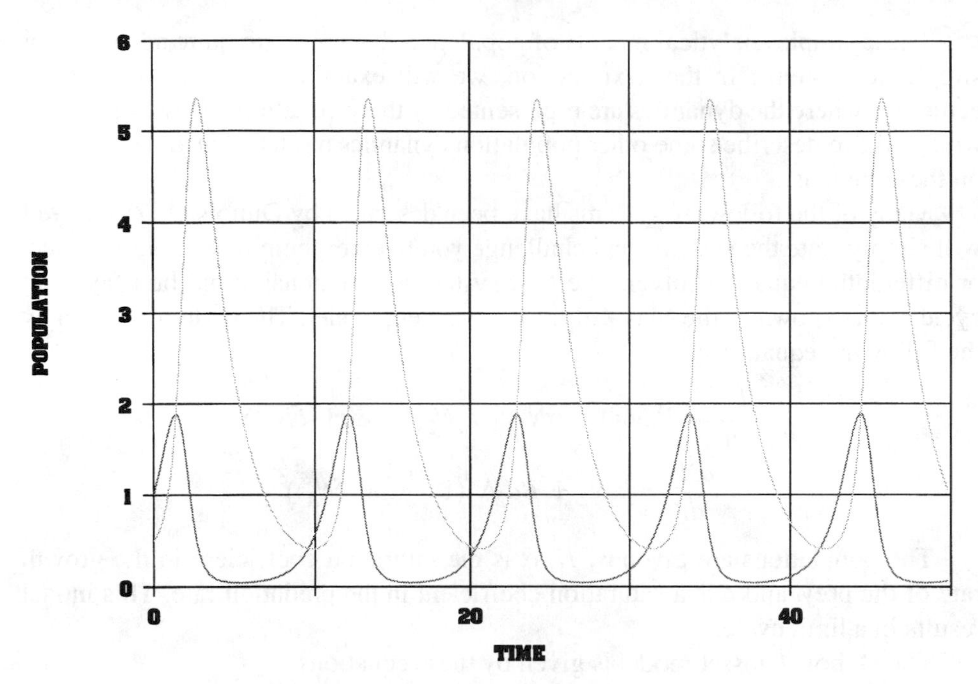

6-13 A solution to the Volterra-Lotka system. $a = 1.0$, $b = 0.5$, $c = 3.0$, $d = 1.5$, $e = 0$, $f = 0.3$, $g = 1.7$, $h = 0.05$, $i = 0.5$.

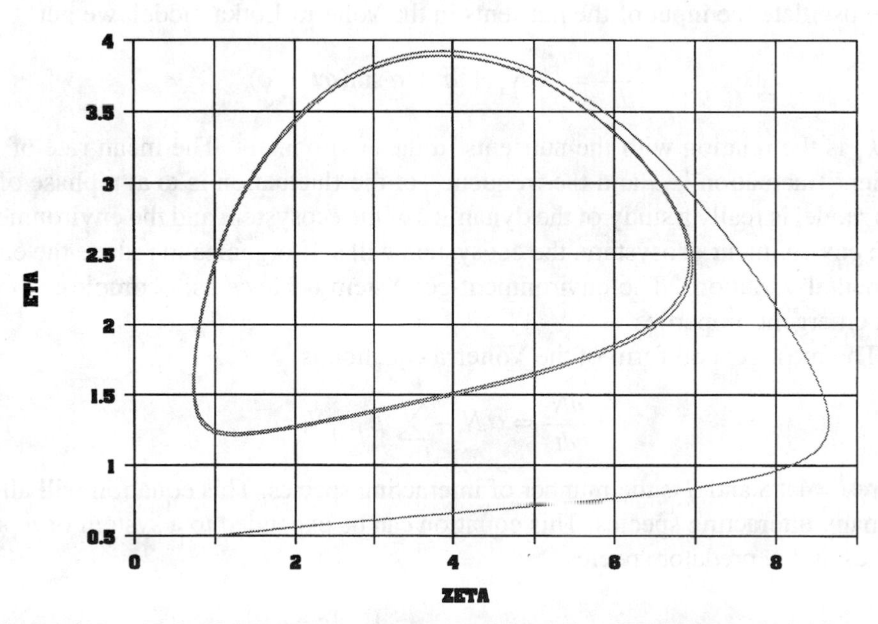

6-14 A solution to the Volterra-Lotka system. $a = 6$, $b = 0.6$, $c = 1$, $d = 1$, $e = 1$, $f = 1$.

These simple analytical models of population dynamics are quite adequate for simple ecosystems. In the next section, we will examine actual closed system ecologies where the dynamics are represented by these equations. However, now I would like to describe some other population dynamics models that are extensions on these models.

Many of the following systems have been described by Dubois (1978). Here I will simply state the systems and challenge you to enter them in your spreadsheet or differential equation solver. The first system is an extension on the May limit cycle and is known as the May-Gilpin-Rosenzweig model. The system is given by the following equations:

$$\frac{dN_1}{dt} = k_1 N_1(1 - \alpha N_1) - k_2 N_2(1 - exp(-\delta N_1))$$

$$\frac{dN_2}{dt} = k_3 N_2 + B d_2 N_2(1 - exp(-\delta N_1))$$

The populations are given by N, α is the saturation coefficient in the growth rate of the prey, and δ is a saturation coefficient in the predation rate. This model results in a limit cycle.

The Dubois-Closset model is given by these equations:

$$\frac{dN_1}{dt} = d_1 N_1(1 - \alpha N_1) - d_2(N_1 N_2)$$

$$\frac{dN_2}{dt} = -d_3 N_2 + B d_2 N_1 N_2$$

If we oscillate the input of the nutrients in the Volterra-Lotka model, we get

$$\frac{dk_1}{dt} = -k_1 N_1 + \bar{a} + a \sin(\omega t + \phi)$$

k_1 is the relation with the nutrients in the environment. The mean rate of the nutrient fluctuation is $\bar{a}$ and the frequency of the fluctuation is ω at a phase of ϕ. This model is really a study of the dynamics of the ecosystem and the environment. In an environment-ecosystem, the ecosystem will self-organize to reduce the environmental variations. The environment-ecosystem behaves as a complex system with emergent properties.

The most general form of the Volterra equation is

$$\frac{dN_i}{dt} = \alpha_i N_i - \sum_{j=1}^{S} \beta_{ij} N_i N_j$$

where $i = 1 \ldots S$ and S is the number of interacting species. This equation will allow for many interacting species. This equation can be expanded to a system of n prey species and m predator species.

$$\frac{dX_i}{dt} = k_{1i}X_i - \frac{1}{m}\sum_{j=1}^{m} k_{2ij}X_iY_j$$
$$\frac{dY_j}{dt} = -k_{3j}Y_j + \frac{1}{n}\sum_{i=1}^{n} k_{4ij}X_iY_j$$

The system is well behaved for $n = m$ and all X and Y should be positive (negative population is meaningless).

As an example of a more complex system, the following equation is for two-preys and one-predator with saturation and competition.

$$\frac{dX_1}{dt} = (k_{11} - k_{71}X_1 - k_{72}X_2)X_1 - k_{211}X_1Y$$
$$\frac{dX_2}{dt} = (k_{12} - k_{81}X_1 - k_{82}X_2)X_2 - k_{221}X_2Y$$
$$\frac{dY}{dt} = -k_3Y + (k_{422}X_1 + k_{421}X_2)Y$$

All of this section so far has been an analytical approach to the study of population dynamics for modeling simple ecosystems. Dubois (1978) also reviews a stochastic approach. You must consider three possible events: the birth rate of the prey, the death rate of the predators, and the attack of a prey by a predator. The probabilities of these three events are:

$$P_1 = Ck_1N_1$$
$$P_1 = Ck_3N_2$$
$$P_3 = Ck_2N_1N_2$$

The constant is adjusted such that the probabilities sum to unity. The simulation starts by generating a random number, R, in the range [0,1]. If $R <= P_1$, then the event is the birth of a prey. If $P_1 < R < P_1 + P_2$, then the next event will be the death of a predator. Continuing on like this and keeping score of the events, you can produce curves of the predator-prey dynamics similar to the ones we have just seen here. The major structural difference in the curves will be the fine structure in the plots due to the random nature of the events.

Just as a side note, besides the analytical dynamics and the stochastic approach, you could also use diffusion equations, as discussed by Okubo (1980).

Maguire (1978) reviews yet another method to study ecosystem dynamics. The advantage of his method is that real-world data could easily be incorporated into the model. The method is known as *statistical response surface modeling*. In this technique, the effects of changing a control variable in a system is observed on the output of the system. Thus the patterns and dynamics of the system are visualized as a hyper surface in some complex space of the control variables and the outputs of the system.

Population growth is the most common output parameter to observe in ecosystem dynamics. A contour plot of the density of algae versus the temperature and nutrient is a simple response surface. The actual effects of the temperature and nutrient concentration on the density of algae can be read off the graph. The data for the contour plot may be theoretically generated or experimental observations. You can observe population oscillations between predators and preys from these response surfaces.

Closed ecosystems: Evolution in a bottle

In this section, we will review the state-of-the-art in closed ecosystems. These ecosystems are materially closed and energetically open. We will review some past and current work on these systems. Following this, I will show you how easy it is to break-into closed ecosystem research and explain possible avenues the research might take. The last part of this section will explore a very large experiment, Biosphere II, that involves humans and may be a prototype for planetary and solar system exploration.

Although microcosms are not the focus of this section, they are the precursors of closed ecosystems. A microcosm example would be a balanced aquarium. In these systems, there are often matter and energy as inputs to the system and, many times, must be cleaned. These systems are often supported by technology (e.g., air pumps). Such systems are very useful in studies of the perturbation and recovery of ecosystems (cf. Ravera, 1991; Polunin, 1986; Lalli, 1990; and Giesy, 1980).

Early work by Tuba and Crow (1980) showed that it is possible to build totally synthetic or artificial ecosystems. These systems were built from artificial lake water, artificial sediments consisting of clays, and algal and Daphnia populations. Oscillations and limit cycles were observed in these artificial systems, and they were well described by the equations of the last section. In this work of Tuba and Crow, you should note that the systems were not closed in the material sense or the energetic sense but were open to the air and had technology support.

Schindler et al. (1980) proposed that a PH-PE phase diagram would describe the dynamics of ecosystems. Waide et al. (1980) demonstrated this. The PH-PE phase diagram is a plot of the hydrogen ion concentration and the electron concentration. The authors did not actually measure electron concentration but rather dissolved oxygen in their experimental aquatic ecosystems. Their phase diagram is sketched in Fig. 6-15. The oxygen and hydrogen isoclines are shown along with the region in which ecosystems exist. Outside this region the ecosystem will collapse.

One of the primary ways to discover the dynamics of ecosystems is by perturbation and response. You can use this method for collection of data to build statistical response surfaces, as explored in the last section. Leffler (1980) demonstrated that the damping ratio and the natural frequency alone did not describe ecosystems. The product of these two parameters are a much better description of the system response to perturbations.

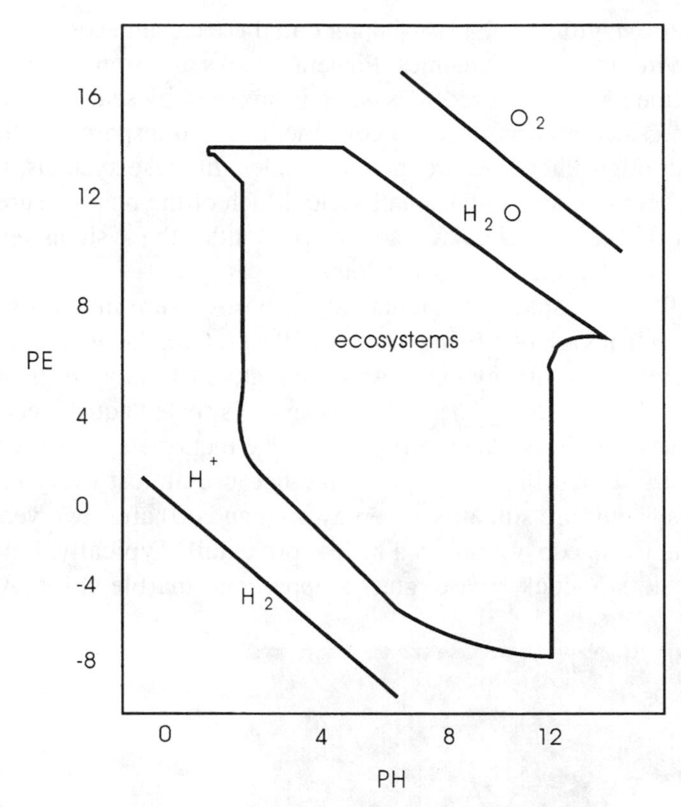

6-15 PH-PE phase diagram for ecosystems (after Schindler et al. 1980).

An example of perturbation-response studies is given by Beyers (1963). He studied the effects of length of light cycle, temperature, nutrients, etc., on the populations of algae; he also monitored oxygen, carbon dioxide, and PH. Similar and more advanced studies have been reported by Giesy (1980), Polunin (1986), Lalli (1990), and Ravera (1991). All these experiments involved either field studies or open laboratory systems supported by any technology necessary for survival.

Ecosystems, as Hill and Wiegert (1980) point out, can be studied from a systems and cybernetics approach. In the open microcosms, examined earlier in this chapter, the energy balance and/or the material balance is often maintained by technology. These systems are excellent for formulating simple ecosystem theory, developing the analytical techniques, and identifying ecosystem structure and function. These studies often go hand-in-hand with the analytical studies explored in the previous section. I will now change the focus and concentrate on closed system research.

Closed ecosystems have no technology support and no material input. These closed ecosystems are just like a planet, having an energy source (such as light) as the only input. As Maguire et al. (1980) point out, these closed ecosystems are excellent for studying the local dynamics. Planetary sinks and sources are removed from consideration. A small closed ecosystem is prepared by sealing a population of organisms and their environment in a container that is transparent to light.

Photo 6-1 through Photo 6-3 are three examples. In these systems, the entire dynamics have been reduced to the small scale. Much of the post-closure dynamics is a result of this small scale. After about two months, the systems settle down to a steady state, limit cycle, or attractor point.

Maguire (1980) has studied the dynamics of post-closure in a design experiment, with the control variables being the size of the system, the water/air ratio, the water/sediment ratio, and energy input. An excellent way to monitor the post-closure dynamics is through the species list. In Maguire's sealed aquatic ecosystems, he concluded that they failed due to the fact that the metazoans (multi-celled animals) population reached 0. This surprises me because almost every one of my fresh water ecosystems are still alive (even by his standard) after two years.

Let's look at these ecosystems in a little more detail. Typically, I go out to a pond with a flask, stopcock grease, glass stopper, and marble chips. After stir-

Photo 6-1 Closed fresh-water ecosystem.

Photo 6-2 Closed fresh-water ecosystem.

ring the bottom mud slightly, I collect an aliquot of muddy water. I then clean off the ground joint, add a few marble chips to the flask, and then seal the flask with the glass stopper and silicon grease. I add the marble chips to buffer the PH of the ecosystem and to act as a carbon source for the carbon-based life. This will keep the system within the boundary of the phase space shown in Fig. 6-15.

Three such ecosystems are shown in Photo 6-1 through Photo 6-3. The first one is a 1000 mL flask and the second is a 2000 mL flask. Each of these are fresh water systems with small tufts of algae, cyclopods, ostracods (cf. Pennik, 1978), and a collection of bacteria in the bottom mud. These are balanced and have reached a limit cycle of steady state. The cyclopods are herbivores, and the ostracods are omnivorous and detritus feeders. These species produce nutrients for the algae and bacteria. The ecosystems receive fluorescent light for 12 hr per day.

I have another similar ecosystem without the cyclopods. It also has been alive and in a limit cycle for over two years. It consists of a small 250 mL flask and sits in a window, with the temperature not being controlled in any way. Wide variations in temperature, from 50 F to 95 F, are seen over the course of a year.

Photo 6-3 Closed marine ecosystem.

The flask shown in Photo 6-3 is a marine system. No metazoians are present, but the system is very much alive. The algae (some of which can be seen on the side of the flask) and the bacteria in the bottom sand is at a limit cycle stage.

All these systems can be perturbed by affecting the energy input. You can easily perturb the systems such that there will be oscillations in the algae and grazers (ostracods and cyclopods). However, these systems will quickly settle to a limit cycle. I want to emphasize that these systems are still alive after two years; Maguire (1980) claims that his systems failed after about 400 days.

Let's review some state-of-the-art in closed system research and what questions are being explored. Maguire (1980) was the first to publish on closed micro-ecosystems. But apparently Folsome, (c. f. Kearns 1981, 1983) started research along this line years before Maguire. Folsome's systems are marine ecosystems with micro algae and microorganisms. These systems have been productive, (indicated by green color) for over 25 years. His early research is reported by Kearns and Folsome (1981), Takano, Folsome, and Karl (1982), and Obenhuber and Folsome (1983).

Flosome's last results, before his untimely death, has been reported by Shaffer (1991). As Maguire pointed out, after the systems are closed, it is difficult to monitor the state of the system. Maguire used species lists to monitor the system, while the Folsome team used septums to remove micro samples that would hopefully perturb the system very little. These workers found that, after closure, the oxygen concentration in the atmosphere above the aqueous phase increases due to photosynthesis. The concentration reaches a stead state slightly above that found on earth. The Folsome team also monitored the ATP production and found an increase over that found in the open ocean. The ATP oscillated with a period of about 50 days. Bacterial counts were also found to oscillate and be in phase with the ATP production.

Although it was not stated in the paper, this 50-day period oscillation between the bacteria indicates a limit cycle. The eucaryote-to-procaryote ratio was found to increase by two orders of magnitude within 30 days after closure. The systems then reached a steady state, with respect to this indicator.

Hanson (1982, 1986), at JPL, working with Folsome, designed marine micro-ecosystems with Halocaridina Rubra (shrimp), blue-green algae, and microorganisms. These systems have been productive for over 15 years. By using carbon-labeling experiments, they discovered that the entire system shows a carbon turnover in 5–6 days and that the eucaryote/procaryote ratio reached 0.05 within 30 days of closure.

What are some of the open research questions on these systems? Folsome and Hanson (1986), point out that these materially-closed energetically-open micro-ecosystems are oxidation-reduction systems in a steady state. The redox state could be studied after closure by monitoring traces of the conductivity and the PH of the system. Previous workers used species lists, oxygen concentration, and ATP as system monitors.

What other monitors are available? What are the ideal gas, liquid, and solid volumes for stability of these closed ecosystems? What is the minimum set of species for stable systems? Folsome and Hanson (1986) have suggested that acoustic spectroscopy would be an ideal technique to monitor the state of the system. It could be used to build up a species list map and could be updated in milliseconds. These closed micro-ecosystems are the only known complex dynamic systems that are thermodynamically closed.

Closed micro-ecosystems are truly miniature worlds. They can undergo their own evolution and adapt to outside environmental (energy flux input) influences. Sagan (1990) points out that we can now create new living worlds. The emergent dynamics of these microworlds (limit cycles) and the dissipative nature of the system indicate that they are essentially artificial living organisms (thus their inclusion in this book). This is the beginning of a new era; the biotechnic era, a synthesis of biology and technology. Further, these closed-system ecology experiments are precursors and prototypes for closed systems in space-technology applications.

Allen and Nelson (1989) and Allen (1991) describe a large scale experiment in which thousands of plant and animal species, including humans, will be enclosed in a sealed glass structure. The Biosphere II, as it is called, will be supported by technology for air conditioning, hot and cold water, irrigation systems, pumps and motors, and timers for automatic shutdown of subsystems. The entire experiment is a prototype experiment for space colonies. Some of the problems of this experiment and contrasts with NASA experiments have been described by Ressmeyer (1992). It is a bold experiment that will clearly produce much data. If this complex assemblage of data is examined carefully by statistical techniques, and perhaps by nonlinear techniques, like neural networks, then we might learn much from this experiment.

As an aside, most of the oscillations, limit cycles, and attractor points observed in natural ecosystems have also been observed and/or predicted for computational ecosystems such as large computer networks. Many of these systems contain an emergent computation just as in ecosystems. (cf. Forrest, 1991, and Huberman, 1988).

Comments on the Gaia hypothesis

The Gaia hypothesis (Lovelock, 1979; Margulis and Lovelock, 1974; Lovelock, 1986 and Lovelock, 1988) is a theory that the biosphere of the planet controls the overall thermal and chemical balance of the earth. This is really quite acceptable. From the previous text, we have seen that closed ecosystems (the earth being one such thing) are stable from a chemical potential perspective. These systems reach steady states and limits cycles in which the carbon and other elements are turned over in the ecosystem by the biota.

A decade ago, while I was an undergraduate student, I often pestered my professors about the chemical differences in the terrestrial planets atmospheres. The nitrogen in the atmosphere could be explained by denitrifying bacteria. The oxygen could be explained by green algae, cyanobacteria, and plants.

But what would the Earth look like without a biosphere, and what if we transplanted a biosphere to Venus or Mars? What effects on the atmospheres would we observe?

I did not get acceptable answers. Table 6-1, after Lovelock (1988), shows the expected effects. This table is worth a close study. The dominant molecular species in the atmosphere of Venus and Mars is carbon dioxide; the same would be expected of earth if there were no biosphere. The biosphere produces the unstable atmosphere we have, of 79% nitrogen and 21% oxygen. The microorganisms are the primary agents for this planetary change (cf. Margulis and Sagan, 1986).

Table 6-1 Planet atmosphere compositions.

Gas	Venus	Abiological Earth	Mars	Biological Earth
Carbon dioxide	96.50%	98.00%	95.00%	0.03%
Nitrogen	3.50%	1.90%	2.70%	79.00%
Oxygen	trace	0.00%	0.13%	21.00%
Argon	trace	0.10%	1.60%	1.00%
Methane	0.00%	0.00%	0.00%	1.7ppm

7 CHAPTER

Robotics
Hardware artificial life

Most of the previous chapters can apply in some way to development of robots or hardware artificial life. For example, genetic algorithms can be used to select the best morphology for the robot from biomorphs. Genetic algorithms can also be used to develop the neural network controllers for the robot. Some of the ecological and population dynamics of Chapter 5 can be applied to populations of robots.

In this chapter, we will examine the details of several robot systems. We will focus on neural network controlled robots—both manipulators and mobile robots. We will also review state-of-the-art in autonomous mobile robots and novel robot locomotion techniques.

Robotics has two fundamental branches. One branch is functional, focusing primarily on getting work done and concentrating on the robot as an automation machine. The second branch of robotics is behavior-based, focusing on autonomous robots that are able to survive in the real world by themselves. The second branch is, almost by definition, at the core of artificial life research. We will examine both branches of robotics, focusing primarily on the behavior branch. To ignore the first branch would be foolish, though; many valuable contributions have been made in functional robotics that can be applied to behavior-based robotics.

First we will review some technical papers on sensors, control, manipulators, mobile robots, and various subsystems. A complete review of the literature would not be possible. Later sections of the chapter will focus on reviews of work by several behavior-based robotics research groups, LEGO robotics, neural controllers, autonomous agents, and robotics for space research and manufacturing. The experimental sections of the chapter will focus on neural control of a robotic manipulator, a simple vision system, LEGO robots, and small autonomous "toy" robots.

Before I start into the subject, I should point out that much philosophical debate occurs concerning AI and Alife. Fetzer (1990) has summarized the capabilities and limits of AI. There are a few good sources of general robotics reviews. Brady (1989) edited a collection of papers by leading researchers. Fu et al., (1987) is an excellent review of the fundamental issues. Holland (1983), Taylor (1990), and Young (1973) explore the fundamental aspects of robotics at a beginners level. The work by Young is somewhat dated, however.

The basics: A literature review

Controllers

The Braitenberg machines (cf. Braitenberg, 1984) examined in Chapter 3 are feedforward controllers. The sensors send a signal directly to the actuators. This feedforward control is represented in Fig. 7-1a. If sensors are used to detect the position of the actuator joints, then the sensor signal can be feedback to the controller in order to improve the precision of the position. This feedback control is shown in Fig. 7-1b and operates like the feedback control of an electronic circuit. If the signal at the output is too high, the controller will reduce the input. If the signal at the output is too low, the

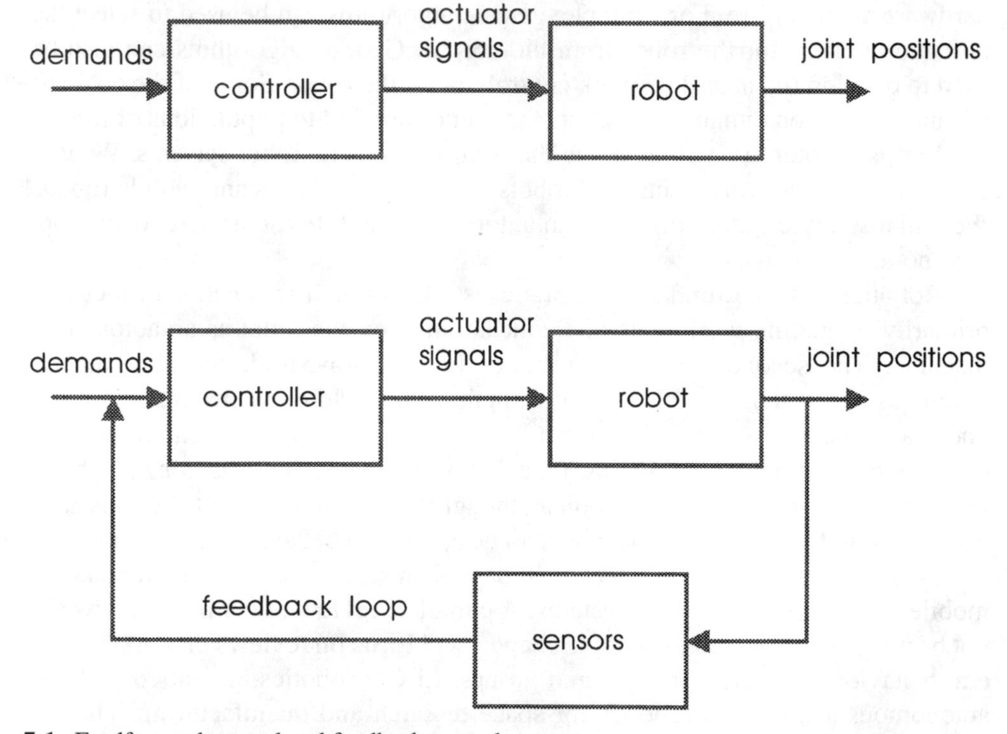

7-1 Feedforward control and feedback control.

controller will increase the input. Many variations of feedback control exist, such as proportional, differential, and integral control. All these, and combinations of them, are used to compensate for the time delay in the feedback loop.

Later in this chapter, we will examine a neural network controller in parallel with a proportional controller. Fu et al. (1987) have written a review of control for robotics. This type of control is generally applicable to functional robotics and therefore of little concern for us. Recall that behavior robotics is like the Braitenberg ballistic control.

All these controllers assume linearity and are developed by a system of differential equations. For legged motion, it is possible to build a memory-based controller, similar to the Heiserman robots examined in Chapter 3. Heiserman (1979) built a wheeled autonomous robot based on this memory-control idea.

If you recall, the memory controllers are self-developed by the robot as it roams around its environment. Contact situations and the direction and speed are stored in a memory matrix to build up a kind of map of the environment. After enough contacts and successful escapes, the robot will have built up a complex map of its environment—a map complex enough for the robot to behave in an intelligent way when it encounters a new object in its environment. (cf. Chapter 3 for computer simulations).

Dawamura and Nakagawa (1990), apparently independent of Heiserman, built two robots based on memory control. One robot was a two-degree-of-freedom robotic fish, and the other was a three-degree-of-freedom walking robot. Both were controlled by the memory-based control technique.

Sensors

Sensors range from the simple to the complex and cover many sensing modes. Visual sensing can range from a simple detection of light and dark to a stereo video system (cf. Faugeras, 1989, Herbert and Kanade, 1989, and Aloimonos and Rosenfeld, 1991). Tactile sensing can range from one pressure-sensitive resistor or capacitor to a whole matrix of these built into an artificial hand with artificial skin (cf. Dario, 1989). A simple tactile sensor is a limit switch to which is attached a stiff wire to act as an artificial insect antenna.

Potentiometers can be used for rotary and linear motion detection. Hall-effect switches and optical encoding can also be used for motion detection. Range sensing is often done with ultrasonics. Crecraft (1983) explores the methods of ultrasonics. Duchesne et al., (1991) and Weinstein, (1981) describe using ultrasonic detectors with personal computers and robotics, respectively. Range sensing can also be done with IR emitters and detectors.

Manipulators

The definitive source for robotic manipulators is Paul (1981). Craig (1989) examines in more detail the control of manipulators. Robotic manipulators can be as simple as an x-y crane, a pick-and-place robot, a revolute-arm robot, or an artificial arm with a multifingered hand (Holland, 1983). Many times the robot manipulator must be placed in a highly structured environment (cf. Middleton and Weston, 1982). More complex robots might be able to deal with an unstructured environment more efficiently (cf. Abidi et al., 1991 and Miller, 1989).

Hashimoto et al. (1985) and Kuribayashi (1986) describe the use of TiNi alloy wire (shape memory alloy) for robotic manipulators. Due to the limited number of thermal cycles the wire can withstand, it will undergo some crystallographic stress and fail. The application of this material is not being perused in many laboratories.

Two unusual robot manipulators have been described by Pierrot et al. (1991) and Shahinpoor (1992). These are both platform manipulators with the rough general configuration of Fig. 7-2. The manipulator of Pierrot et al. is two platforms: one the waist, and one the wrist. The manipulator of Shahinpoor is more than two platforms. Both these manipulator designs can operate at very high speed.

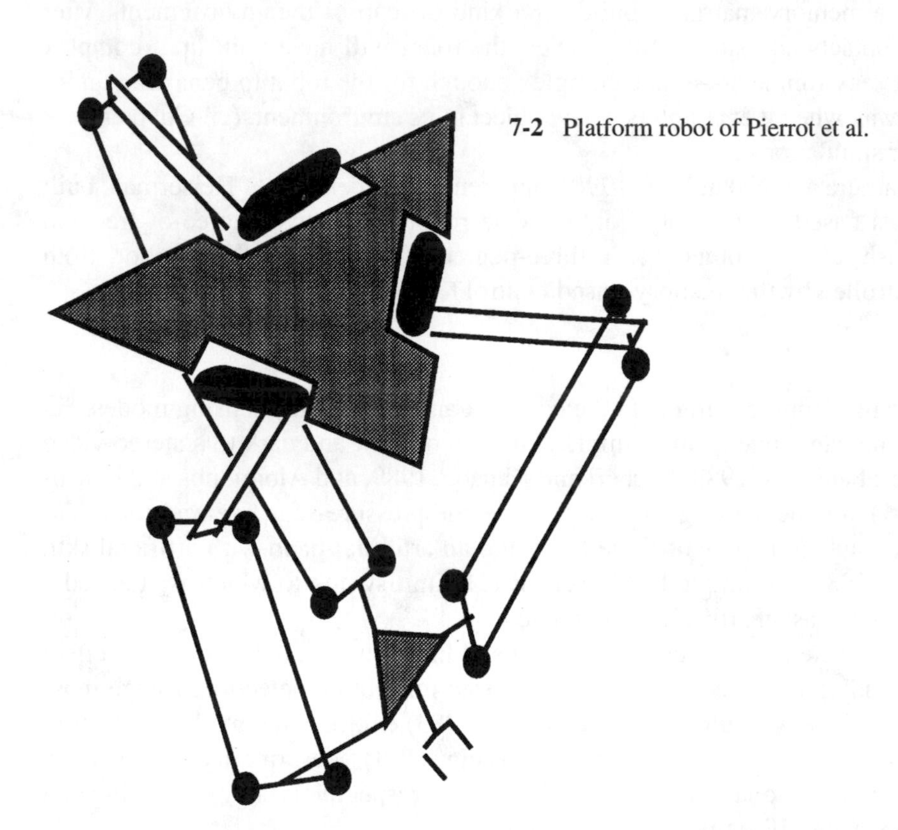

7-2 Platform robot of Pierrot et al.

You must consider several more issues when a robot hand must pick up an object. Unless the object is exactly centered, there will be some degree of error during grasping for a two-fingered robot. If one of the fingers strikes the object first, some rotation will likely occur. Depending on the mass distribution, friction, and initial misalignment, for example, the object might not be gripped firmly (cf. Rivetta, 1986).

Craig (1989) describes two important concepts for robot manipulators: kinematics, and inverse kinematics or dynamics. Robot kinematics refers to the mechanics of the robot joints without consideration of the causes of the motion. Inverse kinematics refers to the joint coordinates, given the end effecter coordinates. The transforms consist of matrix manipulation of Jacobian determinates. The amount of mathematical manipulation can be quite awesome. Later in this chapter we will explore a neural network solution that avoids all that.

Robot mobility

Mobile robots provide a greater range or working capability than fixed robots. This greater range requires greater sensibility to the environment, as well as the ability to behave intelligently with unexpected obstacles. Mobile robots might have one wheel or scores of wheels. They might have only one leg and hop around, or they might have scores of legs like an artificial centipede. Holland (1983) and Weinstein (1981) discuss the basics of robot mobility and should be consulted for the relevant engineering designs. A full review is beyond the scope of this work. Meystel (1991) and Cox and Wilfong (1990) reviewed the entire field of autonomous mobile robots or vehicles with cognitive control. Legged robots are reviewed by Raibert (1986).

Mechanical stability is the most important aspect of mobile robots. The center of gravity must be close to the ground, and large shifts of the center of gravity must not occur when the robot is traveling up or down a slope. Wheeled robots are the most simple mobile robots but prove less adaptable on rough ground. Tractor treads are a good option for robots in slightly rough terrain. Legged robots, like animals, can usually cover almost any terrain.

Besides the stability issues and the method of locomotion, one should be concerned with sensor, control, and navigation subsystems. Mobile robots are the ultimate of robots: they are the most complex of all robotic systems. In the remainder of this subsection, I will examine several mobile robots of novel configuration. We will disregard, for the moment, many of the subsystems and focus on the leg mechanics.

There are two types of walking machines: statically balanced and dynamically balanced. The one-legged hopping robot, to be reviewed shortly, is an example of the dynamically balanced robot; if it stops, it falls over. It is stable only while walking. Contrastingly, the statically balanced robot is stable only while standing in one place. An analysis of the statically balanced robot shows that, while it is walking, the stability to keep from falling over is by pausing momentarily during the individual steps. This pausing might only be a fraction of a second and not even be noticed by observers. The small windup walking toys are usually dynamically stable walkers and often fall over at the slightest perturbation while standing. Analysis of these little walking toys is given by Jameson (1985), and the analysis of biped walking is given by Takanishi et al. (1990).

Quadruped walking robots are usually the static stability type. Three feet must be touching the ground at all times. Adachi et al. (1988) and Hirose and Kunieda (1991) explore quadruped leg mechanisms and the stability of the systems.

Raibert (1986) examined the theoretical and experimental dynamics of a one-legged hopping robot. The body contained the actuators and electronics for the operation, and the one leg could telescope for different lengths and contained sensors measuring the angle of the body to the leg. The basic control mechanism for balance consisted of the equivalent of that necessary for broom balancing on one finger.

Koditschek and Buhler (1991) and Vakakis et al. (1991) studied the dynamics of the Raibert robots. Koditshek and Buhler concentrated on analytical solutions. Vakakis et al. concentrated on discrete dynamical solutions. They discovered period doubling leading to a chaotic behavior in the dynamics. The resulting strange attractor could be controlled by a parameter corresponding to the duration of the applied hopping thrust. The strange attractor could collapse to a period one cycle, which is the stable hopping motion.

Pipeline inspection is often difficult in chemical plants and nuclear plants because of the huge number of pipes and often narrow spaces. It is inconvenient to shut down systems in order to use robots that crawl along the inside of a pipe. Furthermore, valves will interfere with the continued progress of the pipe inspection. Fukuda et al. (1990) designed a robot with articulated body and suction feet to crawl on the outside of a pipe for inspection purposes. The robot can adapt to any radius pipe or tank surface. It can crawl straight up a pipe or tank, and it can pass over valves, flanges, hangers, etc. L and T-joints are not an impediment. It can even jump to an adjacent pipeline if necessary. This almost sounds too much like science fiction. The authors have built four models of this robot and include photographs in their paper.

Other robots with articulated bodies have been built and described by Hirose and Morishima (1990). Like a snake, these robots have 3-dimensional mobility. Figure 7-3 is a rough sketch of one such robot. The body segments are connected by rack and pinion gears, and each body segment is a tractor-tread drive. The robots can easily climb stairs.

Two more unusual robots are described by Iwamoto and Yamamoto (1985) and by Kawamura and Nakagawa (1990). The first is sketched in Fig. 7-4. It is a tractor-tread driven robot able to climb stairs. The second robot has only three degrees of freedom (three joints) and is able to walk on mostly smooth terrain.

Neural control

In this section, we will review a few papers on neural control of robots; a complete review, however, is beyond the scope of this book. Later in this chapter, we will describe our own work on neural control of a robot arm. Eckmiller (1989) and Horn et al. (1990) survey the application of neural networks for robotics.

One major advantage in neural network control of actuators for robots is that solutions to the dynamical equations don't need to be explicitly computed. To train a

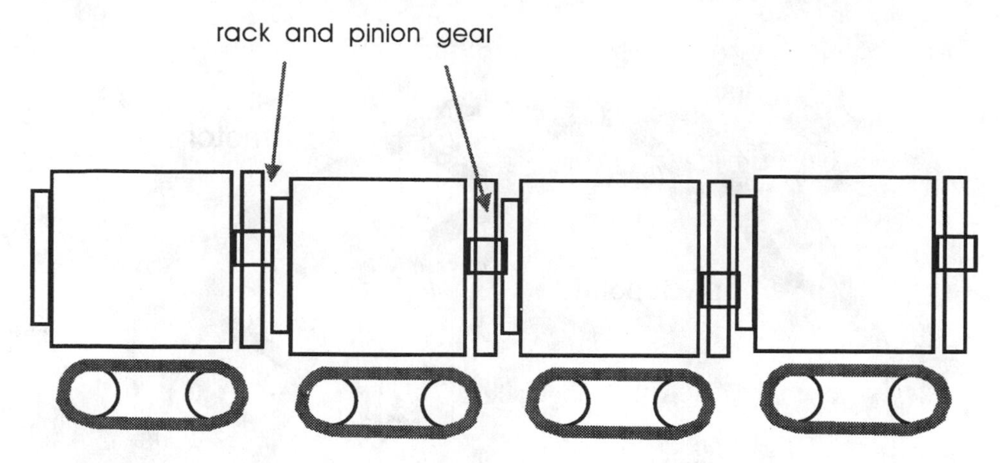

7-3 Articulated body mobile robot of Morishima.

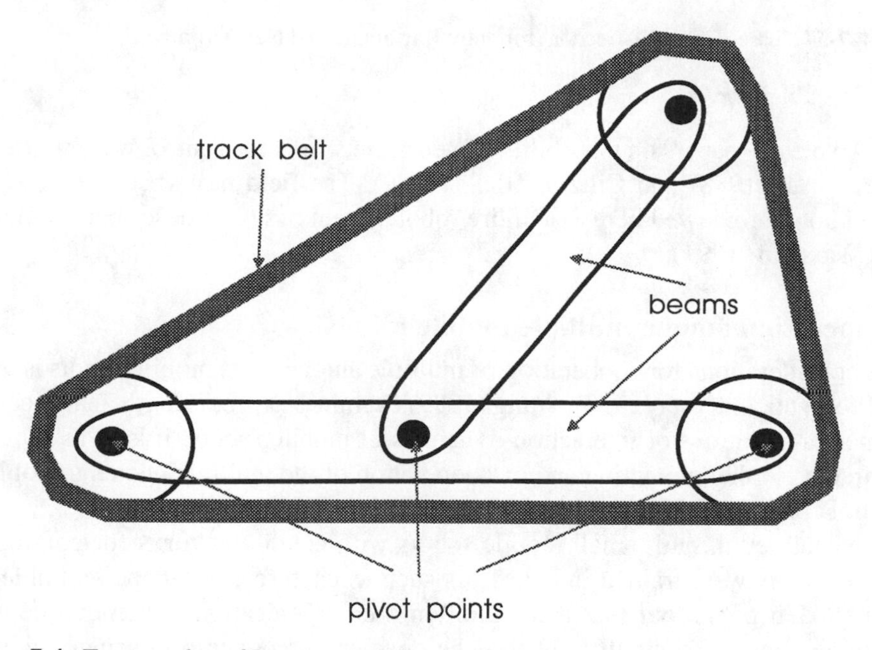

7-4 Tractor robot of Iwamoto and Yamamoto.

network to operate a pair of legs for walking, all you must do is minimize some error measure. This approach was taken by Beer (1990) and by de Garis (1991). Taga et al. (1991) study the neural dynamics of bipedal locomotion and discover that only one parameter needs to be changed to move from walking to running. Their work is similar to Vakakis et al. (1991) in the discovery of single parameter control.

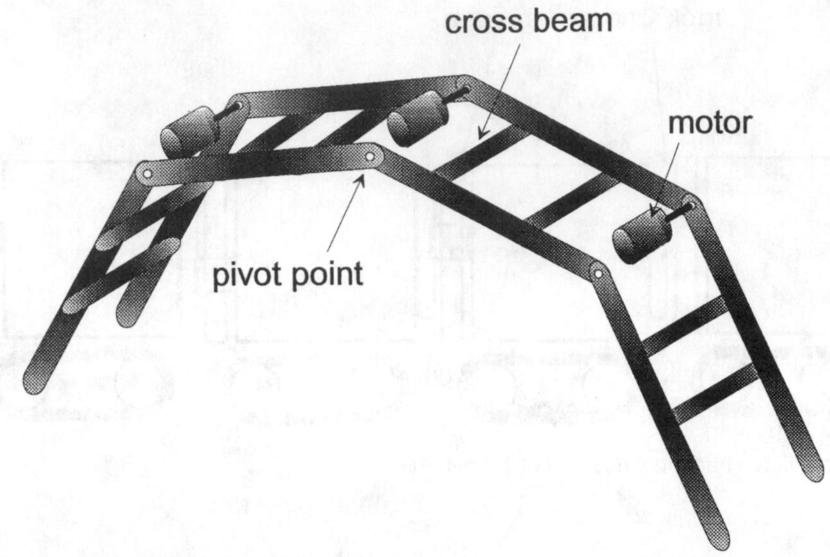

7-5 Three-degree-of-freedom robot by Kawamura and Nakagawa.

Two early papers on possibility of neural networks for control were written by Guez et al. (1988) and Ciliz and Isik (1989). The field has since exploded and could not be reviewed here. An entire robot system has been described by Baloch and Waxman (1991).

Cooperation among multiple mobile robots

The considerations for cooperation of multiple autonomous mobile robots is given by Premvuti and Yuta (1990). Wang (1989) examines at great length analytical and numerical solutions of interaction dynamics of mobile robots. It is in many cases desirable for the interaction and/or cooperation of the multiple agents to not have central control (cf. Nagata et al. 1990, and Arkin 1992).

Nagata et al. built small mobile robots with neural networks for controllers. These robots were trained in behaviors such as capture and escape and hide and seek. Arkin points out that there is often a communication bottleneck between multiple agents. To circumvent this, he develops a cooperation schema without communication between the agents. When central control is required, Noreils (1990) uses petri nets for demonstrating the control of the multiple agents. Fukuda et al. (1991) suggest a central controller sending out messages to the other robots. And finally Lee and Bien (1990) describe a neural network to optimize the path for mobile robots. The objective is to have the robots in some final configuration and start from some known configuration. The question they study is the mapping from one of these states to the other using a neural network optimization circuit similar to the Tank and Hopfiled (1986) circuit.

Robots for space operations

Heretofore we have been concerned with actual system and subsystem research. In this section, we will begin to speculate on the capabilities of robots for extraterrestrial activities.

Freitas (1980) proposed a self-reproducing probe for interstellar exploration. Interstellar distances are so great that communication with robotic probes and/or extraterrestrial entities would be essentially impossible. However, autonomous self-repairing and self-reproducing robotic probes could explore most of the galaxy within about ten centuries. A simple robotic seed probe sent to a Jovian moon, able to utilize space resources, could reproduce and send similar seeds out to explore the galaxy. The atmospheres of the gas giants could be mined for fuel and/or cometary capture. There are many comets in an inner cloud at 40 AU and the Oort Cloud at 100,000 AU. Volatiles from these bodies could be used as fuel and minerals for construction purposes.

These ideas are not just idle speculations. NASA has sponsored a workshop on the subject of advanced automation for space missions (cf. Freitas and Gilbreath, 1982). (Of course, we would expect uncomprehending politicians to cause a major funding bottleneck.) Most of the theoretical background has been known for decades since von Neumann worked out the details. There are some technical questions on chemistry, materials processing, and artificial intelligence, but these should be solvable.

On a much smaller scale, Miller (1990) has written about behavior-controlled robots for planetary surface missions. He is proposing the use of vast numbers of microrobots working in cooperation. Brooks and Flynn (1989), Angle and Brooks (1990) and Brooks et al. (1990) discuss lunar base construction using microrobots and the use of microrobots for planetary rovers. Each robot, weighing about 1 kg, will be a complete autonomous mobile system. A team of about 20 small bulldozers, about the size of a Tonka Toy bulldozer, each powered by solar cells could build a lunar base in a few years. The little bulldozers, or moon-units, could dig trenches and stockpile loose lunar soil to cover habitation units. The robots could "communicate" with each other by infrared beacons. The cooperative activity could easily be programmed with the following rules:

- Maintain same distance from neighbor robots.
- Match velocity with neighbor robots.
- Move toward center of mass of robots in neighborhood.
- Relate its velocity to the number of big rocks in its neighborhood.

These rules will result in simple flocking behavior. And like termites, these robots can collect loose soil. If a wandering robot finds loose soil, it scoops it up. If the robot finds a small pile of soil, it drops its scoopful of dirt. Over the course of time, following this simple algorithm, the piles of soil will cluster together into a few large piles. None of the robots need to have a great level of intelligence.

Comments on conventional robotics

Much of this chapter so far, except for the section on neural network control, has been concerned with the conventional approach to robotics. In that approach, the signals from sensors must be integrated into some robotic mental map of the world in order for the robot to plan a task execution. The task execution requires massive computation of the inverse kinematics to apply the correct voltages to the actuators. The mental map of the world is internal to the robot and is often simplified because of the difficulty of programming a real world map. The robots with these control maps often require very structured environments; and if a new element is placed in the environment, the robot is often not able to respond in an intelligent fashion.

A new approach to robotics is more stimulus/response-based. If a mobile robot detects an obstacle, it moves. Quite simple. There is no need for the robot to identify the object as a wall or a chair. A roach walking on the floor would not be able to identify a chair. What is a human's chair to a roach? Just an obstacle. So the first rule of the new robotics paradigm is simply to wander and avoid objects. Complex maps for this activity are not needed. The objective is simply to build a system that can survive in the real world (cf. Wallich, 1991).

Robot toys

In many cases, the most hi-tech innovations are first tried out in toys. Children don't ask for much. If the robot wanders around and avoids obstacles, then it is successful. Functional application of the toy isn't necessary.

Hobbyist and toy robots are often lacking in functionality. In many cases, the toy market is at the cutting edge of technology. If a new product, barely past the prototype stage, fails in a critical system, it could cost a human life. But if a system fails in a toy, the biggest result will be a few tears from a child. The toy market is often the proving ground for prototype systems. It is surprising that the long-term real-world survivability of many toy robots is much higher than systems built at research institutes and universities to study conventional robotics (although there are a few exceptions). In this section, we will review the work of some home-built robot systems and some commercial robot toys.

Large and complex systems

Pennisi (1991) reports on some work of a high school student's artificial insect. Christopher Stone, a student in Bloomington, Ind., built an artificial insect powered with old syringes for pistons and junk parts. This artificial insect was then controlled by his microcomputer.

He next studied the way a crayfish walks and probed the nervous system for an understanding of the firing dynamics. Armed with this understanding, he built a computer interface to the live crayfish (cyborg) and used his artificial insect program to cause the live crayfish to walk across the table. He won an invitation to attend an Office of Naval Research workshop on robotics.

Several amateur robots have been built. I already referred to the Heiserman (1976, 1979) robots. These gave rise to his experiments with computer simulations similar to those of Chapter 3. His hardware experiments were rather ambitious. His first mobile robot included bumper switches, line tracking, and self-detection of low battery and plugging itself in for a battery recharge. His second robot included the self-programming algorithms, reviewed in Chapter 3, and included all the features of the first mobile robot.

Weinstein (1981) explores at length many of the subsystems for a mobile robot, including obstacle avoidance, mechanical drive, stairway climbing, the chassis, motor drive circuits, battery selection, cooling of subsystems, collision avoidance, fingers and arms, vision systems, world mapping, speech synthesis, and computer interfacing. Holland (1983) and Robillard (1983, 1984) discuss many of the same subsystems. Robillard also covers, leg mechanics, feedback control, hand mechanics, tactile, motion, and attitude sensing, and robot control languages for microcomputer interfacing with the system. The Hero robot, from the now defunct Heathkit Co., is also discussed by Robillard (1984). Robotics Age was an excellent journal for amateur roboticists. Many of the original articles have been reproduced in a book edited by Helmers (1983).

Filo (1979) built a robot vision-arm system that corresponds to the eye-tongue system of a fly. The system operates by reaching out to grasp an object just like the frog tongue. When a frog detects a fly (or any small flying object), the tongue reaches out to catch it. Again, biology shows a simple algorithm. The system includes an articulated arm, hand, and a compound eye.

Two even more ambitious amateur robot projects have been described by Hollis (1977) and by Madison (1986). The Hollis robot is a wheel-based system with a three-degree-of-freedom arm, a vision system, and multiple sensors. The subsystems are controlled by 8080 microprocessors. The robot is based on the conventional approach where the robot is confined to a structured environment, and it builds a world model with classification of objects in its world. The system described by Madison is much like the Hollis robot except that it is a bipedal walking machine that drags its feet. The feet dragging is assisted by tractor treads. Madison claims that the cost to build the machine was about the same as for a family automobile.

Small systems

Just because a robot is made of plastic doesn't mean that it's junk. Trends in the costs of robots indicate that they will continue to drop. Hydrolic systems could be built from injection molded parts and operate by pressurized air or water. Hydrolic pistons and motors can easily be constructed from plastic. Hypodermic syringes are an example of low-cost pistons. Gears and gear boxes can easily be made of plastic. For high strength applications, the plastic parts could be composites of fiberglass or carbon fibers. Low-cost mass-produced robots will change the world just as microprocessors have.

There are 14 small robots listed in the Edmond Scientific Catalog, including robots that are programmable by interfacing with a personal computer, hexapods, quadrupeds activated by sound, sound-following, containing line-tracking systems, having a robotic arm, and having IR sensors for avoidance. All these subsystems are on separate robots designed for educational and experimental purposes.

Photo 7-1 shows the line-tracking system. This robot will follow a line drawn on the floor. Photo 7-2 shows a quadruped robot. When the robot detects a loud noise, it will begin to walk. Photo 7-3 presents a hexapod robot controlled by switches and having a power supply attached via the cable. Both the quadruped and the hexapod are not very good at walking. The quadruped is barely stable on standing; and, while the hexapod walks well, it doesn't pick its feet up far enough to get over small obstacles.

Photo 7-4 shows a robot that will turn away from loud noises and detects obstacles by bumping a microphone pickup into the obstacle. This robot shows the importance of morphological design of the robot. Although it is able to function quite well in a structured environment, it gets stuck easily in cluttered environments. The robots are quite cheap and, if nothing else, they have ready-built circuits that can be cannibalized for parts. These small robots are similar to the Amarobot kits (cf. Brown, 1987) that are popular among hobbyists.

Photo 7-1 Line-tracking robot.

Photo 7-2 Quadruped sound-activated robot.

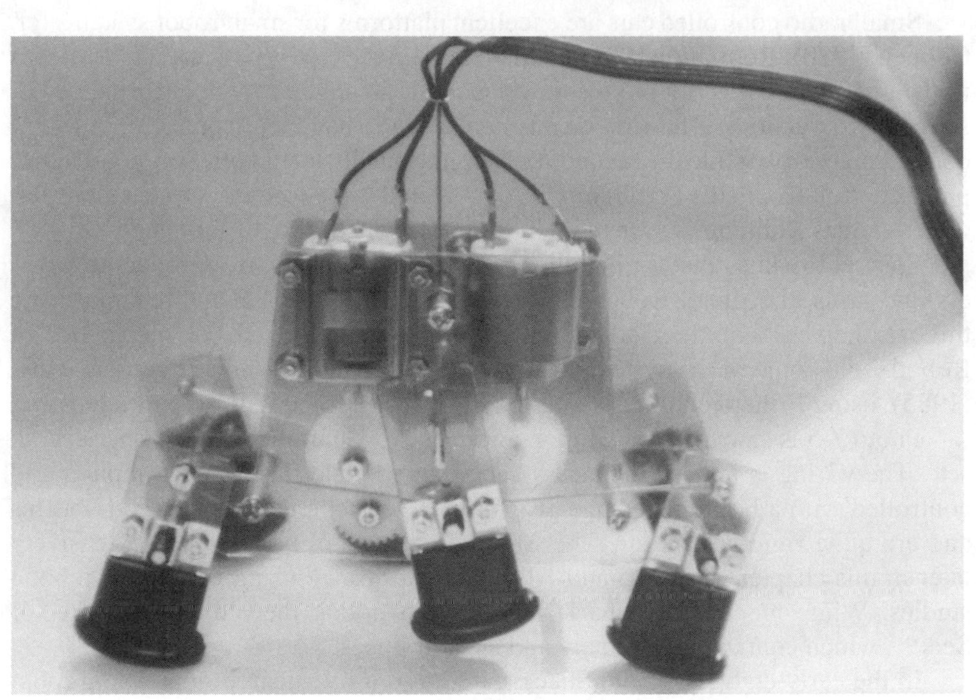

Photo 7-3 Simple hexapod robot.

Photo 7-4 Microphone bumper robot.

Small radio controlled cars are excellent platforms for small robot systems (cf. Connell, 1988). Robillard (1984), in his book *Advanced Robot Systems*, describes in detail how to convert a radio controlled car to a mobile robot platform. He describes two versions: The first simply replaces the handheld control unit with a computer control, while the second replaces the handheld unit with on-board computer control. Connell's (1988) mobil robot includes photosensors to detect obstacles, whereas Robillard's system includes microswitches in the bumpers.

Several toy kits exist for building many types of robots, from rovers to walkers and arms and more. Robotix™, Fishertechnik™, and LEGO Technic™ are three examples. Robillard (1984, 1985, 1986) has explored the mechanics of Robotix and some of the systems that can be built with it. Wrainwright and Moss (1985), using Fishertechnik, have built a huge system modeling an entire factory.

Photo 7-5 is an example of a pick-and-place robot arm from the Fishertechnik set. (The wiring is not yet in place.) The system is interfaced to a computer and controlled with a Fourth language program. The structural elements are H-beams and are quite rigid and strong. The LEGO systems will be reviewed extensively later in this chapter. Many AI and Alife laboratories are using LEGO for robotic studies. While mentioning kits, I should not overlook the old standby Erector Sets™, which contain large motors and steel beams.

In the early eighties, Radio Shack introduced a five-degree-of-freedom robot arm: The Armatron. This robot arm was made from rigid plastic. The joints were

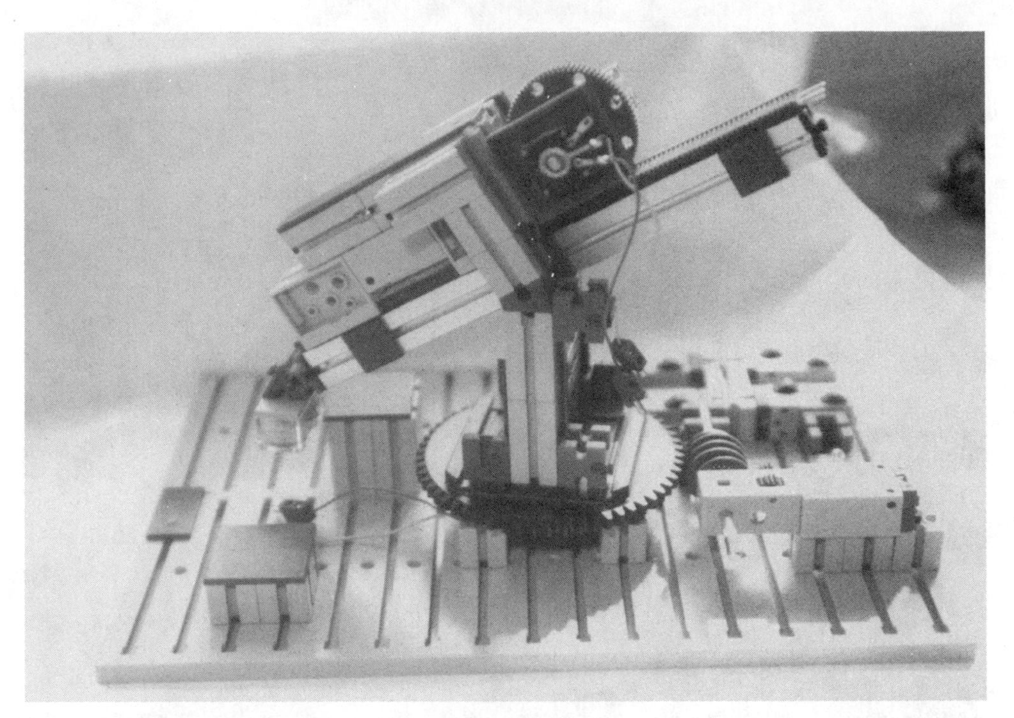

Photo 7-5 Fishertechnik pick-and-place robot.

activated by one motor (joystick controlled). The presence of a large number of plastic gears and levers allowed the one motor to operate all the joints. Only one joint could be activated at any given moment. Schiavone et al. (1984) replaced many of the levers with solenoids and Hall-effect sensors. The entire system was then controlled by a personal computer.

In 1986, Radio Shack introduced the Mobile Armatron. Barbarello (1987) and Rietman (1986) dissected this mobile robot arm and built a computer-control interface. Photo 7-6 and Photo 7-7 show the robot.

This mobile arm provides 100 degrees of motion at the shoulder. The elbow is not electrically controlled but can be rotated 100 degrees in 20-degree steps. The wrist provides 180 degrees of motion up and down and can rotate 360 degrees. The hand opens 2.25". Power is provided by four D type cells, which produce +3 volts and –3 volts for robot control. The control module is connected to the robot by a seven-wire color-coded cable. Push-button contacts on the control module make contact between a control function and + or – voltage. The batteries are in the base of the robot.

The control functions are shown in Table 7-1. The black wire is 3 volts with respect to a function wire, and the brown wire is –3 volts. The potential between the black and brown wire is 6 volts, with black +. Table 7-1 is easy to follow and is included here, along with a circuit, as an example of how to control simple robot actuators.

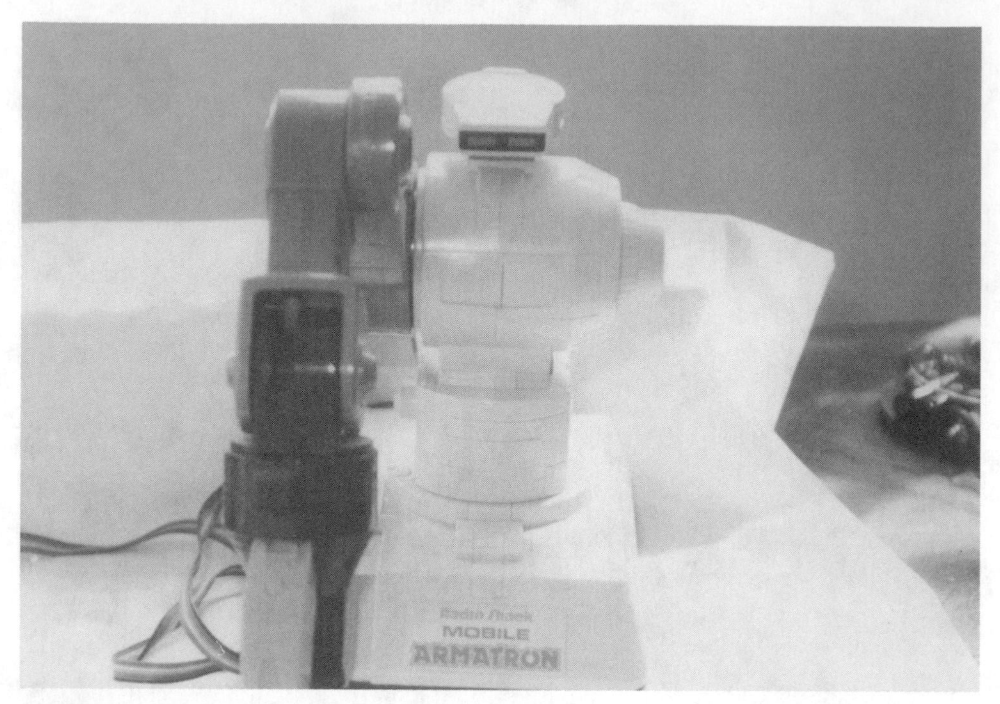

Photo 7-6 Front view of Mobile Armatron.

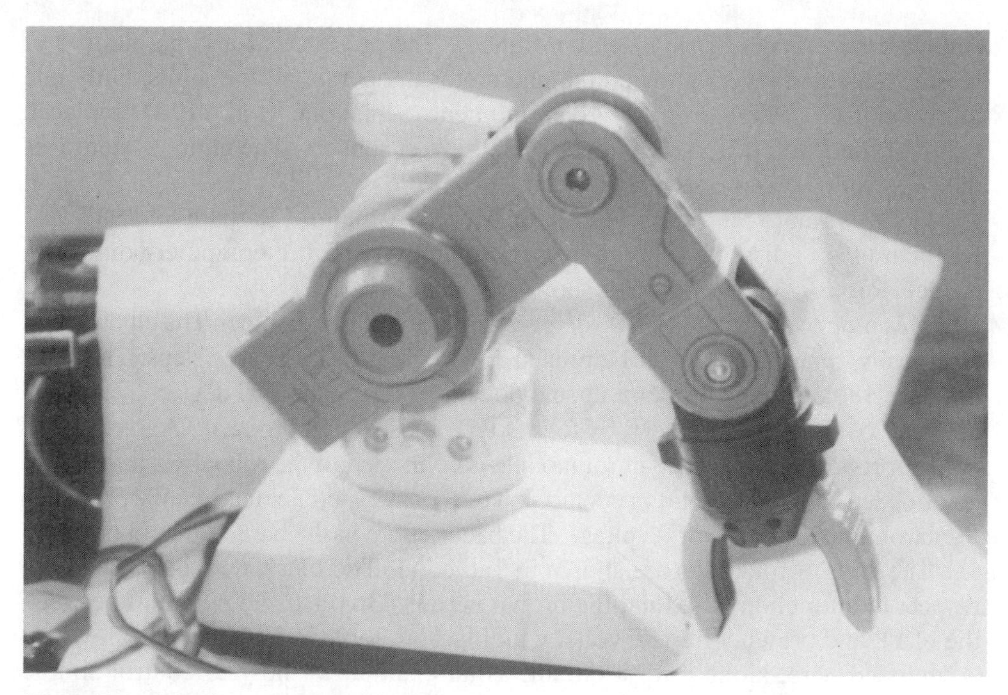

Photo 7-7 Side view of Mobile Armatron.

Table 7-1
Color codes to control robot movements.

Color code connections	Function
Black-red	Hand open/close
Black-orange	Wrist down
Black-yellow	Move left
Black-green	Move right
Black-blue	Arm down
Brown-red	Rotate wrist
Brown-orange	Wrist up
Brown-yellow	Move right
Brown-green	Move left
Brown-blue	Arm up
Black-yellow & green	Move backward
Brown-yellow & green	Move forward

From the table, notice that by making contact to close a switch connecting the black wire and the red wire, the hand will open/close. Forward motion is effected by connecting black with yellow and green. This suggests an obvious control circuit. Figure 7-6 is an example of controlling one of the lines. The lines A & B for robot control would connect to the lines shown in the table; for example, A = black and B = red.

The circuit can easily be controlled by the I/O ports from an 8255 PPI chip. Interface circuits for this chip and software have been described by Goldsbrough

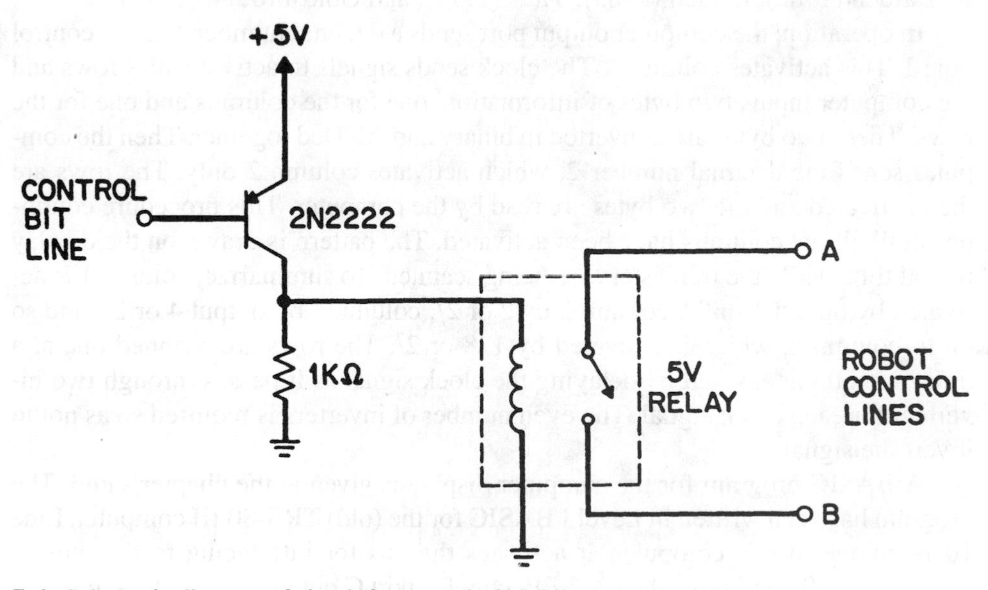

7-6 Schematic diagram of circuit for control of Mobile Armatron.

(1979). A robot control language has been described by Robillard (1984) and would provide easy command level communication with the robot. The language Forth would also be an excellent control language.

In early 1985, I built a simplified pattern recognition system that could be mounted on the front of the Mobile Armatron (Rietman, 1985). The system has poor resolution and recognizes only basic geometric shapes, but the system can be expanded by experimenters. The main part of the system is an 8×8 phototransistor array. When geometric objects are placed over the array in the presence of light, certain phototransistors send signals to the input channels in the computer. The computer then decodes this information and draws a picture on the display.

Let's examine the circuit. The PNP phototransistors are type GE L14C1212 and are connected in an 8×8 array. All the emitters are connected in each column and all the collectors are connected in each row. The base connections were cut off the transistors. The 8 emitter leads, one for each column, are connected to +5V via a CMOS 4066 analog switch. The emitter leads are also connected to one of the two eight-bit input ports for the computer. The 8 collector leads, one for each row, are connected to ground via a CMOS 4066 switch and a low pass filter. Connections just before the filters are also made to a computer input port. The "emitter switches" are controlled by an output port from the computer. The "collector switches" are controlled by a clock signal. The clock is two CMOS 4049 NOT gates used as an oscillator. The 8 clock lines are taken off a simple delay line made up of 4049 NOT gates.

Figure 7-7 and Fig. 7-8 show the circuit. The input ports and output port are made from an Intel 8255 programmable peripheral interface chip (PPI) described by Barden (1982), O'Dell (1983), Titus (1979), and Goldsbrough (1979).

In operation, the computer output port sends a decimal number 1 to the control board. This activates column 1. The clock sends signals to activate all 8 rows and the computer inputs two bytes of information, one for the columns and one for the rows. These two bytes are converted to binary and ANDed together. Then the computer sends the decimal number 2, which activates column 2 only. The rows are then activated and the two bytes are read by the computer. This procedure continues until all the columns have been activated. The pattern is drawn on the display in real time while the transistors are being scanned. To summarize, column 1 is activated by output 1 or 2^0, column 2 by 2 or 2^1, column 3 by output 4 or 2^2, and so on to column 8, which is activated by 128 or 2^7. The rows are scanned one at a time, something ensured by delaying the clock signal as it passes through two inverters for each clock signal. The even number of inverters is required so as not to invert the signal.

A BASIC program for the computer display is given at the chapter's end. The program has been written in Level I BASIC for the (old) TRS-80 III computer. Line 10 is unique to this computer; it activates the bus for interfacing to the outside world. Line 20 configures the 8255 PPI chip for port C output, and ports A and B as input. Line 30 begins the row activation loop, while line 40 activates the appropri-

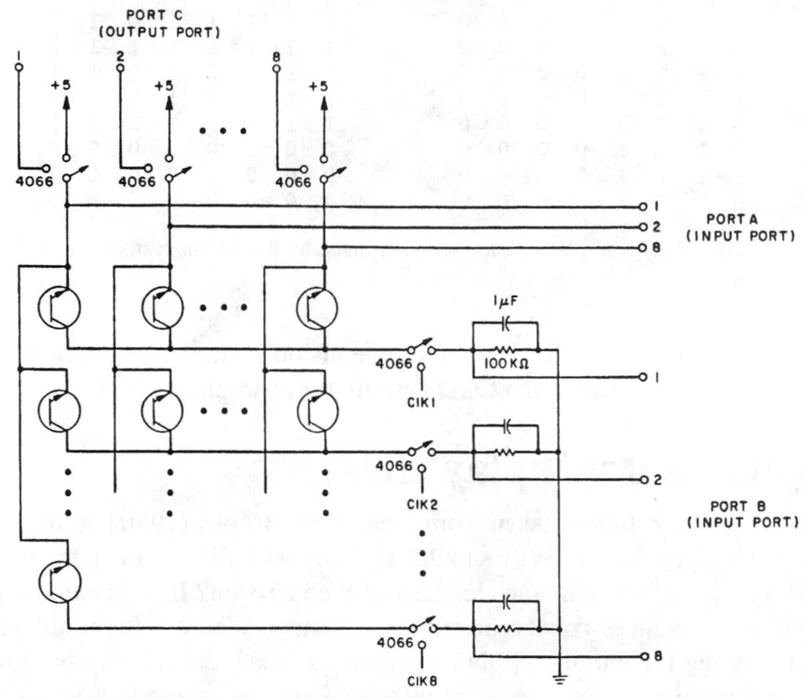

7-7 Circuit diagram for pattern recognition system.

ate row. Lines 50 and 60 input one byte each. These are converted to binary in lines 70–420 and then ANDed in lines 430–450 and displayed in lines 460–465. Line 475 provides a time delay before the screen is cleared and the process started over.

Figure 7-9 is two displays of rectangles placed over the phototransistor array. Notice the image consists of 1's and 0's. This could easily be converted to light and

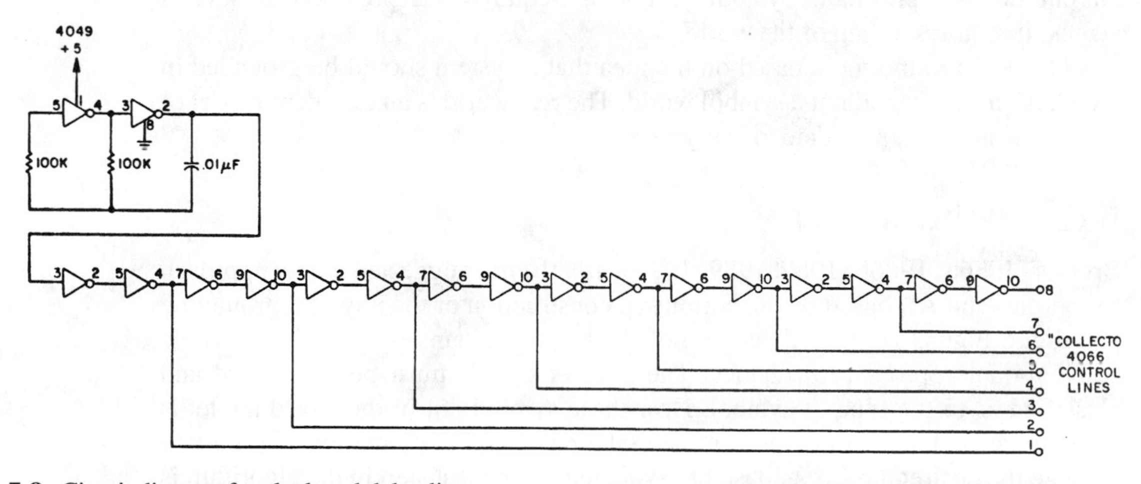

7-8 Circuit diagram for clock and delay lines.

```
0 0 0 0 0 0 0 0     1 1 1 1 1 0 1 1
1 1 1 1 1 0 0 0     1 1 1 1 1 1 1 1
1 1 1 1 0 0 0 0     1 1 1 1 1 1 1 1
1 1 1 1 0 0 0 0     1 1 1 1 1 1 1 1
1 1 1 1 0 0 0 0     1 1 1 1 1 1 1 1
1 1 1 1 0 0 0 0     0 0 0 0 0 0 0 0
1 1 1 1 0 0 0 0     0 0 0 0 0 0 0 0
1 1 1 1 0 0 0 0     0 0 0 0 0 0 0 0
```

7-9 Two examples of patterns recognized by the vision system.

dark pixels. Although this is a rather simple vision system, it contains all the fundamental circuit elements for a CCD system, for example.

Elephants don't play chess

The title of this section is taken from a paper by Brooks (1990) on the problems with AI. The question is "Why should AI focus on building a machine as intelligent as a human?" Why not an elephant or even a roach? Insects can out-perform conventional advanced robots and super computers. Where is the problem?

The problem lies in the symbol grounding of classical AI. Symbol grounding means that the perception and motor interfaces are sets of symbols operated on by a central processor or intelligence—symbols that represent actual physical objects, properties, or whatever, in the real world. An intelligent system must interpret the symbol in order to operate on it with its central processor. All real intelligent systems, except laboratory systems, are embedded in the real world. By the symbol grounding hypothesis, the perception system must convert the information to symbols for the central processor. For the central processor to utilize the information, it must have a world map that includes the properties of the objects in the world. Herein lies the problem. Symbol grounding requires interpretation at several stages; it requires a map of the world.

Physical grounding is based on the idea that a system should be grounded in the physical world and not a symbol world. The real world is an excellent model of itself. It is always up-to-date in every detail.

MIT robots

Brooks (1986a, 1986b, 1987, 1989, 1991a, 1991b) has suggested a new approach to robotics and AI, based on the bottom-up construction of the physical grounding paradigm. Figure 7-10a is a diagram of the symbol grounding paradigm and the conventional approach to robotics. The sensors signals must be interpreted and classified as to the type of symbol. How these symbols fit in the world model is then deduced. The action plan can then take place and computations for how to achieve the desired task (e.g. obstacle avoidance). The last step in the algorithm is

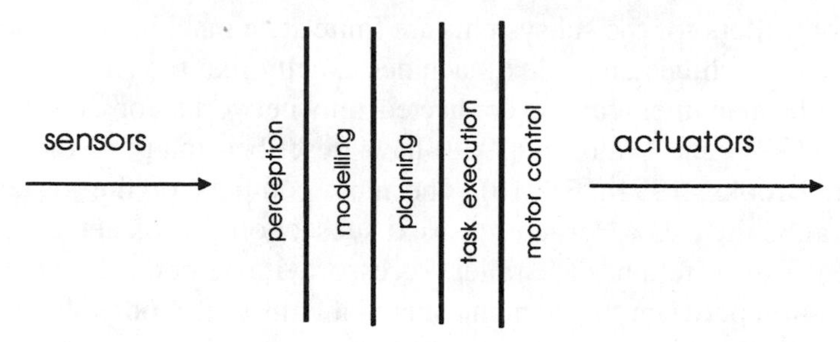

7-10a Conventional robotics solutions to real-world problems.

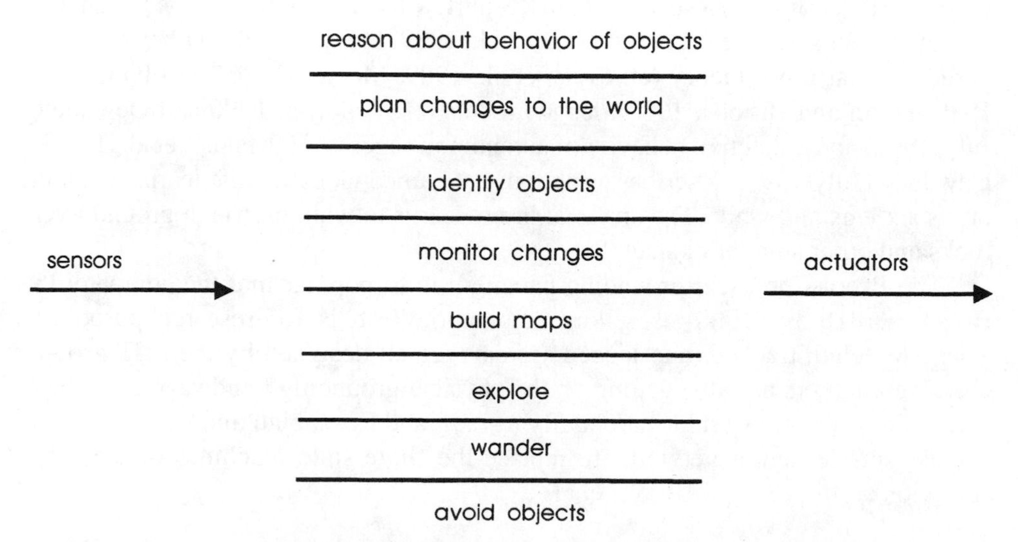

7-10b Brooks' robotics, subsumption solution to real-world problems.

to compute the voltages for the actuators and send the appropriate signals to activate them.

The physical grounding approach is shown in Fig. 7-10b and is called the *subsumption architecture*. The lowest level of activity is to avoid objects. Sensors are hardwired to actuators or perhaps though amplifiers to the actuators (Braitenberg vehicles operate at this level). If there are no obstacles in the current path of the robot, then it wanders or explores the world. By relating contact situations with motion vectors, a robot can build a map of the world, although a strange map (cf. Heiserman 1976, 1979). This idea is the same as the Heiserman robots we simulated in Chapter 3. (I would like to point out that Heiserman was ten years ahead of Brooks. Heiserman clearly pointed out the flaws with conventional robotics and then went on to build systems based on what he called Evolutionary Adaptive Machine Intelligence. The fact that Heiserman didn't publish his ideas in a mainstream research journal is beside the point.)

The controllers for the subsystems are finite state machines (cf. Brooks 1989). These simple machines are reflex machines exactly like the Braitenberg vehicles. These simple machines can be connected into networks for complex behavior. Wolkomir (1991) and Waldrop (1990) have described many of the Brooks et al. robots (cf. Brooks, 1991a, 1991b). The most complex of the systems built by Brooks et al. is the robot Herbert (cf. Next subsection, Brooks et al. 1986, Brooks et al. 1988). The system has a parallel processor with 32 nodes. There is little to no communication between them and no shared memory. The only shared resource is the power supply.

After the Herbert project, the focus of the MIT mobile robot laboratory (Brooks et al.) shifted to smaller robots. Squirt, a 1-cubic-inch robot, is the smallest autonomous mobile robot (cf. Flynn et al., 1989, Memo #1120). Other work focused on gnat robot theory (cf. Flynn et al., 1989, Memo #1126 and Flynn et al. 1989, Flynn and Brooks, 1988), legged robots (cf. Waldrop, 1990 and Dewdney, July, 1991) and collective behavior of autonomous agents (cf. Mataric et al., 1992). Dewdney (July 1991) describes a robot insect named Genghis that learns to stand on its six legs and walk. The artificial insect learns to walk on rough ground over rocks and other smaller obstacles.

The Brooks physical grounding paradigm is so popular that the company IS Robotics (818-597-1900) is selling autonomous robots for research purposes. These wheeled, tracked, and legged systems are all designed by the MIT group. Other researchers are also getting on the physical grounding bandwagon. Hartley and Pipitone (1991) simulate a robot flying craft and Rosenblatt and Payton (1989) discuss simple neural networks to replace the finite state machines of the subsumption architecture of Brooks et al.

Herbert

Herbert is a subsumption architecture mobile robot, like all the Brooks MIT robots. The subsumption architecture on this robot is built from a collection of 8-bit microprocessors; the only shared resource is the power supply. Each processor has its own memory and clock; there is no shared back plane or global clock. Each processor has three serial input lines and three serial output lines, each serial line having a control line and a data line. Messages are self-clocking; the clocks of the sending and receiving processors need to only be within 20% of a common frequency for such a scheme to work. Each processor also has a reset line and an inhibit line, with the inhibiting line inhibiting output messages for a specified period of time. The time period is set by an on-board potentiometer and can range from fractions of seconds to hundreds of seconds.

If new state machines must be added to the subsumption architecture, then a new processor board is just plugged in to the power supply and a few other connections on the patch panel are made for communications. The system is then essentially ready for real-world processing.

The robot Herbert has mobility and a two-degree-of-freedom arm with a parallel jaw gripper. The arm is a simple pick and place device that can reach from table top to the floor covering about 40" high and 18" deep. Each of the motors in the joints has an analog position detector, which is interfaced to finite state machines in the subsumption architecture of parallel processors. These can be used to read the position of the joint angles and the gripper separation. The processors also send output signals to the motors for repositioning the arm and hand.

Infrared sensors are on board for fast obstacle detection. The hand contains touch sensors for the object it is picking up. The infrared sensors are used, rather than ultrasonics, because it is much faster for a high-speed mobile robot. The robot also has a laser range scanner configured for detection of specific objects (soda cans).

In operation, the robot wanders from some home position and searches for empty soda cans. While wandering, it keeps a record of the path it has traversed by keeping a record of the distance and angles of the various motor drives. If a can is found, then the robot positions itself in front and reaches out to snatch the can. The can is then transported back to the robot's nest at the home position by retracing its "steps." In its wandering mode, if the infrared sensors detect an obstacle, the robot will move out of the way and keep a record of the move.

Herbert's entire existence is dedicated to collecting soda cans. It has been called the Collection Machine and has been extensively examined by Connell (1990).

BEAM robots

Mark Tilden (1992) has extended the work of Brooks to include total self-autonomy, small size, and light weight (cf. Smit and Tilden, 1991). Tilden's approach to robotics is quite novel. He has coined the term BEAM robotics where BEAM is an acronym for *Biology Electronics Aesthetics Mechanics*, *Building Evolution Anarchy Modularity*, and *Biotech Ethology Analogy Morphology*. He aptly calls the subject Robobiology. The robotic-bio creatures do not have to serve any practical or industrial purpose; they only need to survive outside in the real world without herd-dependence. Creatures have been built to mimic everything from self-folding flowers to window washers.

There are three aspects to the control theory of the robot creatures: decentralized control, stimulus engines, and biological mimicry. Together, these components result in autonomous robot creatures. All the BEAM robots are solar powered, with no on/off switch. Once they are built, they live their lives—for whatever it's worth—free of human intervention. Tilden's Laws of Robotics are as follows (quoted with permission of Mark Tilden):

Law 1. A robot must protect its existence at all costs.

Law 2. A robot must obtain and maintain access to a power source (with conditional and weighted deference to the First and Third Law).

Law 3. A robot must continually search for better power sources (with conditional and weighted deference to the First and Second law).

These laws are otherwise known as

Law 1. Protect thy "ass."
Law 2. Feed thy "ass."
Law 3. Look for better real-estate.

The most important principle of robobiology, pointed out by Tilden, is that robots do not have to obey any biological laws. They do not have to resemble any biological lifeform. They do not have to behave like any biological form. Survivability is the only essential. Thus Tilden builds unorthodox, selfish, self-centered robot pests and runs competitions. Some of the robots include Solarollar, Photovores, Ropeclimbers, Followers, Jumpers, Legged Racers, Tumblers, and Nanomice. The article by Smit and Tilden (1991) describes how to get started building these little creatures and also includes instructions on how to enter the BEAM Robotics contest.

Let's look at some of the beasts. Photo 7-8 shows solar engines. A schematic is also shown in Fig. 7-11. (A 1-megaohme resistor in parallel with the diode will improve the efficiency by about 25%. This circuit will then be adequate for small toy motors.)

Photo 7-8 Solar engine. (Mark Tilden)

Photo 7-9 shows two solarollers. Photo 7-10 is a Solar Flower in the folded position, and Photo 7-11 is the Solar Flower in the open position.

Photo 7-12 is a solar insect. Photo 7-13 is a solar mouse running a maze. Photo 7-14 is a solar hopper robot.

Photo 7-15 is a robot with three legs. Photo 7-16 is a two-legged walking robot. Photo 7-17 is a solar rope climber, and Photo 7-18 is two views of walking robots.

Readers interested in entering robots in the BEAM contest should contact

Mark Tilden
c/o Mathematics Dept.
University of Waterloo
Waterloo, Ontario, Canada N2L 3G1.

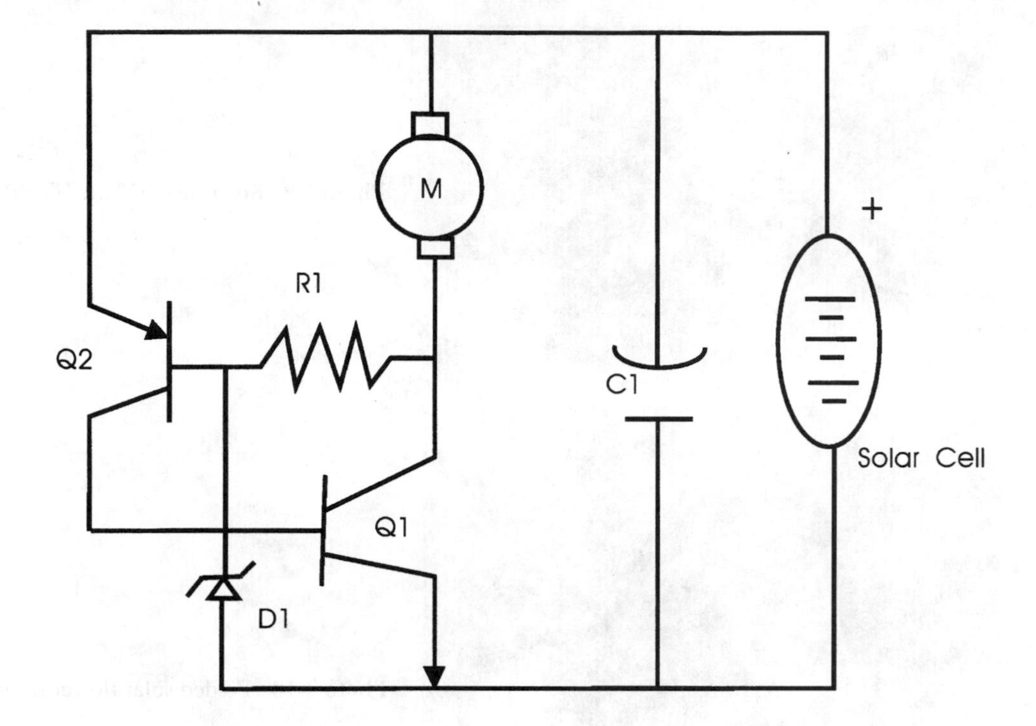

R1 2K to 15K
C1 2,2000 to 47,000 micro F
D1 1N4728 to 1N4731
Q1 2N2222
Q2 3906

7-11 Circuit diagram for solar engine.

TAG robots

TAG is a small company in England that has been making breakthroughs in neural-controlled robots. None of their controlled systems use the conventional analytical approach to finding the inverse kinematics. One of their early systems was based on a radio-controlled car with simple neural circuits wired like a Braitenberg machine. More complex controllers consist of neural networks with the input as the state vector of the robot. The output is one simple behavior (cf. Snath, 1989; Snath and Holland, 1990; Holland and Snath, 1991a). To avoid conflicts of the behaviors, they are then fed into a neural n-flop circuit to determine the winner. All the other behaviors are inhibited.

This simple robotic platform proved an adequate research tool for initial experiments, but it soon became obvious that a vehicle more like a breadboard was

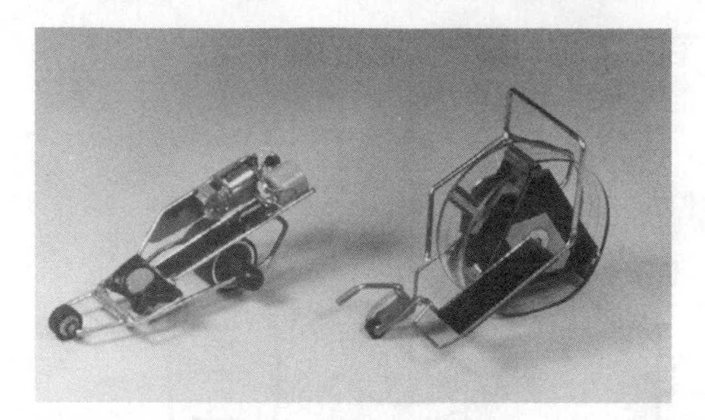

Photo 7-9 Solarollers (Mark Tilden)

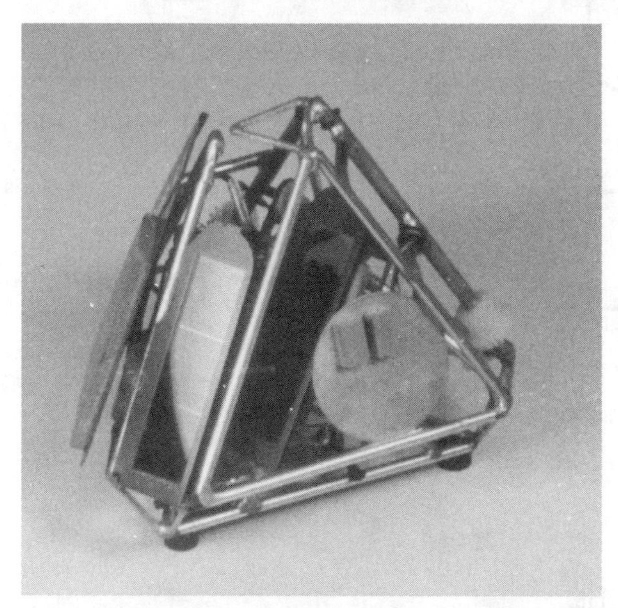

Photo 7-10 Folded solar flower (Mark Tilden).

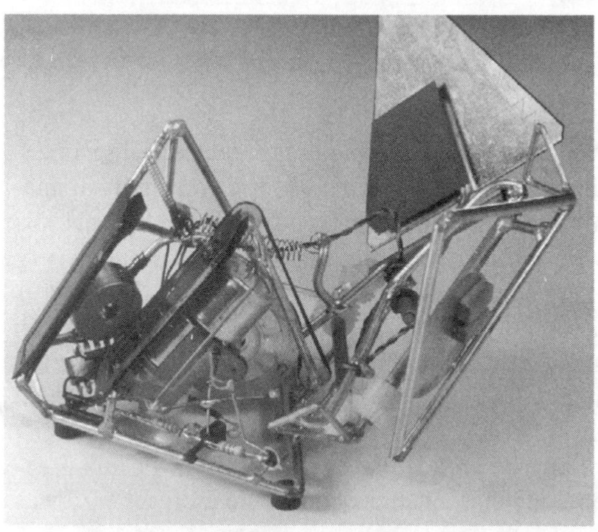

Photo 7-11 Closeup of an open solar flower (Mark Tilden).

Photo 7-12 Solar insect (Mark Tilden).

Photo 7-13 Solar mouse (Mark Tilden).

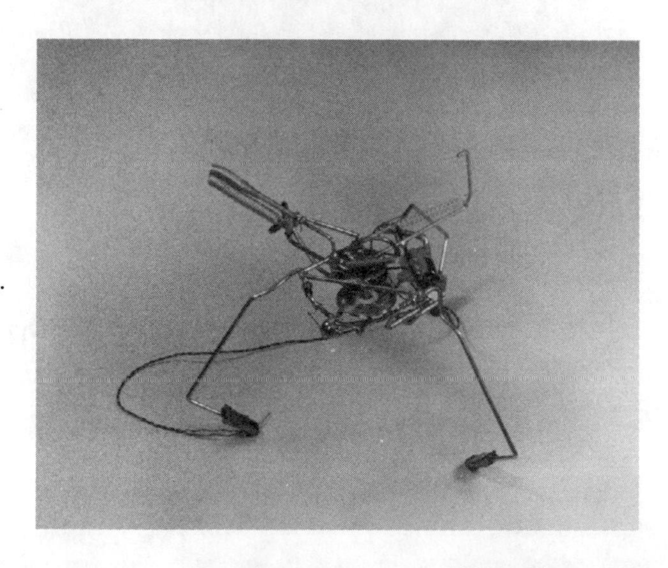

Photo 7-14 Solar hopper (Mark Tilden).

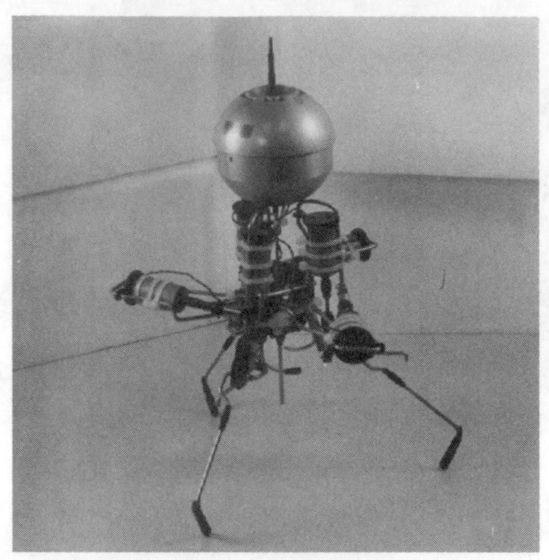

Photo 7-15 Solar hopper—three legs, (Mark Tilden).

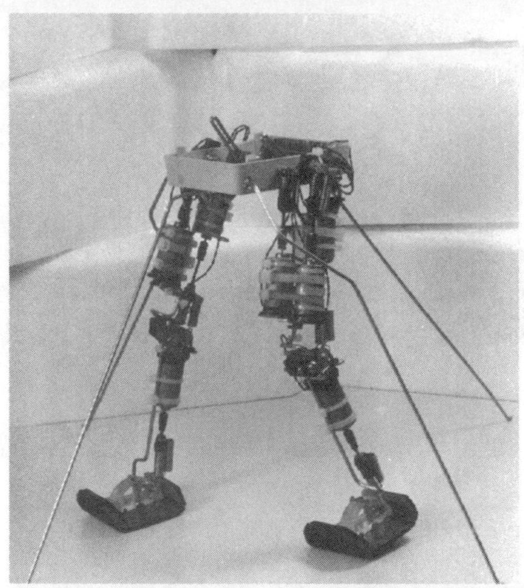

Photo 7-16 Solar walker (Mark Tilden).

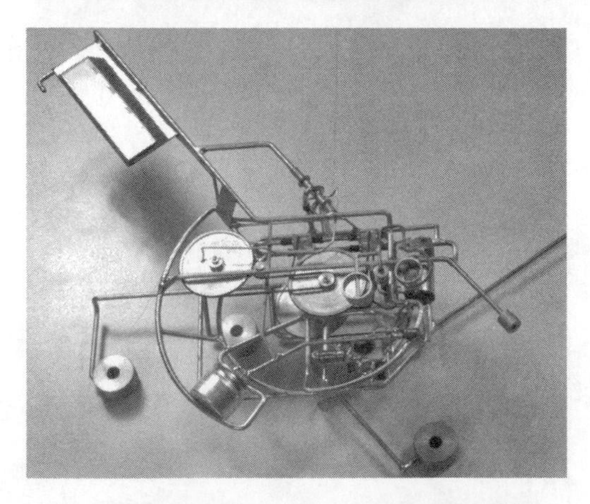

Photo 7-17 Solar rope climber (Mark Tilden).

Photo 7-18 Walking robots (Mark Tilden).

needed for more versatile experimentation. Photo 7-19 is a photograph of their answer to the problem. This robot includes a back plane for plug-in boards that contain various electronics. Photo 7-20 is a photograph of a neural network board. Notice that the board contains many connection points for wiring the network.

Later work has also focused on neural control of pneumatic actuators for a quadrupedal walking robotic platform (cf. Holland and Snath, 1991b; Holland and

Photo 7-19 Mobile experimental robot (TAG Robotics).

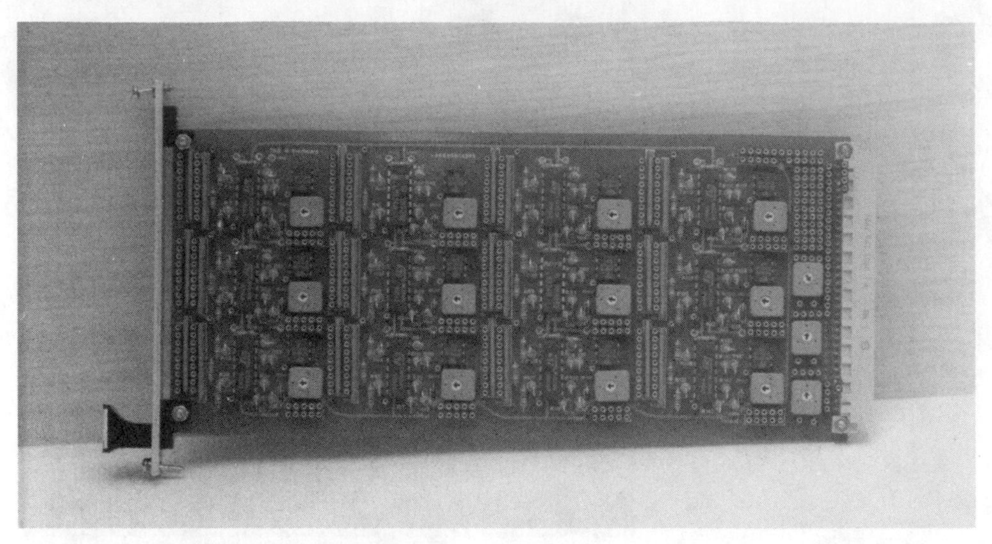

Photo 7-20 Neural circuit board (TAG).

Snath, 1991c; and Snath and Holland, 1991). They conclude, based on combinatorial calculations, that a constrained morphology derived from some type of recursion algorithm would be most efficient for real hardware robots. Snath and Holland have used the Dawkins ("Blind Watch Maker," 1987) constrained morphology to study the advantages of a symmetrical nervous system. They have also used a genetic algorithm to evolve the neural networks for their mobile robots.

Snath and Holland further conclude that the capability of n-flops as connection matrices for behaviors should not be underestimated. These circuits are capable of some hysteresis and transfer distortion. These emergent properties give rise to a memory effect and might show adaptability (cf. Rietman, 1988).

LEGO robots

LEGOs come in the standard blocks for children and high-tech kits with everything from rack and pinion steering and differential drives to motors and optical encoders. Many artificial intelligence and artificial life researchers are using these high-tech building blocks for assembly of prototype robots and biorobots.

Photo 7-21 is a photograph of a small collection of the construction beams, gears, and other parts that come with the LEGO Technic kits. Photo 7-22 is a simple LEGO vehicle with rack and pinion steering. Also evident in this photograph is some of the construction beams for vehicle assembly. Photo 7-23 is a motorized LEGO vehicle powered by a battery pack.

In this section we will review literature from researchers using LEGOs. Later we will look at the LEGOs in more detail and examine two vehicles with breadboards, on board, for circuit development.

Photo 7-21 Examples of LEGO parts.

Photo 7-22 LEGO vehicle with rack and pinion steering.

Photo 7-23 Motorized LEGO vehicle.

One of the earliest workers with LEGO robots was Resnick (1989, 1991) at MIT. His objective was to use LEGOs with LOGO programming and teach children the elements of control theory and programming. The objective was to use the LEGO-LOGO as a medium for interactive teaching. He was so successful that the LEGO has become almost a standard Universal Constructor for artificial life. MIT now offers a course in LEGO vehicle assembly and control for the undergraduate and graduate students (Martin, 1992).

Donnett and Smithers (1991) have described work that they and their students are doing on studies of sensing, control, animat morphology, and cooperative behavior with LEGO vehicles. Some of their discoveries are that the robustness and reliability of the vehicle is dependent on the placement of the sensors, as well as on the shape of the vehicle and its subsystems. Thus, the robot morphology determines the survivability of the robot. The problems encountered in developing a successful LEGO robot are vastly different than those for a successful computer simulation. The LEGO robot exists in the real world with all the messiness of bad sensors, poor morphology, fussy motors, and circuits, fragile mechanics, etc. Much of this is forgotten by people who build simulations.

The bulk of the published literature on LEGO robots has come from the work of the Department of Artificial Intelligence at Edinburgh University, and the work, cited previously, from the Artificial Intelligence Laboratory at MIT by Martin and Resnick. Adrin (1990) studied simple sensors for optokinetic and tactile competences in mobile LEGO robots. He discovered—something obvious in hindsight and never observed in simulations—that vehicle morphology plays a key part in the success of the robot. A robot with sensors placed too high might manage to get itself stuck in a low tight space. A robot with sensors too low might also find itself in positions from which it cannot extract itself.

Daskalakis (1990) studied sensor-action behavior of LEGO robots. The robots were able to learn wall following and obstacle avoidance using self-organizing associative memories on board the mobile robot. Pebody (1991) built LEGO robots for studies of the subsumption architecture of Brooks.

Dallis (1990) built several LEGO robots in order to study the cooperative search behavior. Each robot was more or less identical. They communicated with each other using pulsed IR signals, which acted as simple beacons for location of neighbors. The goal was for the robots to locate a bright light and group there. The first robot to locate the source sent a signal to the others to attract them. This strategy proved to be successful.

Mein (1991) built several distinctly different LEGO robots in order to study the cooperative behavior in search and task performance. The robots had to locate a light hidden behind a shield. The robots also had to knock over the shield in order to find the light. There was a total of six lights to be found by four robots. One of the lights had a shield. The robots communicated by pulsed IR to keep count of the number of lights found. When all six lights were found, the robots came to a stop.

As pointed out earlier, LEGOs are a universal construction kit with many bricks, beams, gears, plates, sprockets, wheels, sensors, motors, etc. In order to construct a LEGO vehicle, one will have to deal with the major problem in adding crossbeams for stability. Figure 7-12 shows the dimensions of a LEGO brick. All the building units have cylindrical bumps on the top and cylindrical structures on their bottom to allow the parts to mate. The positions of the cylindrical bumps dictate the positions of other components.

This point is most important for fitting non-LEGO parts to the robot. One should consider the dimensions of the components and be sure that they are commensurate with LEGO if nuts and bolts are to be used. Of course, glue is far less demanding in dimensions.

Photo 7-24 is a small LEGO tractor vehicle with an experimenter breadboard attached. The robot is designed for simple studies of sensors and actuators. The battery power is off the robot. In this robot, the LEGO dimensions were considered for the breadboard. Notice that the breadboard is bolted to LEGO 1/3 plates, and these are then attached to the LEGO vehicle. Once a design is finalized, the vehicle can be made more robust by gluing the components together. This particular robot can only travel forward or backward as determined by the touch switches shown on each end as bumpers. If two of these tractor vehicles are attached, it will be possible to steer the robot.

Photo 7-25 to Photo 7-27 show the side, end, and bottom views of a robot built with two LEGO tractor vehicles. This robot has eight times as much breadboard space and three sets of three AA rechargeable batteries. The batteries are supported

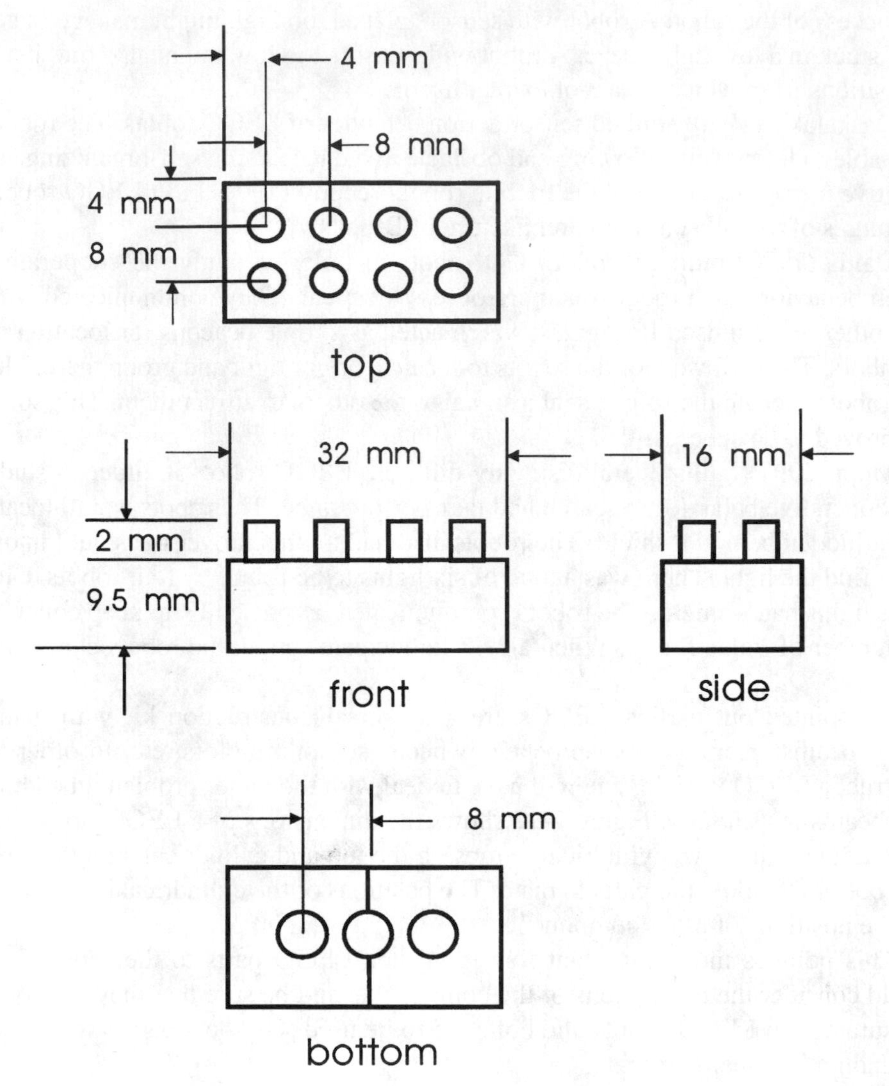

7-12 A LEGO brick.

between the two vehicles and just below the breadboard. Two crossbeams couple the two vehicles together (see Photo 7-27) and two further crossbeams hold the battery packs. These battery packs provide 4.5 V for TTL logic and 4.5 V for each of the tractor motors. The robot has four bumper switches, two on each end (see Photo 7-26).

The advantages of this robot are the large breadboard space, on-board power supplies, and zero-turning radius. Zero-turning radius means that the robot can turn around without applying any forward or backward vector to the motion. I am planning to use this robot for sensor-actuator studies, Braitenberg vehicle studies,

Photo 7-24 Small LEGO tractor vehicle with experimenter breadboard.

Photo 7-25 Side view of a large LEGO tractor vehicle.

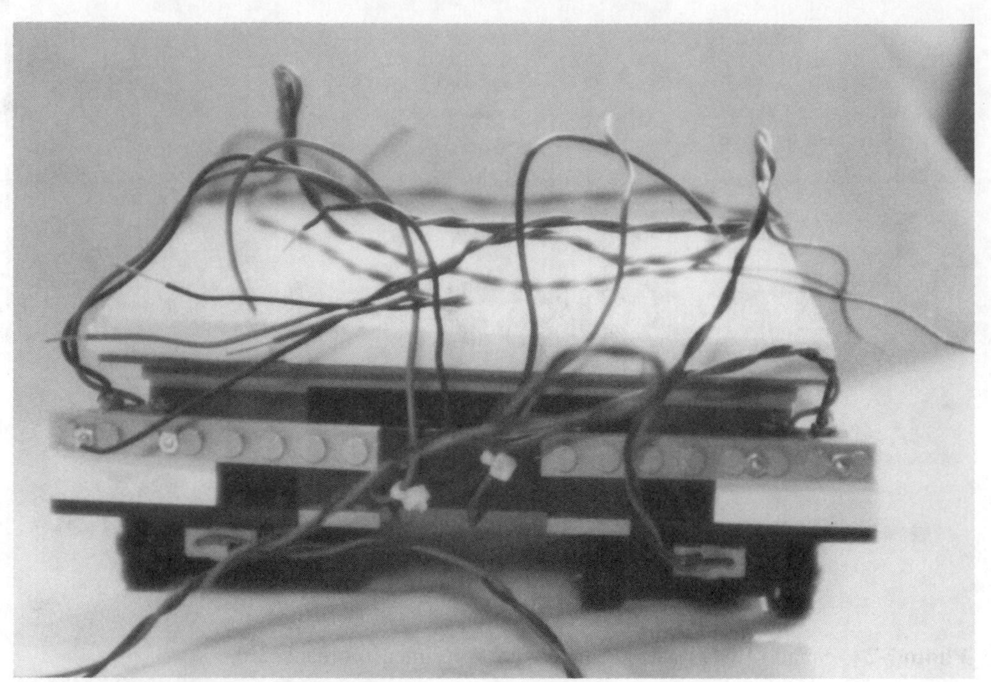

Photo 7-26 Front/Back view of a LEGO tractor vehicle. Notice large electrical breadboard and bumper switches.

Photo 7-27 Bottom view of a large LEGO tractor vehicle. Notice the cross beams holding the two vehicles together and the two cross beams holding the battery packs underneath the breadboard.

and neural control of autonomous mobile robots. Perhaps these results will be reported in a later book.

Neural controllers for robot arms

In this section, we will explore the use of neural networks in controlling a robot arm and examine two such systems. The first is a robot arm with a vision system, and the second is a robot arm with inherent time delays.

Murphy

Mel (1990) describes in some length a robot arm/vision system called Murphy that is controlled by a neural network. The robot arm has three degrees of freedom and is constrained to operate in a plane. The system also includes an autofocus camera. As we have seen earlier, the conventional approach to robot arm trajectory planning consists of either using a teaching program, in which the exact trajectory is known ahead of time (e.g., robots on an assembly line) and the environment is highly structured, or massive computations of the dynamics and inverse kinematics.

Murphy was trained to reach for objects within its field of view. The system has two networks: one for forward dynamics, and one for differential inverse kinematics. The forward dynamics network consists of a population of neurons that view the visual field. These are connected to a joint angle layer of neurons and then to the joint actuators. The differential inverse kinematics (derivative of the inverse kinematics) is a linear network and consists of camera input of the hand location. This layer is connected for feedback to the joint angle computations. The layers are not fully connected but rather connected with nearest neighbors in a Gaussian neighborhood. The problem of reaching for a visual target by means of constructing an inverse kinematic model is avoided by a simple heuristic search on a forward model. Murphy thus learns the inverse map from a visual image of its arm/hand and the visual target.

The learning is a Hebb-type construction of look-up associational tables. The output unit computes an activation as a weighted sum over a set of input units with appropriately placed receptive fields

$$y = \sum_i w_i g(x - \zeta)$$

where g is a Gaussian radial basis function and w_i is its weight. The objective of the learning is to minimize the squared difference between the actual function value and the one produced by the output, averaged over the set of training examples. The system learns by a gradient descent to the target. The networks have a total of 2.5 million connections and were trained with about one billion vectors.

The configuration of the robot, video camera, target, and obstacles as follows. The robot arm is an actual electromechanical system with five degrees of freedom. Only three of these degrees of freedom are used to confine the robot to a planer orientation. The visual target and the obstacles are painted on a background on which the camera focuses. The robot joint angles are visualized by white marks painted on the joints. So the robot arm moves parallel to the plane with the target and obstacles.

This heuristic method for visually guided reaching proved quite successful. The system learns to push its hand directly toward the object in the visual field. If no obstacles are in the path, very little flailing about is needed. Each time the robot arm moves, with respect to its visual field, it generates its own new training vector. Mel points out that this is the same way human infants learn to reach for objects. In humans, this learning occurs during the first 15 weeks after birth. The algorithm for humans is called "learning by doing" and is quite successful.

Neural tracking controller

This section describes in-depth some work that I and a colleague did on developing a neural controller for a robot arm (cf. Rietman and Frye, 1991). Conventional computer control of machines and processes (cf. Bollinger and Duffie, 1989) and adaptive control (cf. Astrom and Wittenmark, 1989) usually start by developing a linear model of the machine or process and then devise a controller that will operate within the linear region of the system's operation. Adaptive stochastic modeling and control techniques also require linearized system representations. Neural networks, on the other hand, are not restricted by such linearity demands and have been shown to be useful in many problems of nonlinear system identification and control (cf. Antsaklis, 1990). Nguyen and Widrow (1990) examined a model nonlinear problem of learning to back a tractor trailer truck to a target location and explore methods of indirect learning to train neural controllers. Josin et al. (1988) explains a neural network to assist a conventional proportional controller, reducing its error. Psaltis et al. (1987, 1988) and Kung and Hwang (1989) have described several neural controllers for robotic manipulators.

In the experiments we describe here, we have used a small robot arm as a prototypical nonlinear system (see Photo 7-28). This arm had three degrees of freedom: horizontal rotation about a central waist, vertical rotation at a shoulder, and vertical rotation at an elbow. Each of these motions was powered by a dc survo-motor, rather than the stepper-motor more commonly found in robotic systems. These motors stall at low drive voltages, leading to nonlinear torque characteristics.

In addition, we used a digital computer controlling the analog robot to simulate varying amounts of delay in the controllers to study the effect this had on the system. In the course of developing adaptive neural controllers for this system, we found that these delays and the inertial effects in the hardware constrain our approach to adaptive control.

There are several conventional approaches to adaptive control. The one that most closely resembles neural networks, known as Model-Reference Adaptive Systems (MRAS), has been described by Astrom and Wittenmark (1989) and is shown in Fig. 7-13.

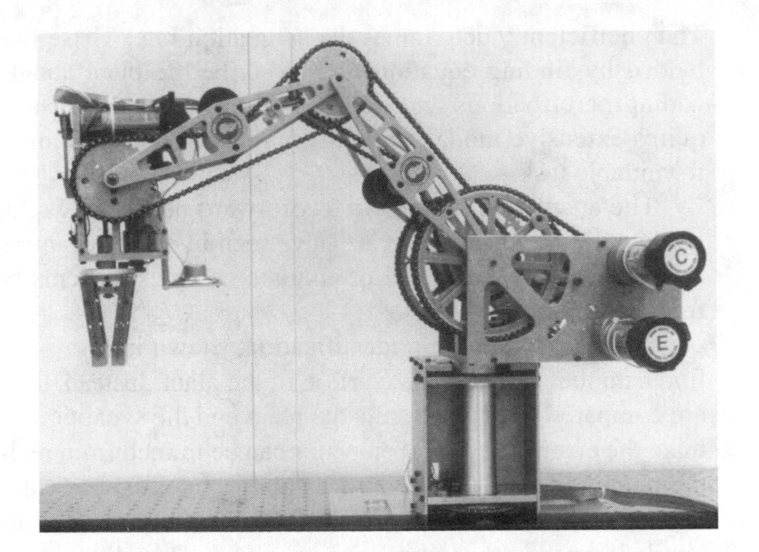

Photo 7-28 Robot arm used in neural network control experiments.

A linear model of the plant (robot) is first specified. This model tells the process output how it should ideally respond. The regulation parameters θ are adjusted to minimize the mean square of the error E between the process output and the model output using a gradient descent method. The parameter adjustment is given by the relation

$$\frac{d\theta}{dt} = -\gamma E \frac{\partial E}{\partial \theta}$$

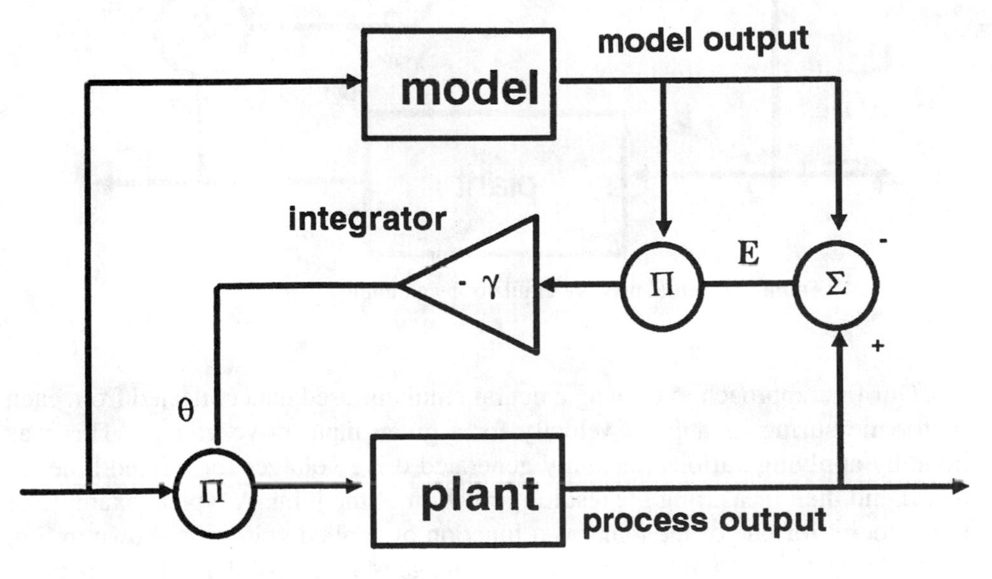

7-13 Model reference adaptive system.

The coefficient γ determines the adaptation rate. These linear models can be extended by finding equations that describe the plant about an equilibrium point, adding perturbations and doing a Taylor expansion. This method, however, requires extensive modeling. In many real physical systems, this can prove to be a formidable task.

The adaptation process in feedforward neural networks is based on a similar gradient descent method as we have seen in earlier chapters. Neural networks are an excellent tool to model or emulate complex systems because they can work from examples.

In the task of system identification, shown in Fig. 7-14, it is not necessary to find a mathematical representation of the plant. Instead, the outputs of the network are compared with those from the plant and the synaptic weights adjusted to minimize the error. The neural emulator can be trained from real-time determination of the plant's responses or from a database. In either case, the advantage of using a neural network for system identification is that the nonlinear description of the system is learned, rather than being specified. This eliminates the often difficult task of modeling the system.

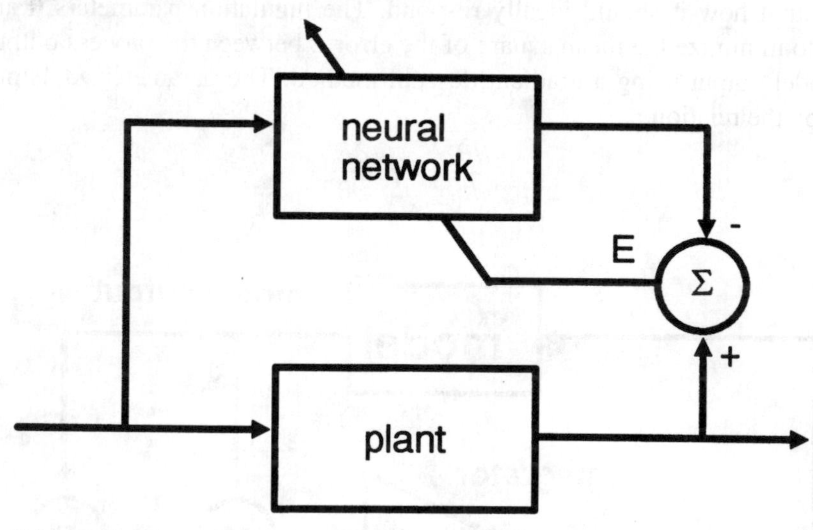

7-14 Training a neural network emulator by example.

Our first approach to training a neural emulator used data collected from each motor measuring its angular velocity for a given input drive voltage. This was done by applying various randomly generated drive voltages for a fixed time interval and then measuring the resulting rotation of the joint. A typical example of the velocity for one of the joints as a function of applied voltage is shown in Fig. 7-15. The flat portion of the curve at low voltage is a result of the stall characteristics of the motor and also of crossover distortion in a bipolar push-pull circuit used

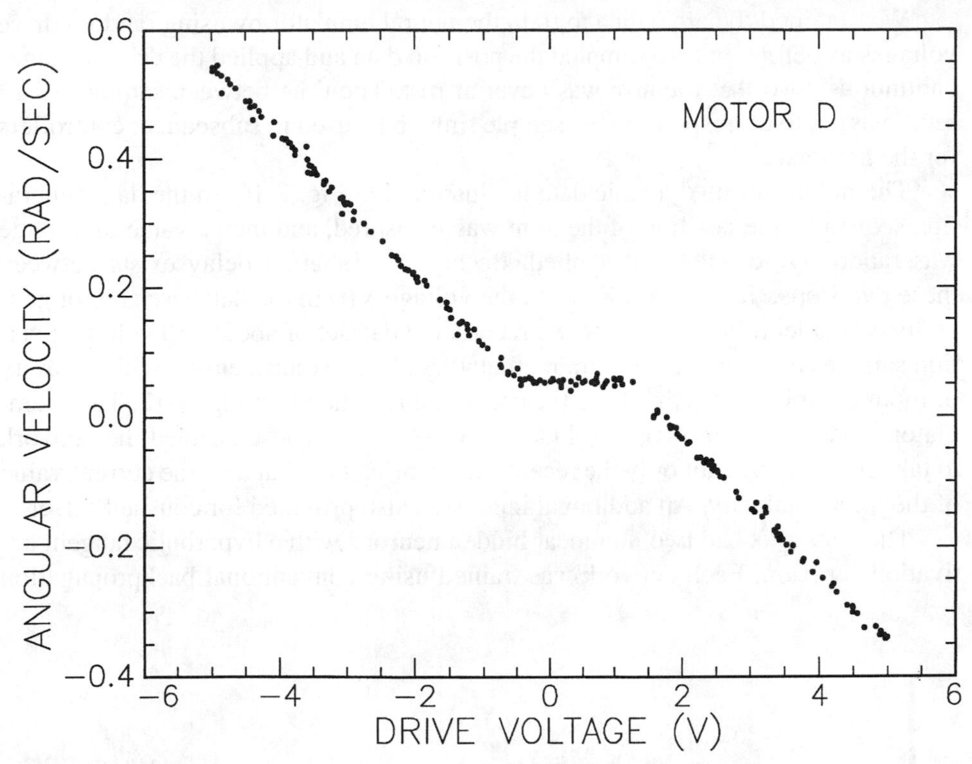

7-15 Servo-motor driven joint velocity vs. drive voltage characteristics.

to buffer the computer-generated analog signals to the motor. Training a neural network to emulate this response was straightforward. The network had two inputs, one being the drive voltage and the second being a constant value 1.0 to allow for threshold shifts. The network had two hidden neurons with a hyperbolic tangent activation function. The single output neuron was linear. Using the method shown in Fig. 7-14, this network learned to emulate the measured characteristics of the motor drive joint to an accuracy of a few percent.

We simulated the operation of the robot arm using neural emulators for each of the three joints in the arm. We then controlled each of the joints by a proportional controller, comparing the behavior of the simulated to that of the robot. Despite the high accuracy of the emulators, the simulation showed poor agreement. The most likely cause for the poor performance of this method is that it is essentially static. Each measurement of the velocity started with the arm in a home position, at rest. This made no provision for previous states of motion (inertia) or for position dependent torque effects (i.e., the arm behaves differently in a horizontal position than in a vertical one). In order to take into account the dynamic behavior of the arm, we need dynamic data, and the network must have inputs that tell it not only the instantaneous drive voltages but also the present position and past states of motion and drive.

We obtained dynamic data to train the neural emulator by using random drive voltages as before, but we sampled the position data and applied the drive voltages continuously, so that the arm was never at rest. The time between samples (0.12 sec.) was set to correspond to the sample time to be used in subsequent controllers for the hardware.

The nature of the dynamic data is illustrated in Fig. 7-16. In the data generation sequence, the position of the joint was measured, and then a value of voltage was randomly selected and applied. Because an inherent delay exists between these two steps, the position X(t) and the voltage V(t) in the database are not generally coincident. For each joint, we recorded a dataset of about 250 voltage-position samples and used them to train emulators. After experimenting with a variety of input combinations, we chose the configuration shown in Fig. 7-17. These emulators had inputs for V(t), V(t–1) and [X(t)–X(t–1)]. These enabled the network to take into account not only the recent values of voltage but also the current value of the joint's velocity. An additional input was also provided for constant offset.

The networks had two nonlinear hidden neurons with a hyperbolic tangent activation function. Each network was trained using conventional backpropagation

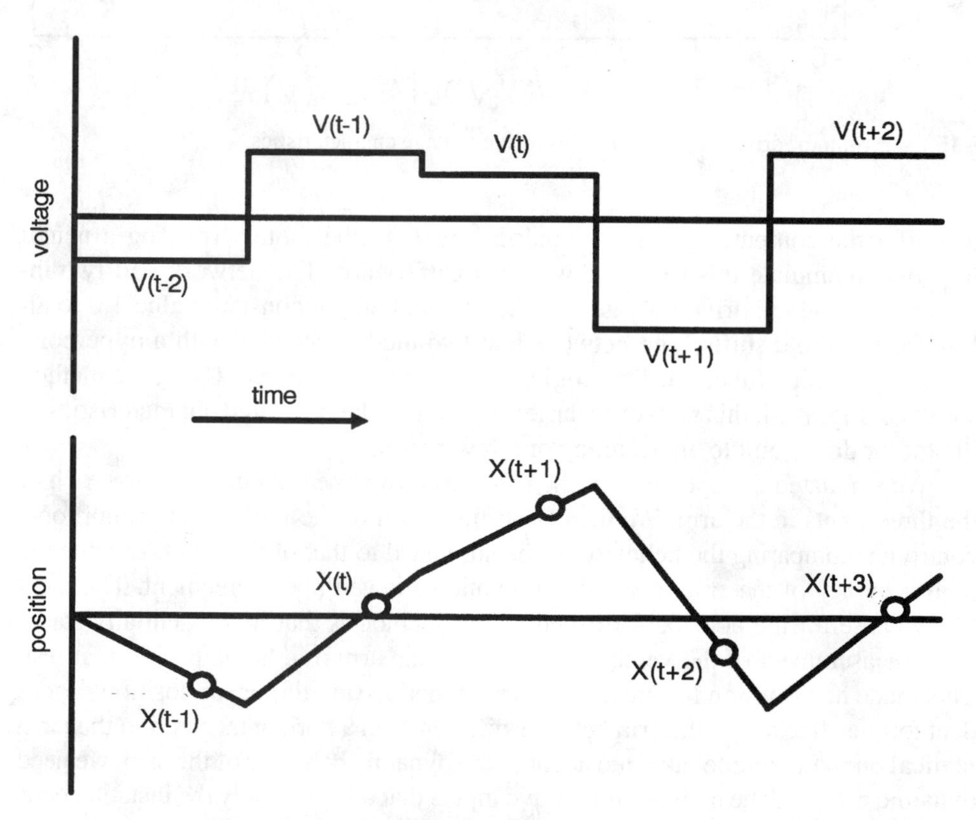

7-16 Time-sampled dynamic position-voltage data.

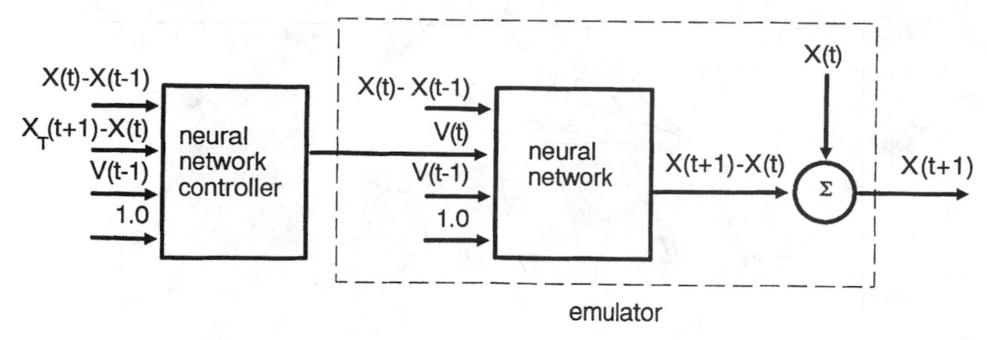

7-17 Neural network emulator configuration.

learning to output the new value of velocity, [X(t+1)–X(t)], from this input information. This output could be added to the current position, X(t), to give a prediction of the next position, X(t+1). These networks learned to do this to an accuracy of about two degrees for the waist joint and about five degrees for the shoulder and elbow.

We tested the neural emulator robot arm by using proportional control and comparing the performance of the emulator with that of the hardware. Agreement between the two was good. Figure 7-18a shows a comparison of the behavior of the neural emulator and the hardware in a non-ideal control situation. In this experiment, we set the gain of the proportional controller to a high value. This, combined with an additional 0.15 seconds in delay in the feedback loop, led to an unstable operation. The dotted line in the figure shows a target path for the arm to trace. The motion called for the coordinated action of all three of the robot's joints.

Figure 7-18b shows the performance of the simulation, using the neural network emulators for each of the joints, under the same control conditions. The qualitative similarity of the behavior, even in these extreme circumstances, is evident. The neural emulators for each of the joints was used for subsequent work on a variety of adaptive controllers.

A basic approach to developing neural controllers has been explored by Nguyen and Widrow (1990). The technique, illustrated in Fig. 7-19a, involves training a neural controller by back-propagating errors through a neural emulator of the plant. Because we do not generally know for any desired trajectory the value of the drive voltage needed to control the plant, we cannot directly train the controller network. Instead, we develop a neural network emulator like the one just described and try to control it using the neural network controller. For a particular value of desired position, the controller will generate an initially erroneous drive signal.

The problem is that we cannot modify the connection strengths in the controller network because we do not know the correct drive signal. Instead we propagate the signal through the emulator, resulting in a position that is also erroneous.

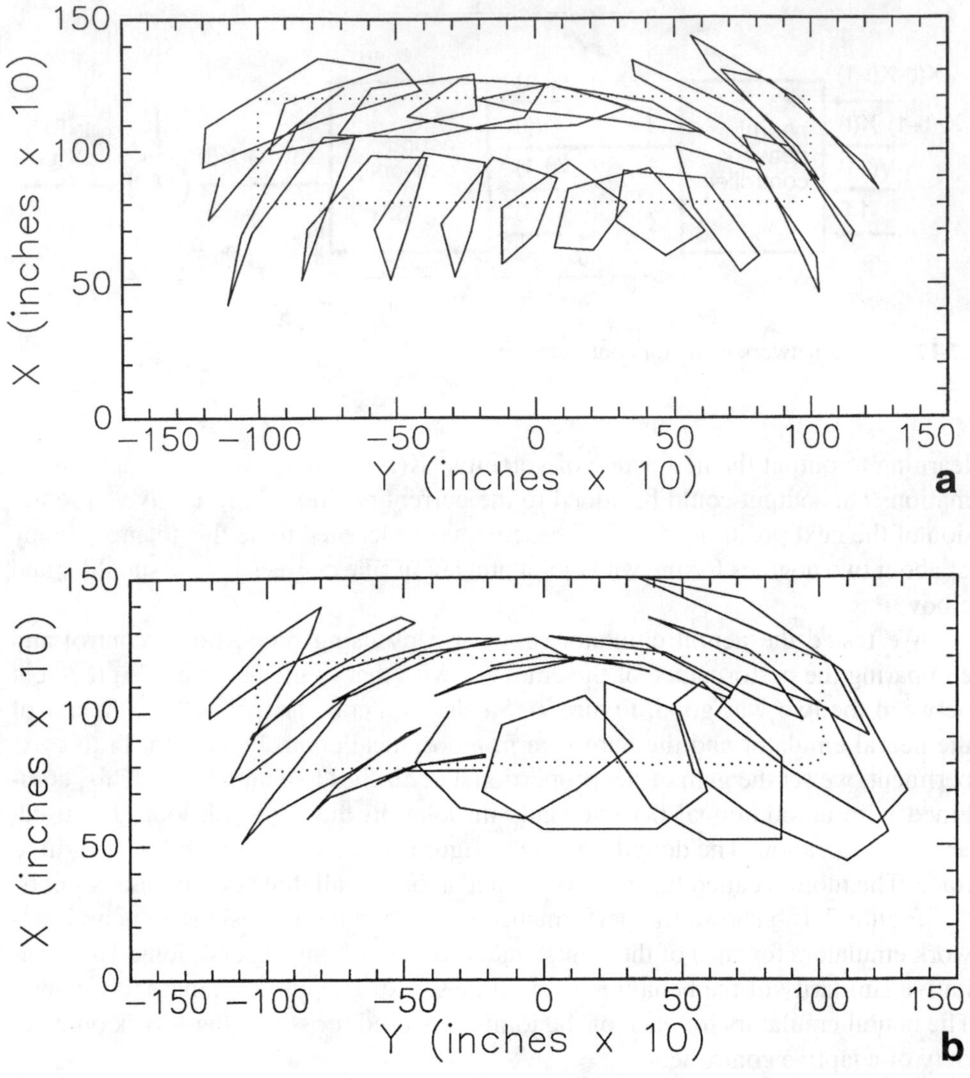

7-18a, b A comparison of the behavior of the hardware (a) and neural-network-based simulation (b). The dotted line is the target trajectory and the solid line is the result, projected on the x,y plane.

We can determine the error at this point, and this error can be back-propagated thought the emulator network to give us an equivalent error at the output of the controller network. This error can then be used to modify the connection strengths in the controller network.

Figure 7-19b shows the implementation of this method for our problem. Note that the emulator had three input variables and one output variable. The controller was the same, but the output variables V(t) and [X(t+1)–X(t)] are switched. For

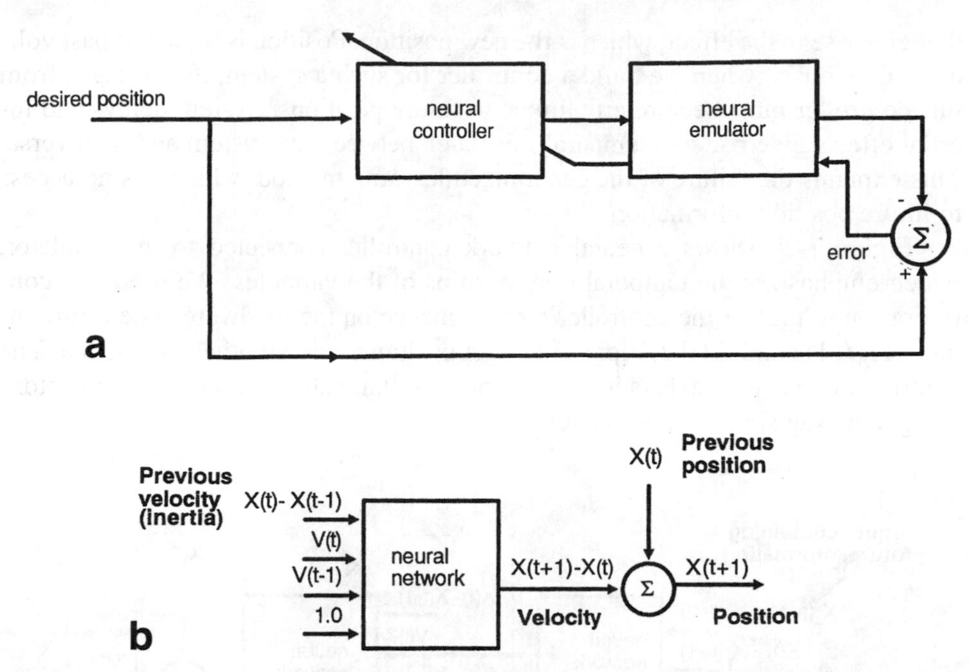

7-19a,b Block diagram of the indirect learning method and its implementation for our system.

training, the target position XT(t+1) was compared to the output of the emulator, and the error back-propagated through the emulator.

All of our attempts to use this indirect learning method to train a controller failed. We varied the number of hidden neuron and the learning rates, but the network showed no tendency to converge. The object of this method it to learn the inverse kinetics of the system, and out subsequent experiments on alternative methods to achieve this have provided some insight into the reasons for this failure.

A more general approach to the inverse kinematic problem is to train the controller directly from a stimulus-response database. In this approach, we simply attempted to train the neural network to invert the map between cause and effect, telling us the input voltage (given information about the position). This is also the basic idea behind the indirect learning method that we described earlier; but in this case, we allow a more general access to the position dataset rather than restricting ourselves to a particular subset of the data. We examined a range of data points about the time t to find position data that would allow us to successfully train the network. We found that, to successfully train the inverse kinematic controller, we had to provide inputs specifying future joint positions.

A fundamental difference between the emulator and the controller becomes apparent if we consider them from the viewpoint of cause and effect. The robot arm is obviously a causal system, something clearly reflected in the nature of the inputs to the emulator. Present and past values of position and voltage are the cause that give rise to the effect, which is the new position. Position is linked to past voltages. Obviously, when we build a controller for such a system, the voltages from this controller must be similarly linked to future positions. System delays and inertial effects give rise to a temporal mismatch between the system and its inverse. This explains the failure of the generalized learning method, which has no access to future position information.

Figure 7-20 shows a neural network controller connected to the emulator, which emphasizes the temporal relationships of the variables. We used this configuration to predict the controller's performance on the hardware. The future inputs XT(t+1) and XT(t+2) represent target positions, and all others are actual. The controller copes with delays in the system by looking ahead at the target trajectory and generating signals appropriately.

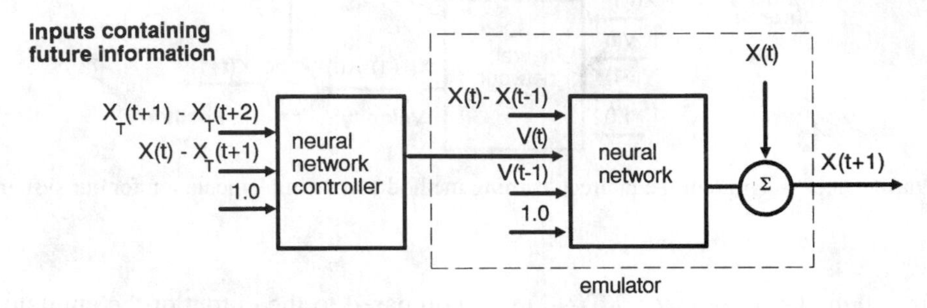

7-20 Feedforward neural controller connected to the neural network emulator. Future positions (denoted by subscript T) are targets.

Despite the fact that they learned the inverse dynamics of the robot joint generation output voltages with about 10% accuracy, the trained neural controllers performed poorly when tested on the emulators. This wasn't entirely unexpected because this method is effectively feedforward control and not robust to system noise or inaccuracies. A weak feedback path is established because the actual position X(t) is present at the input, but this did not appear to be sufficient to give good control of either the emulator or the hardware. In operation, the arm was stable but still tended to drift in a random way from its target trajectory. Therefore, we sought a method that would be more tolerant of perturbations to the system.

Figure 7-21a shows a method suggested by Miyamoto et al. (1988) and Kawato et al. (1988). In this method, the feedforward neural controller performs the inverse kinematic transformation just as it does in the previous case; but in addition, there is a feedback path from the output to a proportional controller. This provides added robustness and reduces sensitivity to system perturbations. An interesting advantage to this method is that the feedback error from the proportional controller is used to adjust the synaptic weights of the feedforward controller net-

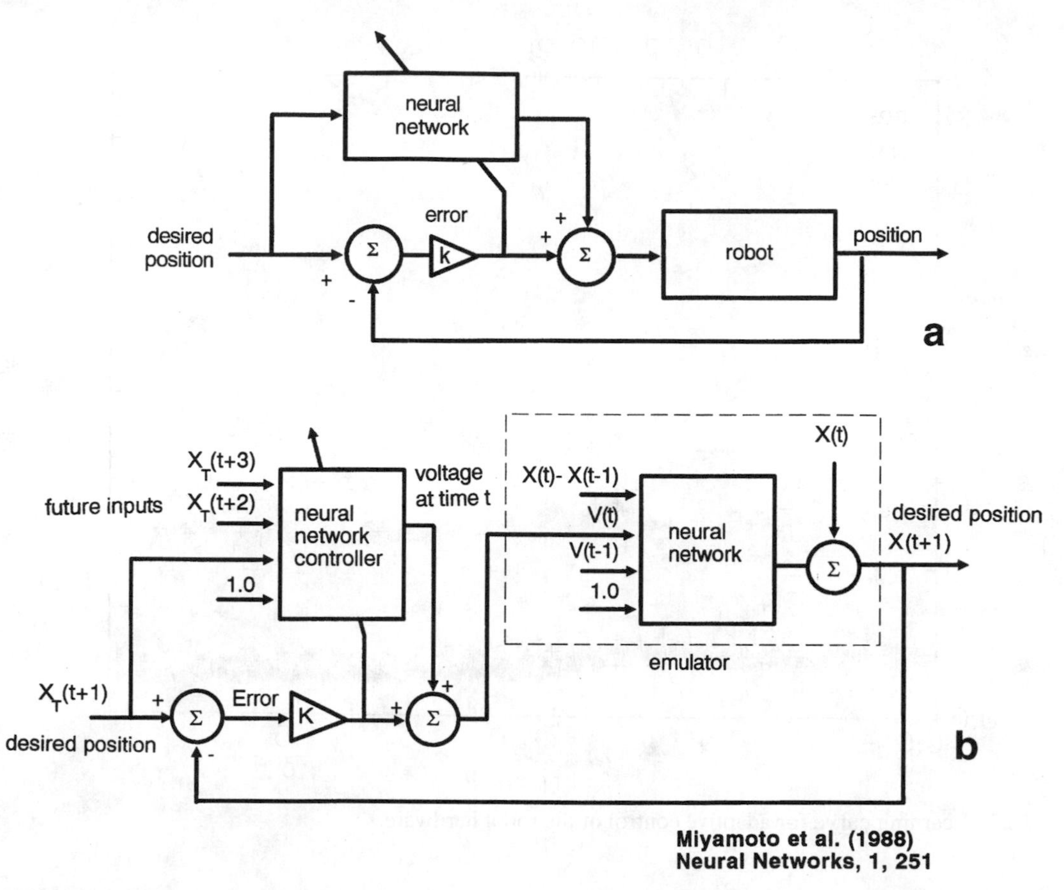

7-21a,b Combined feedforward and proportional feedback controller (a), and its training using the neural emulator (b).

work. In the initial stages of learning, when the weights in the network and its output are small, the overall system behaves like a simple proportional controller. As the network learns, it gradually assumes feedforward control. However, if there are any major departures from the target trajectory, the feedback proportional controller will restore the system.

We used the configuration shown in Fig. 7-21b to train the feedforward controller. The controller network incorporated the same look ahead capability as in the previous experiment, and future values of the position were the target trajectory. For training the network, we did not use random walks. Instead, we trained the network to trace out a rectangular box, similar to that described as the test exercise for the emulator in Fig. 7-44. We obtained good learning for each of the motor controllers using the emulator and found similarly good results for the hardware.

The learning curve for the waist joint motor (hardware) is shown in Fig. 7-22. In this figure, the time is an arbitrary unit, and each data point is the average error

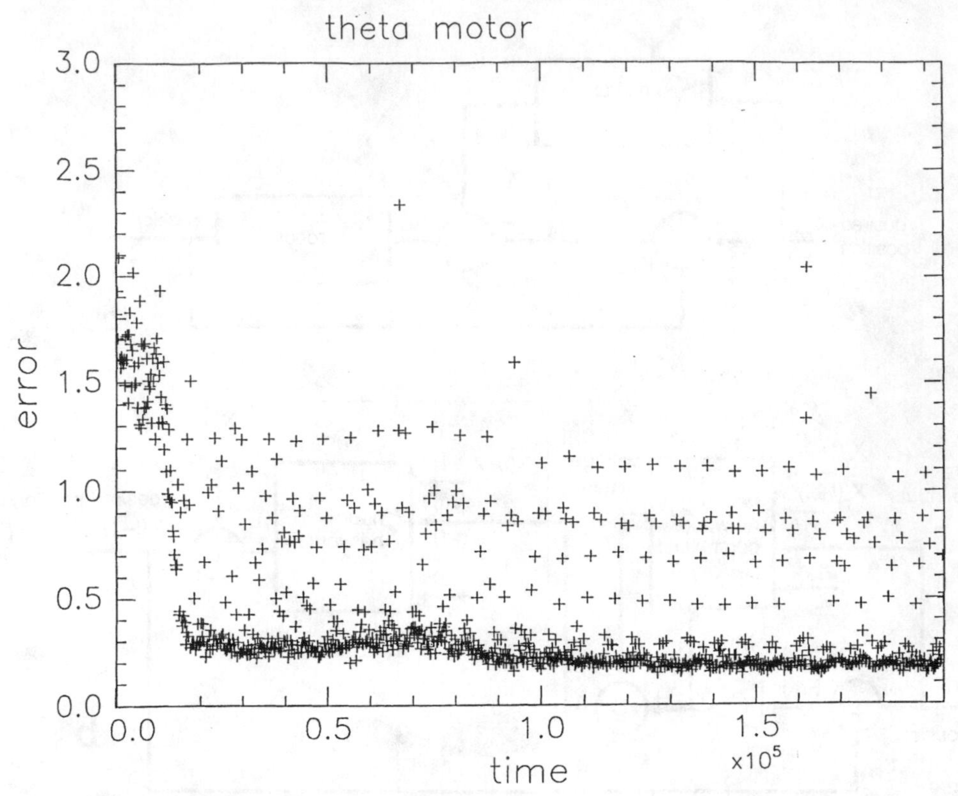

7-22 Learning curve for adaptive control of the robot hardware.

from 100 time samples. The actual training sequence was 8-10 hours. The sporadic noisy data points are a result of bit-error events in the position sensing hardware, which was periodically reset. These results demonstrate the robustness of the system. It is also important to note that the level of error at the start of this learning cycle is that of the proportional controller, and the subsequent decrease in the error demonstrates improvement arising from the adaptive control.

Because the robot arm was repeating a periodic motion, we could take a representative time-window of one cycle of the angular position. Figure 7-23a compares one of these cycles for proportional control of the emulator with neural control of the emulator and neural control of the hardware, all for the waist motor. The solid curve represents the target position of the joint, and the dashed curve is the actual result. The proportional delay is clearly seen here. In some systems, this delay can be reduced by increasing the loop gain in the proportional controller; but in ours, this resulted in severe oscillation.

Figure 7-23b shows the results for the neural controller operating the neural emulator. The results are clearly better than with proportional control alone. Figure 7-23c is the results for the neural controller learning to control the hardware arm.

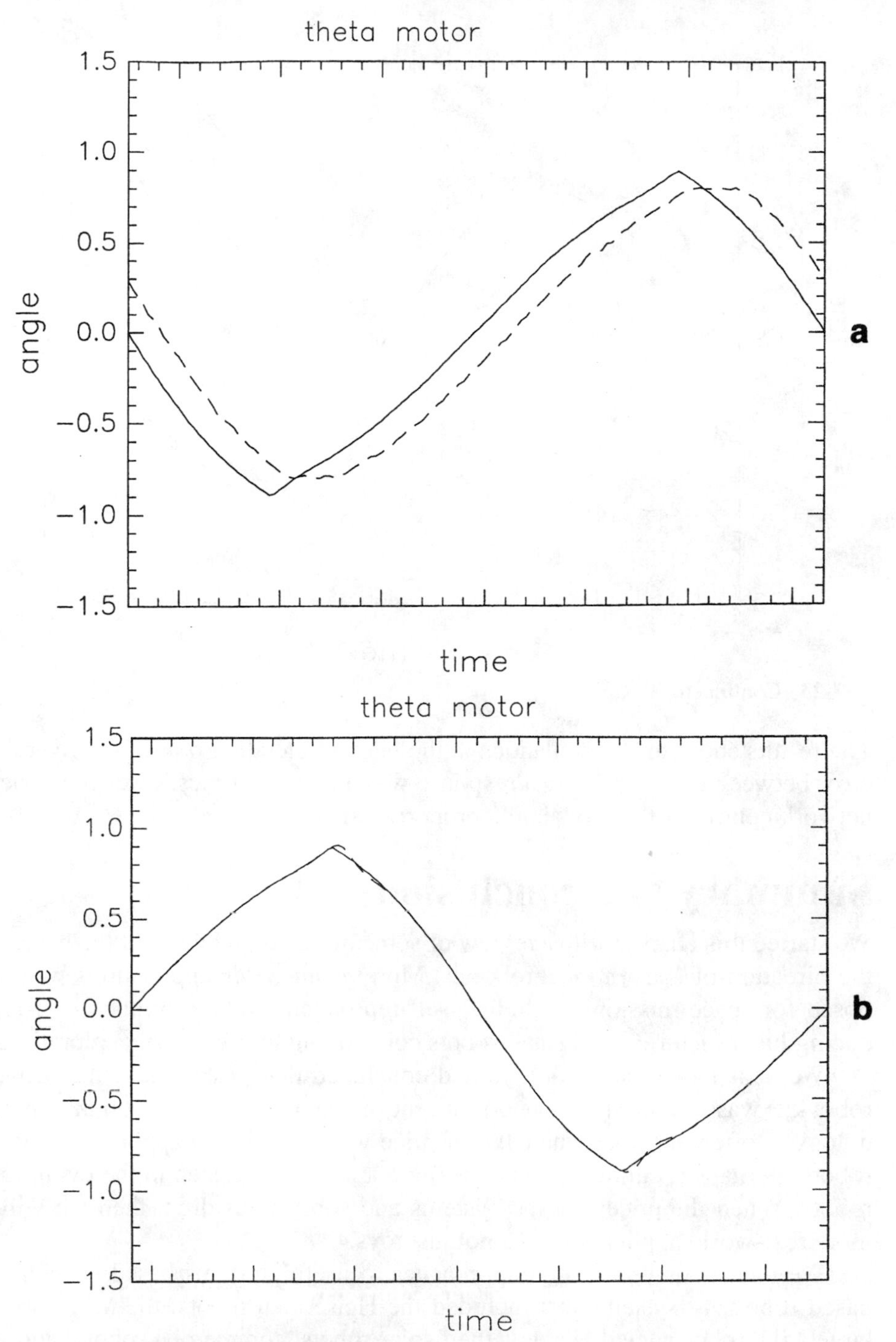

7-23a,b,c Comparison of target (solid line) and actual (dashed line) position for the waist motor of the robot arm with proportional control of the emulator (a), adaptive control of the emulator (b), and adaptive control of the hardware (c).

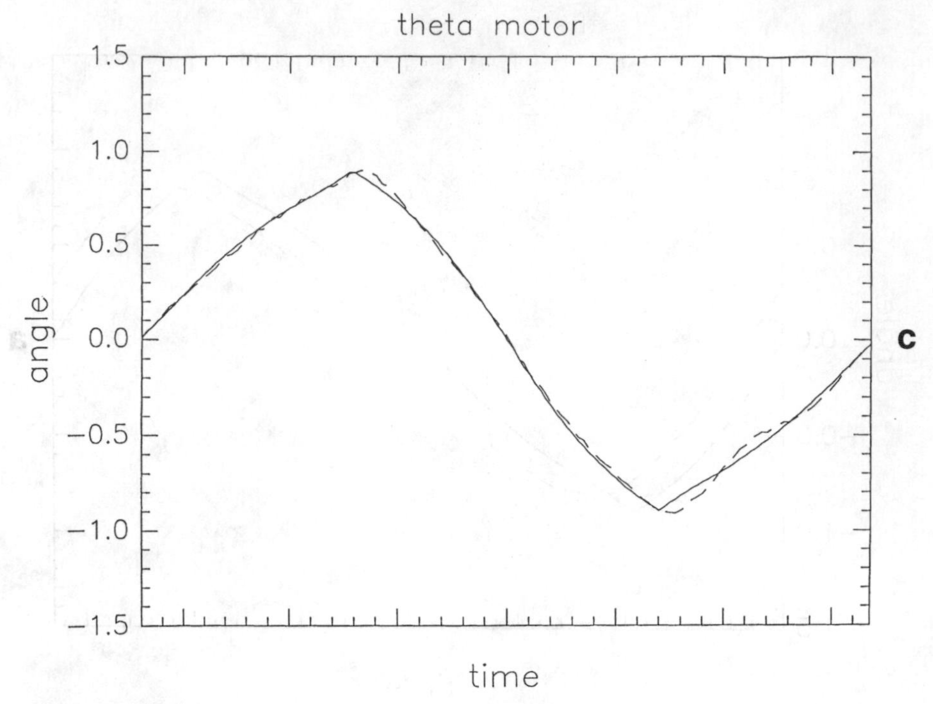

7-23 Continued.

The results show some oscillations at the high acceleration points, but overall the error between target and actual response was about five times lower in the neural network controller than for simple proportional control.

Summary and conclusion

We started this chapter with a review of some literature to highlight the basics and the direction of conventional robotics. More *avant-garde* applications were purposed for space missions, including self-reproducing robotic probes, self-reproducing lunar factories, and small robots cooperating for planetary explorations.

We next looked at robot toys and concluded that just because the robot (or robot kit) was sold in a toy store doesn't mean that it is useless or is junk. Toy technology is often at the cutting edge of price wars for hi-tech applications. Plastic robots (perhaps reinforced composites) are a next logical step in the evolution of robots. When the prices for the systems and subsystems drop, then we will see more real-world applications and not just toys.

We then reviewed several approaches to useful systems based on behavior-based robotics—systems that included the Heiserman robots, the MIT robots, a large MIT robot named Herbert, then solar robots, commercial robots, and a hi-tech toy (LEGO Technic) for assembly of simple artificial life-forms.

We concluded the chapter by examining the topic of neural networks for visually guided reaching and neural networks for robot arm controllers. In the control problems, we discovered that, with systems accompanied by delay or by inertia, the methods commonly used to adaptively learn inverse feedforward controllers do not work. In such systems, the controllers must necessarily anticipate the future demands on the system. We have presented neural networks suitable for this task and demonstrated their use on a real-time hardware system.

Program for Pattern Recognition

```
  2 CLS
  5 DIM A(10,B(10),C(10)
 10 OUT 236,16
 20 OUT 7,146
 30 FOR ROW=0 TO 7
 40 OUT 6,2^ROW
 50 A=INP(4)
 60 B=INP(5)
 70 A1=0:A2=0:A3=0:A4=0:A5=0:A6=0:A7=0:A8=0
 80 IF A-2^7<0 THEN GOTO 100
 90 A=A-2^7:A8=1
100 IF A-2^6<0 THEN GOTO 120
110 A=A-2^6:A7=1
120 IF A-2^5<0 THEN GOTO 140
130 A=A-2^5:A6=1
140 IF A-2^4<0 THEN GOTO 160
150 A=A-2^4:A5=1
160 IF A-2^3<0 THEN GOTO 180
170 A=A-2^3:A4=1
180 IF A-2^2<0 THEN GOTO 200
190 A=A-2^2:A3=1
200 IF A-2^1<0 THEN GOTO 220
210 A=A-2^1:A2=1
220 IF A-2^0<0 THEN GOTO 240
230 A=A-2^0:A1=1
240 A(8)=A8:A(7)=A7:A(6)=A6:A(5)=A5:A(4)=A4:A(3)=A3:A(2)=A2:A(1)=A1
250 B8=0:B7=0:B6=0:B5=0:B4=0:B3=0:B2=0:B1=0
260 IF B-2^7<0 THEN GOTO 280
270 B=B-2^7:B8=1
280 IF B-2^6<0 THEN GOTO 300
290 B=B-2^6:B7=1
300 IF B-2^5<0 THEN GOTO 320
310 B=B-2^5:B6=1
320 IF B-2^4<0 THEN GOTO 340
330 B=B-2^4:B5=1
340 IF B-2^3<0 THEN GOTO 360
350 B=B-2^3:B4=1
360 IF B-2^2<0 THEN GOTO 380
370 B=B-2^2:B3=1
380 IF B-2^1<0 THEN GOTO 400
390 B=B-2^1:B2=1
400 IF B-2^0<0 THEN GOTO 420
410 B=B-2^0:B1=1
420 B(8)=B8:B(7)=B7:B(6)=B6:B(5)=B5:B(4)=B4:B(3)=B3:B(2)=B2:B(1)=B1
430 FOR I=1 TO 8
440 C(I)=A(I) AND B(I)
450 NEXT I
460 FOR I=8 TO 1 STEP -1:PRINT C(I);:NEXT I
465 PRINT
470 NEXT ROW
475 FOR I=1 TO 500:NEXT I
480 CLS
490 GOTO 10
500 END
```

Bibliography

Aarts, E. and Korst, J. *Simulated Annealing and Boltzmann Machines: A Stochastic Approach to Combinatorial Optimization and Neural Computing*. New York: John Wiley, 1989.

Abidi, M. A., Eason, R. O., and Gonzalez, R. C., "Autonomous Robotic Inspection and Manipulation Using Multisensor Feedback." *Computer*, April 1991. pp. 17–31.

Ackley, D. and Littmen, M. "Interaction Between Learning and Evolution" in Langton, C. G.; Taylor, C.; Farmer, J. D.; and Rasmussen, S., Eds. *Artificial Life II*. Redwood City: Addison-Wesley, 1992. p.487.

Adachi, H., Koyachi, N., and Nakano, E. "Mechanism and Control of a Quadruped Walking Robot." *IEEE Control Systems Magazine*. October 1988, pp. 14–19.

Allen, J. *Biosphere 2: The Human Experiment*. New York: Penguin Books, 1991.

Allen, J. and Nelson, M. *Space Biospheres*. Oracle, Arizona: Synergetic Press, 1989.

Aloimonos, Y. and Rosenfeld, A. "Computer Vision." *Science* 253, 1991. pp. 1249–1254.

Anglc, C. M. and Brooks, R. A. "Small Planetary Rovers." Proceedings IROS '90, IEEE Int. Workshop on Intelligent Robots and Systems '90, Towards a New Frontier of Applications, July 3–6, 1990. pp. 383–388,

Antsaklis, P. J. "Neural Networks in Control Systems," *IEEE Control Systems Magazine* 10(3), 1990

Ardin, P. *The Development of Tactile and Optokinetic Competences in a Mobile Robot*. MSc. Thesis, Edinburgh University, 1990.

Arkin, R. C. "Cooperation Without Communication: Multiagent Schema-Based Robot Navigation." *Journal of Robotic Systems* 9(3), 1992. pp. 351–364.

Arganbright, D. E. *Mathematical Applications of Electronic Spreadsheets*. New York: Byte Books, McGraw-Hill, 1985.

Astrom, K. J., and Wittenmark, B. *Adaptive Control*. Addison Wesley, 1989.

Baloch, A. A. and Waxman, A. M., "Visual Learning, Adaptive Expectations, and Behavioral Conditioning of the Mobile Robot MAVIN." *Neural Networks* 4, 1991. pp. 271–302.

Banks, E. R. "Universality in Cellular Automata." IEEE 11th Ann. Symp. Switching and Automata Theory. Santa Monica, CA, 1970.

Barbarello, J. "Computer-controlled Robot." *Radio-Electronics*, May 1987. pp. 144–147

Barden, W., *BYTE*, #291, 1982.

Barnsley, M. F. and Demko, S. "Iterated Function Systems and the Global Construction of Fractals." Proc. Royal Soc. London, A399, 1985. pp. 243–275

Barnsley, M. F. *Fractals Everywhere*. New York: Academic Press, 1988.

Batten, G. L. *Design and Application of Linear Computational Circuits*. Blue Ridge Summit: TAB Books, 1987.

Becker, K. and Dorfler, M. *Dynamical Systems and Fractals: Computer Graphics Experiments in Pascal*. Cambridge: Cambridge University Press, 1989.

Beer, R. D. *Intelligence as Adaptive Behavior*. New York: Academic Press, 1990.

Belew , R. K., and Booker, L. B., Eds. *Genetic Algorithms*. Proceedings of the Fourth International Conference on Genetic Algorithms, Morgan Kaufman Publishers, 1991.

Beni, G. "The Concept of Cellular Robotic Systems." Proceedings of IEEE Int. Symp. on Intelligent Control, 1989. pp. 57–62.

Bentsman, J. and Hannon B. "Cyclic Control in Ecosystems." Eds. Thoma, M., and Wyner, A. *Lecture Notes I Control and Information Sciences*. Berlin: Springer-Verlan, 1988. pp. 423–436.

Berlekamp, E.R.; Conway, J.H.; and Guy, R.K. *Winning Ways for your Mathematical Plays. Vol. 2: Games in Particular*. New York: Academic Press, 1982.

Berge, P., Pomeau, Y., and Christian, V. *Order Within Chaos*. New York: John Wiley, 1984.

Beyers, R. J. "The Metabolism of Twelve Aquatic Laboratory Microecosystems", *Ecological Monographs* 33(4), 1963. pp. 282–306.

Bollinger, J. G., and Duffie, N. A. *Computer Control of Machines and Processes*. Addison Wesley, 1989.

Bonner, J. T. *The Evolution of Complexity by Means of Natural Selection*. Princeton: Princeton University Press, NJ, 1988.

Booker, L. B.; Goldberg, D. E.; and Holland, J. H. "Classifier Systems and Genetic Algorithms." *Artificial Intelligence* 40, 1989. pp. 235–282.

Brady, M. *Robotics Science*. Cambridge, MA: MIT, 1989.

Braintenberg, V. *Vehicles: Experiments in Synthetic Psychology*. Cambridge, MA: MIT Press, 1984

Braumlette, M. F. and Bouchard, E. E. "Genetic Algorithms in Parametric Design of Aircraft" Davis, L., Ed. *Handbook of Genetic Algorithms*. New York: Von Nostrand Reinhold, 1991. pp. 109–123.

Brooks, R. A.; Connell, J. H.; and Flynn, A. "A Mobile Robot with Onboard Parallel Processor and Large Workspace Arm." Proceedings AAAI-86, Fifth National Conference on Artificial Intelligence. Philadelphia, PA, August 1986. pp. 1096–1100

Brooks, R. A. and Connell, J. H. "Asynchronous Distributed Control System for a Mobile Robot." *SPIE* 727, 1986. pp. 77–84

Brooks, R. A. "A Robust Layered Control System for a Mobile Robot." *IEEE Journal of Robotics and Automation*. RA2 1, 1986. pp. 14–23.

Brooks, R. A.; Connell, J. H.; and Ning, P. "Herbert: A Second Generation Mobile Robot." *MIT AI Memo,* January 1988. p.1016

Brooks, R. A. and Flynn, A. M. "Fast, Cheap, and Out of Control: A Robot Invasion of the Solar System." *Journal of the British Interplanetary Society* 42, 1989. pp. 478–485

Brooks, R. A. "A Robot that Walks: Emergent Behaviors from a Carefully Evolved Network." *Neural Computation* 1, 1989. pp. 253–262

Brooks, R. A.; Maes, P.; Mataric, M. J.; and More, G. "Lunar Base Construction Robots." Proceedings IROS '90, IEEE International Workshop on Intelligent Robots and Systems '90, Towards a New Frontier of Applications, July 3–6, 1990. pp. 389–392

Brooks, R. A. "New Approaches to Robotics." *Science* 253, 1991. pp. 1227–1232.

__________., "Intelligence Without Representation." *Artificial Intelligence* 47, 1991. pp. 139–159

Brown, K. "Build BERT, The Basic Educational Robot Trainer, Part I." *BYTE*, April 1987. pp. 113–122

Caderre, W. "Modeling Behavior in Petworld" in Langton, C. G., Ed., *Artificial Life*. New York: Addison-Wesley, 1989. pp. 407–420

Calder, N. *Spaceships of the Mind*. New York: Viking Press, 1978.

Carter, F.L., "The Molecular Device Computer: Point of Departure for Large Scale Cellular Automata." *Physica* 10D, 1984. pp. 175–194.

Chorafas, D. "Machine Learning: The Next Ten Years." Forsyth, R., Ed. *Machine Learning: Principles and Techniques*. London: Chapman and Hall, 1989. pp. 238–249.

Ciliz, M. K. and Isik, C. "Time Optimal Control of Mobile Robot Motion Using Neural Nets." Proceedings of IEEE International Symposium on Intelligent Control, 1989. pp. 368–373.

Codd, E.F. *Cellular Automata*. New York: Academic Press, 1968.

Connell, J. H. "The Omni Photovore: How to Build a Robot that Thinks like a Roach." *OMNI*, October 1988. pp. 201–212

_________. *Minimalist Mobile Robotics: A Colony-Style Architecture for an Artificial Creature*. San Diego: Academic Press, 1990.

Cox, Jr., L. A.; Davis, L.; and Qiu, Y. "Dynamic Anticipatory Routing in Circuit-Switched Telecommunications Networks" in Davis, L., Ed., *Handbook of Genetic Algorithms*. New York: Von Nostrand Reinhold, 1991. pp. 124–143

Craig, J. J. *Introduction to Robotics Mechanics and Control, 2nd Ed.* New York: Addison Wesley, 1989.

Crecraft, D. I., "Ultrasonic Instrumentation: Principles, Methods and Applications." *Journal of Physics, E: Scientific Instruments* 16, 1983. pp. 181–189.

Dallas, J. *Co-Operative Search Behavior in a Group of LOGO Robots*. MSc Thesis, Edinburgh University, 1990.

Danby, J. M. A. *Computing Applications to Differential Equations*. Englewood Cliffs: Prentice-Hall, 1985.

Dario, P. "Contact Sensing for Robot Active Touch" in Brady, M., Ed., *Robotics Science*. MIT Press, 1989. pp. 137–163.

Daskalakis, N. *Learning Sensor-Action Coupling in LEGO Robots*. MSc Thesis, Edinburgh University, 1991.

Davidor, Y. "A Genetic Algorithm Applied to Robot Trajectory Generation" in Davis, L., Ed. *Handbook of Genetic Algorithms*. New York: Von Nostrand Reinhold, 1991. pp. 144–169.

_________. *Genetic Algorithms and Robotics: A Heuristic Strategy for Optimization*. New Jersey: World Scientific Press, 1991.

Davis, L., Ed. *Handbook of Genetic Algorithms*. New York: Von Nostrand Reinhold, 1991.

Dawkins, R. *The Blind Watchmaker*. New York: W. W. Norton, 1987.

_________. "The Evolution of Evolvability" in Langton, C., Ed., *Artificial Life*. New York: Addison Wesley, 1989.

Dayhoff, J.E. *Neural Network Architectures: An Introduction*. New York: Van Nostrand Reinhold, 1990.

Debaney, R. L. *Chaos, Fractals and Dynamics: Computer Experiments in Mathematics*. New York: Addison-Wesley Pub. Co., 1990.

Deken, J. *Silico Sapiens: The Fundamentals and Future of Robots*. New York: Bantam Books, 1986.

Deneubourg, J. L.; Pasteels, J. M.; and Verhaeghe, J. C. "Probabilistic Behavior in Ants: A Strategy of Errors." *Journal of Theoretical Biology* 105, 1983. pp. 259–271.

Dewdney, A. K. *Scientific American*. May 1989.

_________. *The Magic Machine: A Handbook of Computer Sorcery*. New York: W. H. Freeman, 1990.

_________. "Computer Adventure: Bring Tur-Mites to Life." *Algorithm* 1.7, November 1990. pp. 8–10.

_________. *Scientific American*, July 1991.

Dominic, S.; Das, R.; Whitley, D.; and Anderson, C. "Genetic Reinforcement Learning for Neural Networks." *International Joint Conference on Neural Networks*. Seattle, II–71, 1991.

Donnett, J. and Smithers, T. "LEGO Vehicles: A Technology for Studying Intelligent Systems" in Meyer, J. A., and Wilson, S. W., Eds. *From Animals to Animats*. Cambridge, MA: MIT Press, 1991. pp. 540–549.

Drexler, K. E. *Proc. Natl. Acad. Sci. USA* 78(9), 1981. pp. 52–75.

Dubois, D. M., "State-of-the-Art of Predator-Prey Systems Modeling" in Jorgensen, S. E., Ed. *State-of-the-Art in Ecological Modelling*. New York: Pergamon Press, 1978. pp. 163–217.

Duchesne, B.; Fischer, C. W.; and Gray, C. G. "Inexpensive and Accurate Position Tracking with an Ultrasonic Ranging Module and a Personal Computer." *American Journal of Physics* 59(11), 1991. pp. 998–1002.

Eberhart, R.C. and Dobbins, R. W., Eds. *Neural Network PC Tools: A Practical Guide*. New York: Academic Press, 1990.

Eckmiller, R. "Neural Nets for Sensory and Motor Trajectories." *IEEE Control Systems Magazine*, April 1989. pp. 53–59.

Eldridge, N. *Macroevolutionary Dynamics Species, Niches and Adaptive Peaks*. New York: McGraw-Hill, 1989.

Enloe, C. L. "Solving Coupled, Nonlinear Differential Equations With Commercial Spreadsheets." *Computers in Physics*, Jan/Feb 1989. pp. 75–76.

Ewert, J. P. and Arbib, M. A. *Visuomotor Coordination: Amphibians, Comparisons, Models, and Robots*. New York: Plenum Press, 1989.

Faugeras, O. D. "A Few Steps Toward Artificial 3-D Vision" in Brady, M., Ed., *Robotics Science*. MIT Press, 1989. pp. 39–137.

Feistel, R. and Ebeling, W. *Evolution of Complex Systems: Self-Organization, Entropy and Development*. Boston: Kluwer Academic Publishers, 1989.

Fetzer, J. H. *Artificial Intelligence: Its Scope and Limits*. Boston: Kluwer Academic Publishers, 1990.

Filo, A. "Designing a Robot From Nature: Part I, Biological Considerations." *BYTE*, February 1979. pp. 12–29.

________. "Designing a Robot From Nature: Part II, Constructing the Eye." *BYTE*, March 1979. pp. 114–123.

Flynn, A. M. and Brooks, R. A. "MIT Mobile Robots—What's Next?", IEEE, Int. Conf. on Robotics and Automation, Philadelphia, PA, April 1988. pp. 611–617.

Flynn, A. M.; Brooks, R. A.; Wells III, W. M.; and Barrett, D. S. "Squirt: The Prototypical Mobile Robot for Autonomous Graduate Students." MIT AI Memo 1120, July 1989.

Flynn, A. M.; Brooks, R. A.; and Tavrow, L. S. "Twilight Zones and Cornerstones: A Gnat Robot Double Feature." MIT AI Memo 1126, July 1989.

Flynn, A. M. and Brooks, R. A. "Battling Reality." MIT AI Memo 1148, October 1989.

Folsome, C. E. and Hanson, J. A. "The Emergence of Materially-Closed-System Ecology." in Polunin, N., *Ecosystem Theory and Application*. New York: John Wiley, 1986. pp. 269–288.

Fontana, W.; Schnabl, N.; and Schuster, P. "Physical Aspects of Evolutionary Optimization and Adaptation." *Physical Review* A40(6), 1989. pp. 3301–3321.

Forrest, S., Ed. *Emergent Computation*. Cambridge: MIT Press, 1991.

Freitas, Jr., R. A. "A Self-Reproducing Interstellar Probe." *J. British Interplanetary Society* 33, 1980. pp. 251–264.

Freitas, Jr., R. A., and Gilbreath, W. P., Eds. *Advanced Automation for Space Missions*. NASA Conference Publication 2255, 1982.

Frye, R. C.; Rietman, E. A.; and Wong, C. C. "Back-Propagation Learning and Nonidealities in Analog Neural Network Hardware." *IEEE Trans. on Neural Networks* 2(1), 1991. pp. 110–117.

Fu, K. S.; Gonzalez, R. C.; and Lee, C. S. G. *Robotics: Control, Sensing, Vision, and Intelligence*. New York: McGraw-Hill, 1987.

Fukuda, T.; Hosokai, H.; and Shimasaka, N. "Autonomous Plant Maintenance Robot (Mechanism of Mark IV and Its Actuator Characteristics." Proceedings IROS '90, IEEE Int. Workshop on Intelligent Robots and Systems '90, Towards a New Frontier of Applications, July 3–6, 1990. pp. 471–478.

Fukuda, T.; Ueyama, T.; and Arai, F. "Control Strategy for a Network of Cellular Robots: Determination of a Master Cell for Cellular Robotic Network Based on a Potential Energy." Proceedings IEEE Int. Conf. on Robotics and Automation, Sacramento, CA, April 1991. pp. 1616–1621.

Garis, H. de. "Lizzy: The Genetic Programming of an Artificial Nervous System" in Kohonen, T.; Kakisara, O.; Simula, O.; and Kangas, J., Eds. *Artificial Neural Networks*. Elsivier Science Publishers, 1991. pp. 1269.

________."Genetic Programming: Building Artificial Nervous Systems with Genetically Programmed Neural Network Modules" in Soucek, B., and the IROS Group, Eds. *Neural and Intelligent Systems Integration*. New York: John Wiley, 1991. pp. 207–234.

Getting, P. A., and Dekin, M. S. "Tritonia Swimming: A Model System for Integration Within Rhythmic Motor Systems" in Silverston, A., Ed. *Model Neural Networks and Behavior*. New York: Plenum Press, 1985. pp. 3–20.

Giesy, Jr., J. P., Ed. *Microcosms in Ecological Research*. U.S. Dept. of Energy, Technical Information Center, 1980.

Goldberg, D. E. *Genetic Algorithms in Search, Optimization, and Machine Learning*. New York: Addison-Wesley, 1989.

Goldsbrough, P. *Microcomputer Interfacing with the 8255 PPI Chip*. Howard W. Sams Publishers, 1979.

Goles, E. and Vichniac, G.Y. "Neural Networks for Computing" in Denker, J.S., Ed. *AIP* 151, American Institute of Physics, New York, 1986.

Guez, A. Eilbert, J. and Kam, M. "Neuromorphic Architectures for fast Adaptive Robot Control." IEEE International Conference on Robotics and Automation, Philadelphia, PA, April 1988. pp. 145–149.

Haberman, R. *Mathematical Models Mechanical Vibrations, Population Dynamics, and Traffic Flow*. Englewood Cliffs, NJ: Prentice Hall, 1977.

Hanson, J. A. *Workshop on Closed System Ecology*. NASA JPL Publication #82–64, 1982.

Hartley, R. and Pipitone, F. "Experiments with the Subsumption Architecture." Proceedings of the 1991 IEEE International Conference on Robotics and Automation, Sacramento, CA, April 1991. pp. 1652–1658.

Hashimoto, M.; Kakeda, M.; Sagawa, H.; Chiba, I.; and Sato, K. "Application of Shape Memory Alloy to Robotic Actuators." *Journal of Robotic Systems* 2(1), 1985. pp. 3–25.

Hebb, D.O. *The Organization of Behavior*. New York: John Wiley, 1949.

Hebert, M. and Kanade, T. "3-D Vision for Outdoor Navigation by an Autonomous Vehicle" in Brady, M., Ed. *Robotics Science*. MIT Press, 1989. pp. 208–224.

Heiserman, D. L. *Build Your Own Working Robot*. Blue Ridge Summit: TAB Books, 1976.

________. *How to Build Your Own Self-Programming Robot*. Blue Ridge Summit: TAB Books, 1979.

________. *Robot Intelligence with Experiments*. Blue Ridge Summit: TAB Books, 1981.

Helmers, C., Ed. *Robotics Age: In the Beginning*. Hasbrouck Heights: Hayden Book Company, 1983.

Hill IV, J. and Wiegert, R. G. "Microcosms in Ecological Modeling" in Giesy Jr., J. P. *Microcosms in Ecological Research*. U.S. Dept of Energy, Technical Information Center, 1980. pp. 138–163.

Hinton, G. E. and Sejnowski, T. J. "Learning and Relearning in Boltzman Machines" in Rumelhart, D. E.; McClelland, J. L.; and the PDP Research Group, Eds. *Parallel Distributed Processing: Explorations in the Microstructure of Cognition Volume 1 Foundations*. Cambridge: MIT Press, 1986.

Hirose, S. and Kunieda, O. "Generalized Standard Foot Trajectory for a Quadruped Walking Vehicle." *International Journal of Robotics Research* 10(4), 1991. pp. 3–12.

Hirose, S. and Morishima, A. "Design and Control of a Mobile Robot with an Articulated Body." *International Journal of Robotics Research* 9(2), 1990. pp. 99–114.

Holland, J. M. *Basic Robotic Concepts*. Indianapolis, IN: Howard W. Sams, Co., 1983.

Holland, O. and Snath, M. "The Use of Recursively Generated Iterated Structures to Constrain Neural Network Architectures." Paper from TAG Robotics, Alnwick, U.K., 1991.

________. "Interrupt-Driven Architectures for Behavioral Robots: A Neural Implementation." Paper from TAG Robotics, Alnwick, U.K., 1991.

_________. "The Blind Neural Network Maker: Can We Use Constrained Embrologies to Design Animat Nervous Systems?" in Kohonen, T.; Makisara, K.; Simula, O.; and Kangas, J., Eds. *Artificial Neural Networks*. North-Holland: Elsiver Science, 1991. pp. 1261–1264.

Hollis, R. "Newt: A Mobile, Cognitive Robot." *BYTE*, June 1977. pp. 30–45

Hopfield, J.J. *Proc. Natl. Acad. Sci. U.S.A.* 79, 1982. pp. 2554.

_________. *Proc. Natl. Acad. Sci. U.S.A.* 81, 1984. pp. 3088.

Horne, B.; Jamshidi, M.; and Vadiee, N. "Neural Networks in Robotics: A Survey." *Journal of Intelligent and Robotic Systems* 3, 1990. pp. 51–66.

Huberman, B. A., Ed. *The Ecology of Computation*. North-Holland, Amsterdam, 1988.

Hughes, G. *BYTE*, December 1986. p.161.

Iwamoto, T. and Yamamoto, H. "Starway Travel of a Mobile Robot with Terrain-Adaptable Crawler Mechanism." *Journal of Robotic Systems* 2(1), 1985. pp. 125–134.

Jameson, J. W. "The Walking Gyro." *Robotics Age*, January 1985. pp. 7–10.

Josin, G., Charney, D., and White, D. "Robot Control Using Neural Networks," Proc. IEEE Int'l. Conf. on Neural Networks, II, 625, San Diego, CA, 1988.

Kampis, G. *Self-Modifying Systems in Biology and Cognitive Science*, New York: Pergamon Press, 1991.

Karr, C. L., "Design of an Adaptive Fuzzy Logic Controller Using a Genetic Algorithm," in Belew , R. K. and Booker, L. B. *Genetic Algorithms*. Proceeding of Fourth Int. Conf. on Genetic Algorithms, Morgan Kaufmann, San Mateo, CA, 1991. pp. 450–457.

Kawamura, S. and Nakagawa, M. "Memory-Based Control For Recognition of Motion Environment and Planning of Effective Locomotion." IEEE International Workshop on Intelligent Robots and Systems, IROS, Tsuchiura, Ibaraki, Japan, July 3–4, 1990. pp. 303–308.

Kearns, E. A. and Folsome, C. E. "Measurement of Biological Activity in Materially Closed Microbial Ecosystems", *Biosystems* 14, 1981. pp. 205–209.

Kearns, E. A. *Materially Closed Microbial Ecosystems*. Master of Science Thesis, University of Hawaii, 1981.

_________. *Efficiency of Energy Utilization in Thermodynamically Closed Ecosystems*. PhD Dissertation, University of Hawaii, 1983.

Kephart, J. O.; Hogg, T.; and Huberman, B. A. "Dynamics of Computational Ecosystems." *Physical Rev.* A40(1). pp. 404.

Kimura, H.; Wang, Z.; and Nakano, E. "Huge Object Manipulation in Space by Vehicles." Proceedings IROS '90, IEEE Int. Workshop on Intelligent Robots and Systems '90, Towards a New Frontier of Applications, July 3–6, 1990. pp. 393–397.

Koditschek, D. E. and Buhler, M. "Analysis for a Simplified Hopping Robot." *International Journal of Robotics Research* 10(6), 1991. pp. 587–605.

Kohonen, T. *Self-Organization and Associative Memory*. Springer-Verlag, 1984.

Koza, J. R. and Rice, J. P. "Genetic Generation of Both the Weights and Architecture for a Neural Network." IJCNN-91-Seattle, II–397, July, 1991.

Kung, S and Hwang, J. "Neural Network Architectures for Robotic Applications," IEEE Trans. Robotics and Automation, 5(641), 1989.

Kuribayashi, K. "A New Actuator of a Joint Mechanism Using TiNi Alloy Wire." *International Journal of Robotics Research* 4(4), 1986. pp. 47–58.

Kuwato, M., Uno, Y., Isobe, M., and Suzuki, R. "A Hierarchical Model for Voluntary Movement and its Application to Robotics", Proc. IEEE First Intl. Conf. on Neural Networks, IV, 573, San Diego, 1987.

Ladd, S. R. "Computer Life: Creating Mutations on Your Very Own Computer." *Micro Cornucopia* #49, September/October 1989. pp. 50–56.

Lalli, C. M., Ed. *Enclosed Experimental Marine Ecosystems: A Review and Recommendations*. New York: Springer-Verlag, 1990.

Langton, C.G. "Self-Reproduction in Cellular Automata." *Physica* 10D, 1984. pp. 135–144.

________. "Studding Artificial Life With Cellular Automata." *Physica* 22D, 1986. pp. 120–149.

________. *Artificial Life*. Proceedings of an Interdisciplinary Workshop on the Synthesis and Simulation of Living Systems. Redwood City: Addison-Wesley Publishing, 1989.

Langton, C. G.; Taylor, C.; Farmer, J. D.; and Rasmussen, S., Eds. *Artificial Life II*. New York: Addison-Wesley, 1992.

LeCun, Y. "Learning Processes in an Asymmetric Threshold Network" in Bienenstock, E.; Fogelman-Soulie, F.; and Weisbuch, G., Eds. *Disordered Systems and Biological Organization*. NATO ASI Series F, Vol. 20. Berlin: Springer-Verlag, 1986.

Lee, J. and Bien, Z., "Collision-Free Trajectory Control for Multiple Robots Based on Neural Optimization Network." *Robotica* 8, 1990. pp. 185–194.

Leffler, J. W. "Microcosmology: Theoretical Applications of Biological Models." in Giesy, Jr., J., Ed. *Microcosms in Ecological Research*, U.S. Department of Energy, Technical Information Center, 1980. pp. 14–29.

Li, W.; Packard, N. H.; and Langton, C. G. "Transition Phenomena in Cellular Automata Rule Space." *Physica* D45, 1990. pp. 77–94.

Louis, S. J. and Rawlins, G. J. E., "Designer Genetic Algorithms: Genetic Algorithms in Structure Design" in Belew, R. K. and Booker, L. B., *Genetic Algorithms*, Proceedings of the Fourth Int. Conf. San Mateo: Morgan Kaufmann, CA, 1991. pp. 53–60.

Lovelock, J. E. *Gaia: A New Look At Life on Earth*. Oxford: Oxford University Press, 1979.

_________. "Geophysiology: A New Look at Earth Science." *Bull. Am. Meterol. Soc.* 67(4), 1986. pp. 392–397.

_________. *The Ages of Gaia*. New York: W. W. Norton, 1988.

Lucasius, C. B.; Blommers, M. J. J.; Buydens, L. M. C.; and Kateman, G. "A Genetic Algorithm for Conformational Analysis of DNA" in Davis, L., Ed. *Handbook of Genetic Algorithms*. New York: Von Nostrand Reinhold, 1991. pp. 251–281.

Madison, T. J. "A PRIME Walking Machine." *Robot Experimenter* 4, 1986. pp. 2–11.

Maguire, B. "Ecosystem Simulation Through Use of Models of Subsystem Response Structures." *Simulation*, November 1975. pp. 149–158.

_________. "Modeling of Ecological Process and Ecosystems with Partial Response Structures: A Review and a new Paradigm for Diagnosis of Emergent Ecosystem Dynamics and Patterns" in Jorgensen, S.E., Ed. *State-of-the-Art in Ecological Modelling*, Volume 7, International Society fo Ecological Modelling. New York: Pergamon Press, 1978. pp. 59–126.

Maguire, Jr., B.; Slobodkin, L. B.; Marowitz, H. J.; Moore III, B.; and Botkin, D. B. "A New Paradigm for the Examination of Closed Ecosystems" in Giesy, Jr., J. P. *Microcosms in Ecological Research*. U.S. Department of Energy, Technical Information Center, 1980. pp. 30–69.

Maguire, Jr., B. "Some Patterns in Post-Closure Ecosystem Dynamics (Failure)", in Giesy, Jr., J. P. *Microcosms in Ecological Research*. U.S. Department of Energy, Technical Information Center, 1980. pp. 319–332.

Mandelbrot, B. B. *The Fractal Geometry of Nature*. San Francisco: W. H. Freeman, 1977.

Mandrick, B. and Moyson, F. "Self-Organization Versus Programming in Massively Parallel Systems: A Case Study" in Eckmiller, R.; Hartmann, G.; and Hauske, G. Eds. *Parallel Processing in Neural Systems and Computers*. New York: North-Holland, 1990. pp. 195–199.

Maricic, B. "Genetically Programmed Neural Network for Solving Pole-Balancing Problems." in Kohonen, T.; Kakisara, O.; Simula, O.; and Kangas, J., Eds. *Artificial Neural Networks*. Elsivier Science Publishers, 1991. pp. 1273.

Margulis, N., "Physics-Like Models of Computation." *Physicia* 10D, 1984. pp. 81–95.

Margulis, L. and Lovelock, J. E. "Biological Modulation of the Earth's Atmosphere." *Icarus* 21, 1974. pp. 471–489.

Margulis, L. and Sagan, D. *Microcosmos: Four Billion Years of Microbial Evolution*. New York: Summit Books, 1986.

Martin, F. Paper presented at Alife III, Santa Fe, June 1992.

Mataric, M. J. "From Interactions to Intelligent Behavior." Paper presented at ALIFE III, Santa Fe, NM, June 1992.

Maturana, H. R. and Varela, F. J. *Autopoiesis and Cognition: The Realization of the Living*. Boston: D. Reidel Pub., 1980.

May, R. M. *Stability and Complexity in Model Ecosystems*. Princeton: Princeton University Press, 1975.

_________. *Nature* 216, 1976. pp. 459.

_________. "The Dynamics of Insect Faunas" in Mound, L. A. and Waloff, N., Eds. *Diversity of Insect Faunas*. Symposia of The Royal Entomological Society of London, 1984. pp. 188–204.

McEliece, R. J.; Posner, E. C.; and Rodemich, E. R. *Twenty-Third Annual Allerton Conference on Communication Control and Computing*. October 1985.

Mein, R. *Cooperative Behavior in Uniformly and Differentially Programmed LEGO Vehicles*. MSc Thesis, Edinburgh University, 1991.

Mel, B. W. *Connectionist Robot Motion Planning: A Neurally-Inspired Approach to Visually-Guided Reaching*. San Diego: Academic Press, 1990.

Metropolis, N.; Rosenbluth, A.N.; Rosenbluth, M.N.; Teller, A.H.; and Teller, E. "Equations of State Calculations by Fast Computing Machines." *Journal of Chemistry and Physics* 21, 1953. pp. 1087–1091.

Meystel, A. *Autonomous Mobile Robots: Vehicles with Cognitive Control*. New Jersey: World Scientific Publishers, 1991.

Middleton, J. and Weston, R. H. "Structured Hardware and Software for Robots." *The Industrial Robot*, June 1982. pp. 92–96.

Miller, D. P. "Multiple Behavior-Controlled Micro-Robots for Planetary Surface Missions." Conference Proceedings, IEEE Int. Conference on Systems, Man, and Cybernetics, November 4–7, 1990. pp. 289–292.

Miller III, W. T. "Real-Time Application of Neural Networks for Sensor-Based Control of Robots with Vision." *IEEE Trans. on Systems, Man, and Cybernetics* 19(4), 1989. pp. 825–831.

Minsky, M. *The Society of Mind*. New York: Simon and Schuster, 1986.

Miyamoto, H., Kuwato, M., Setoyama, T., and Suzuki, R., "Feedback-Error_ Learning Neural Network for Trajectory Control of a Robotic Manipulator", Neural Networks, 1(251), 1988.

Moravec, H. *Mind Children: The Future of Robots and Human Intelligence*. Cambridge, MA: Harvard University Press, 1988.

Moreno, A.; Fernandez, J.; and Etxeberria, A. "Cybernetics, Autopoiesis, and Definition of Life" in Trappl, R., Ed. *Cybernetics and Systems '90*. New Jersey: World Scientific Pub., 1990. pp. 357–364.

Muhlenbein, H.; Gorges-Schleuter, M.; and Kramer, O. "Evolution Algorithms in Combinatorial Optimization." *Parallel Computing* 7, 1988. pp. 65–85.

Murray, J. D. *Mathematical Biology*. New York: Springer Verlag, 1989.

Nagata, S.; Sekiguchi, M.; and Asakawa, K. "Mobile Robot Control by a Structured Hierarchical Neural Network." *IEEE Control Systems Magazine*, April 1990. pp. 69–76.

Nguyen, D. H., and Widrow, B., "Neural Networks for Self-Learning Control Systems", IEEE Control Systems Magazine, 10(18), 1990.

Noreils, F. R. "Integrating Multirobot Coordination in a Mobile Robot Control System.", Proceedings IROS '90, IEEE Int. Workshop on Intelligent Robots and Systems '90, Towards a New Frontier of Applications, July 3–6, 1990. pp. 43–49.

Obenhuber, D. C. and Folsome, C. E., "Eucaryote/Procaryote Ratio as an Indicator of Stability for Closed Ecological Systems." *Biosystems* 16, 1984. pp. 291–296.

O'dell, J. *TRS-80 as Controller: An Introduction to Interfacing*. Wayne Green Books, 1983.

Odum, H. T. "Self-Organization, Transformity, and Information." *Science* 242, 1988. pp. 1132–1139.

Okubo, A. *Diffusion and Ecological Problems: Mathematical Models*. New York: Springer-Verlag, 1980.

Orvis, W. J. *1-2-3 For Scientists & Engineers*. San Francisco: Sybex, 1987.

Oster, G. and Ipaktchi, A. "Population Cycles" in Eyring, H. and Henderson, D. *Theoretical Chemistry Periodicities in Chemistry and Biology*. Vol.4. New York: Academic Press, 1978. pp. 111–132.

Paul, R. P. *Robot Manipulator Mathematics, Programming, and Control*. Cambridge: MIT Press, 1981.

Pebody, M. *How to Make a LEGO Vehicle Do the Right Thing*. MSc Thesis, Edinburgh University, 1991.

Pennik, R. W. *Fresh-Water Invertebrates of the United States, 2nd Ed.* New York: John Wiley, 1978.

Pickover, C. A. "Biomorphs: Computer Displays of Biological Forms Generated from Mathematical Feedback Loops." *Computer Graphics Forum* 5(4), 1987. pp. 313–316.

Pickover, C. A. *Computers, Pattern Chaos and Beauty: Graphics from an Unseen World*. New York: St. Martins Press, 1990.

Pierrot, F.; Fournier, A.; and Dauchez, P. "Towards A Fully-Parallel 6 DOF Robot for High Speed Applications." International Conference on Robotics and Automation, Sacramento, California, April 1991. pp. 1288–1293.

Polunin, N., Ed. *Ecosystem Theory and Application*. New York: John Wiley, 1986.

Posch, H. A.; Narnhofer, H.; and Thirring, W. "Dynamics of Unstable Systems." *Physical Review* A42(4), 1990. pp. 1880–1890.

Poundstone, W. *The Recursive Universe: Cosmic Complexity and the Limits of Scientific Knowledge*. Chicago: Contemporary Books, 1985.

Premvuti, S. and Yuta, S. "Consideration on the Cooperation of Multiple Autonomous Mobile Robots." Proceedings IROS '90, IEEE Int. Workshop on Intelligent Robots and Systems '90, Towards a New Frontier of Applications, July 3–6, 1990. pp. 59–63.

Press, W. H.; Flannery B. P.; Teukolsky, S. A.; and Vetterling, W. T. *Numerical Recipes in C: The Art of Scientific Computing*. Cambridge: Cambridge University Press, 1988.

Prusinkiewicz, P. and Hanan, J. *Lindenmayer Systems, Fractals, and Plants*. New York: Springer-Verlag, 1980.

Prusinkiewicz, P and Lindenmayer, A. *The Algorithmic Beauty of Plants*. New York: Springer-Verlag, 1990.

Psaltis, D., Sidiris, A., and Yamamura, A, "Neural Controllers", Proc. IEEE First Intl. Conf. on Neural Networks, IV, 551, San Diego, 1987.

_________., "A Multilayered Neural Network Controller", *IEEE Control Systems Magazine* 8, 1988.

Rai, V.; Kumar, V; and Pande, L. K. "A New Prey-Predator Model." *IEEE Transactions on Systems, Man, and Cybernetics* 21(1), 1991. pp. 261–263.

Raibert, M. H. "Legged Robots." *Comm. of the ACM* 29(6), 1986. pp. 499–514.

Ravera, O., Ed. *Terrestrial and Aquatic Ecosystems: Perturbation and Recovery*. New York: Ellis Horwood, 1991.

Ray, T. S. "An Approach to the Synthesis of Life" in Langton, C. G.; Taylor, C.; Farmer, J. D.; and Rasmussen, S., Eds. *Artificial Life II*. Redwood City: Addison-Wesley, 1992. pp. 371–408.

________. "Is it Alive or is it a GA?" in Belew , R. K. and Booker, L. B., Eds. *Genetic Algorithms*. Proceedings of the Fourth International Conference on Genetic Algorithms. Morgan Kaufman Publishers, 1991. pp. 527–534.

________. Notes from the ALIFE network on INTERNET, Alife Digens #066, 1991.

Resnick, M. "LEGO, Logo, and Life" in Langton, C. G., Ed. *Artificial Life*. Redwood City, CA: Addison-Wesley Publishers, 1989. pp. 397–406.

________. "Animal Simulations with *Logo: Massive Parallelism for the Masses" in Meyer, J.A. and Wilson, S.W., Eds. *From Animals to Animats*. Cambridge: MIT Press, 1991. pp. 534–539.

Ressmeyer, R. H. "Trouble in Paradise." *Air & Space*, December '91/January '92, 1992. pp. 55–65.

Rietman, E. A. "A Simplified Pattern Recognition System Using a Personal Computer." AT&T Bell Labs. Memo #11535-851120-48TM, 1985.

________. "A Simple Microprocessor Controlled Robot Arm." AT&T Bell Labs. Memo #11535-860910-27TM, 1986.

________. *Experiments in Artificial Neural Networks*. Blue Ridge Summit, PA: TAB Books, 1988.

________. *Exploring the Geometry of Nature: Computer Modeling of Chaos, Fractals, Cellular Automata and Neural Networks*. Blue Ridge Summit, Pa: TAB Books, 1989.

________. *Exploring Parallel Processing*. Blue Ridge Summit, PA: TAB Books, 1990

________. "Experiments on Simulation of an Artificial Life Form Using a Neural Network." *Journal of British American Scientific Research Association* XXXI(6), June 1991. pp. 34–46.

Rietman, E. A. and Frye, R. C. "Neural Control of a Nonlinear System with Inherent Time Delays." ACM Conference, Analysis of Neural Network Applications, ANNA-91, May 1991.

Rivetta, A., "Design and Control of Mechanical Hands for Robots", *Computers in Industry* 7, 1986. pp. 275–282.

Robillard, M. J. *Microprocessor-Based Robotics*. Indianapolis: Howard W. Sams, 1983.

_________. *Microcomputing* 42, August 1984.

_________. *Advanced Robot Systems*. Indianapolis: Howard W. Sams, 1984.

_________. "Robotix: Part I, Examining the Pieces." *Robotics Age*, December 1984. pp. 20–23.

_________. "Robotix: Part II, External Control." *Robotics Age*, January 1985. pp. 20–21.

_________. "Inexpensive Robotics Arms." *Radio-Electronics*, July 1986. pp. 80–81.

Rosen, R. *Life Itself: A Comprehensive Inquiry into the Nature, Origin and Fabrication of Life*. New York: Columbia University Press, 1991.

Rosenblatt, J. K. and Payton, D. W. "A Fine-Grained Alternative to the Subsumption Architecture for Mobile Robot Control." International Joint Conference Neural Networks, Washington D.C. II–317, June 1989.

Rumelhart, D. E.; Hinton, G. E.; and Williams, R. J. "Learning Internal Representations by Error Propagations" in Rumelhart, D. E.; McClelland, J. L.; and the PDP Research Group, Eds. *Parallel Distributed Processing: Explorations in the Microstructure of Cognition, Vol. 1*. Cambridge: MIT Press, 1986.

_________., "Learning Internal Representations by Backpropagation Errors", Nature, 323(533), 1986.

Sagan, D. *Biospheres: Reproducing Planet Earth*. New York: Bantam Books, 1990.

Sampson, J. P. *Biological Information Processing: Current Theory and Computer Simulation*. New York: John Wiley, 1984.

Schaffer, J. D.; Caruana, R. A.; and Eshelman, L. J. "Using Genetic Search to Exploit the Emergent behavior of Neural Networks." *Physica* D42, 1990. pp. 244–248.

Schiavone, J. J., Dawson, M.; and Brandeberry, J. E. "Super Armatron: An Inexpensive, Microprocessor-controlled Robot Arm." *Robotics Age*, January 1984. pp. 20–28.

Schindler, J. E.; Waide, J. B.; Waldron, M. C.; Hains, J. J.; Schreiner, S. P.; Freedman, M. L.; Benz, S. L.; Pettigrew, D. R.; Schissel, L. A; and Clark, P. J. "A Microcosm Approach to the Study of Biogeochemical Systems. 1. Theoretical Rational" in Giesy, Jr., J. P., Ed. *Microcosms in Ecological Research*. U.S. Department of Energy, Technical Information Center, 1980. pp. 192–203.

Schindler, J. M. *1989 Yearbook of Developmental Biology*. Boca Raton: CRC Press, 1989.

Schneider, S. H., and Londer, R. *The Coevolution of Climate and Life*. San Francisco: Sierra Club Books, 1984.

Selverston, A. I., Ed. *Model Neural Networks and Behavior*. New York: Plenum Press, 1985.

Shaffer, J. *Stable Organization in Closed Ecological Systems: An Examination of Various Parameters*. Preprint of PhD Dissertation, University of Hawaii, 1991.

Shahinpoor, M. "Kinematics of a Parallel-Serial (Hybrid) Manipulator." *Journal of Robotic Systems* 9(1), 1992. pp. 17–36.

Shonkwiler, R., Mendivil, F., and Deliu, A., "Genetic Algorithms for the 1-D Fractal Inverse Problem", in Belew, R. K. and Booker, L. B., *Genetic Algorithms*, Fourth Int. Conf., San Mateo, CA: Morgan Kaufmann Pub., 1991. pp. 495–501.

Smil, V. *General Energetics: Energy in the Biosphere and Civilization*. New York: John Wiley, 1991.

Smit, M. C. and Tilden, M. W. "Beam Robotics." *Algorithm* 2.2, March 1991. pp. 15–19.

Smith, A. R. "Real-Time Language Recognition by One-Dimensional Cellular Automata." *Journal of Computer & System Science* 6(3), 1972. pp. 233–253.

________. "Plants, Fractals, and Formal Languages." *Computer Graphics* 18(3), 1984. pp. 1–10.

________. "Graftal Formalisms Notes." Lucasfilm Ltd. Computer Graphics Dept. Tech. Memo. No. 114, 1985.

________. "Formal Geometric Languages for Natural Phenomena." Pixar Technical Memo. No. 182, San Rafael, CA, 1987.

________. "Designing Biomorphs With Interactive Genetic Algorithms" in Belew, R. K. and Booker, L. B., Eds. *Genetic Algorithms*, Proceedings of the Fourth International Conference on Genetic Algorithms, Morgan Kaufman, 1991. pp. 535–538.

Snath, M. "The Use of Simple Behavioral and Neural Modelling in the Development of Adaptive Control Systems for Autonomous Robots." Paper from TAG Robotics, Alnwick, U.K., 1989.

Snath, M. and Holland, O. "Quadrapedal Walking Using Trained and Untrained Neural Models" in Kohonen, T.; Makisara, K.; Simula, O.; and Kangas, J., Eds. *Artificial Neural Networks*. North-Holland: Elsiver Science, 1991. pp. 1257–1260.

_________. "The Application of Temporal Difference Learning to the Neural Control of Quadruped Locomotion." Paper from TAG Robotics, Alnwick, U.K., 1991.

_________. "A Feature-Based Navigator for Mobile Robots." Paper from TAG Robotics, Alnwick, U.K., 1990.

Stark, J. "Iterated Function Systems as Neural Networks" *Neural Networks* 4, 1991. pp. 679–690.

Stevens, R. T. *Fractal Programming in C*. Redwood City: M & T Books, 1989.

Taga, G.; Yamaguchi, Y.; and Shimizu, H. "Self-Organized Control of Bipedal Locomotion by Neural Oscillators in Unpredictable Environment." *Biol. Cybernetics* 65, 1991. pp. 147–159.

Takanishi, A.; Lim, H.O.; Tsuda, M.; and Kato, I. "Realization of Dynamic Biped Walking Stabilized by Trunk Motion on a Segittally Uneven Surface." Proceedings IROS '90, IEEE Int. Workshop on Intelligent Robots and Systems '90, Towards a New Frontier of Applications, July 3–6, 1990. pp. 323–330.

Takano, C. T.; Folsome, C. E.; and Karl, D. M. "ATP as a Biomass Indicator for Closed Ecosystems." *Biosystems* 16, 1983. pp. 75–78.

Tank, D. W. and Hopfield, J. J. "Simple Neural Optimization Networks: An A/D converter, Signal Decision Circuit, and a Linear Programming Circuit." IEEE Trans. Circuits Systems CAS-33 (5), 1986. pp. 533–541.

Taub, F. B. and Crow, M. E. "Synthesizing Aquatic Microcosms" in Giesy, Jr., J. P. *Microcosms in Ecological Research*. U.S. Department of Energy, Technical Information Center, 1980. pp. 69–104.

Taylor, M. Private communication, 1991.

Taylor, P. M. *Understanding Robotics*. Boca Raton: CRC Press, 1990.

Tedeschini-Lallin, L. *Journal of Stat. Physics* 27, 1982. p.365.

Titus, J. *TRS-80 Interfacing Book 1*. Howard W. Sams Publishers, 1979.

Tilden, M. W. Paper presented at ALIFE III, Santa Fe, June 1992.

Todd, P. M. and Miller, G. F., "On the Sympatric Origin of Species: Mercurial Mating in the Quicksilver Model" in Belew, R. K. and Booker, L. B. *Genetic Algorithms*. Proceeding of the Fourth Int. Conf. San Mateo, CA: Morgan Kaufmann Pub., 1991.

Travers, M. "Animal Construction Kits" in Langton, C. G., Ed. *Artificial Life*. New York: Addison-Wesley, 1989. pp. 421–442.

Ural, S. "Functions of Complex Variables." *Circuit Celler Ink: Computer Applications Journal* #17, October/November 1990. pp. 16–22.

Vakakis, A. F.; Burkick, J. W.; and Caughey, T. K. "An Interesting Strange Attractor in the Dynamics of a Hopping Robot." *International Journal of Robotics Research* 10(6), 1991. pp. 606–618.

Van Neumann, J. *Theory of Self-Reproducing Automata*. Burks, A.W., Ed. Urbana: University of Illinois Press, 1966.

Varela, F. G.; Maturana, H. R.; and Uribe, R. "Autopoiesis: The Organization of Living Systems, Its Characterization, And A Model." *Biosystems* 5, 1974. pp. 187–196.

Waide, J. B.; Schindler, J. E.; Waldron, M. C.; Hains, J. J.; Schreiner, S. P.; Freedman, M. L.; Benz, S. L.; Pettigrew, D. R.; Schissel, L. A.; and Clark, P. J. "A Microcosm Approach to the Study of Biogeochemical Systems. 2. Responses of Aquatic Laboratory Microcosms to Physical, Chemical, and Biological Perturbations" in Giesy, Jr., J. P., Ed. *Microcosms in Ecological Research*. U.S. Department of Energy, Technical Information Center, 1980. pp. 204–223.

Wainwrght, R. E. and Moss, R. H. "A Microcomputer-based Model Robot System with Pulse-Width Modulation Control." *IEEE Micro*, February 1985. pp. 7–21.

Waldrop, M. M. "Fast, Cheap, and Out of Control." *Science* 248, 1990. pp. 959–961.

Wallich, P. "Silicon Babies." *Scientific American*, December 1991. pp. 124–134.

Walter, W. G. "An Imitation of Life" Scientific American, May 1950. pp. 42–45.

_________. "A Machine That Learns." Scientific American, August 1951. pp. 60–63.

Wang, P. K. C. "Interaction Dynamics of Multiple Mobile Robots with Simple Navigation Strategies." *Journal of Robotic Systems* 6(1), 1989. pp. 77–101.

Wasserman, P.D. *Neural Computing: Theory and Practice*. New York: Von Nostrand Reinhold, 1989.

Weinstein, M. B. *Android Design: Practical Approaches for Robot Builders*. Hasbrouck Heights: Hayden Books, 1981.

Weisbuch, B. *Complex Systems Dynamics: An Introduction to Automata Networks*. Lecture Notes Volume II, Santa Fe Institute Studies in the Sciences of Complexity. New York: Addison-Wesley, 1991.

Werner, N. *Cybernetics*. New York: John Wiley, 1948.

Wesley, J. P. *Ecophysics: The Application of Physics to Ecology*. Springfield: C. C. Thomas Pub., 1974.

Whitley, D.; Starkwather, T.; and Bogart, C. "Genetic Algorithms and Neural Networks: Optimizing Connections and Connectivity." *Parallel Computing* 14, 1990. pp. 347–361.

Whitley, D.; Starkwather, T.; and Shaner, D. "The Traveling Salesman and Sequence Scheduling: Quality Solutions Using Genetic Edge Recombination" in Davis, L., Ed. *Handbook of Genetic Algorithms*. New York: Von Nostrand Reinhold, 1991. pp. 350–372.

Widrow, B. and Hoff, M. E. "Adaptive Switching Circuits." *Ire Wescon, Convention Record*. New York, 1960. pp. 96–104.

________. "Associative Storage and Retrieval of Digital Information in Networks of Adaptive Neurons." Bernard, E. E. and Kare, M. R., Eds. *Biological Prototypes and Synthetic Systems Vol. 1*. New York: Plenum Press, 1962.

Wilson, S. "The Genetic Algorithm and Biological Development." *Genetic Algorithms and Their Applications*. Proceedings of the Second International Conference on Genetic Algorithms. Laurence Erlbaum Associates, 1987.

Wolfram, S. "Statistical Mechanics of Cellular Automata." *Rev. Mod. Phys.* 55(3), 1983. pp. 601–644.

________. "Some Recent Results and Questions about Cellular Automata" in Demongeot, J.; Goles, E.; and Tchuente, M., Eds. *Dynamical Systems and Cellular Automata*. New York: Academic Press, 1985.

Wolkomir, R. "Working the Bugs Out of a new Breed of Insect Robots." *Smithsonian* 22 (3), June 1991. pp. 64–73.

Wootters, W. K. and Langton, C. G. "Is There a Sharp Phase Transition for Deterministic Cellular Automata?" *Physica* D45, 1990. pp. 95–104.

Young, J. F. *Robotics*. London: Butterworths, 1973.

Zeleny, M. "Self-Organization of Living Systems: A Formal Model of Autopoiesis." *International Journal of General Systems* 4, 1977. pp. 13–28.

Index

Order Form for Readers Requiring a Single 5.25" Disk

This Windcrest/McGraw-Hill software product is also available on a 5.25"/360K disk. If you need the software in 5.25" format, simply follow these instructions:

- Complete the order form below. Be sure to include the exact title of the Windcrest/McGraw-Hill book for which you are requesting a replacement disk.
- Make check or money order made payable to *Glossbrenner's Choice*. The cost is **$5.00** (**$8.00** for shipments outside the U.S.) to cover media, postage, and handling. Pennsylvania residents, please add 6% sales tax.
- Foreign orders: please send an international money order or a check drawn on a bank with a U.S. clearing branch. We cannot accept foreign checks.
- Mail order form and payment to:

  Glossbrenner's Choice
  Attn: Windcrest/McGraw-Hill Disk Replacement
  699 River Road
  Yardley, PA 19067-1965

Your disks will be shipped via First Class Mail. Please allow one to two weeks for delivery.

✂..

Windcrest/McGraw-Hill Disk Replacement

Please send me a replacement disk in 5.25"/360K format for the following Windcrest/McGraw-Hill book:

Book Title ____

Name ____

Address ____

City/State/ZIP ____

DISK WARRANTY

This software is protected by both United States copyright law and international copyright treaty provision. You must treat this software just like a book, except that you may copy it into a computer in order to be used and you may make archival copies of the software for the sole purpose of backing up our software and protecting your investment from loss.

By saying "just like a book," McGraw-Hill means, for example, that this software may be used by any number of people and may be freely moved from one computer location to another, so long as there is no possibility of its being used at one location or on one computer while it also is being used at another. Just as a book cannot be read by two different people in two different places at the same time, neither can the software be used by two different people in two different places at the same time (unless, of course, McGraw-Hill's copyright is being violated).

LIMITED WARRANTY

Windcrest/McGraw-Hill takes great care to provide you with top-quality software, thoroughly checked to prevent virus infections. McGraw-Hill warrants the physical diskette(s) contained herein to be free of defects in materials and workmanship for a period of sixty days from the purchase date. If McGraw-Hill receives written notification within the warranty period of defects in materials or workmanship, and such notification is determined by McGraw-Hill to be correct, McGraw-Hill will replace the defective diskette(s). Send requests to:

Customer Service
Windcrest/McGraw-Hill
13311 Monterey Lane
Blue Ridge Summit, PA 17294-0850

The entire and exclusive liability and remedy for breach of this Limited Warranty shall be limited to replacement of defective diskette(s) and shall not include or extend to any claim for or right to cover any other damages, including but not limited to, loss of profit, data, or use of the software, or special, incidental, or consequential damages or other similar claims, even if McGraw-Hill has been specifically advised of the possibility of such damages. In no event will McGraw-Hill's liability for any damages to you or any other person ever exceed the lower of suggested list price or actual price paid for the license to use the software, regardless of any form of the claim.

MCGRAW-HILL, INC. SPECIFICALLY DISCLAIMS ALL OTHER WARRANTIES, EXPRESS OR IMPLIED, INCLUDING, BUT NOT LIMITED TO, ANY IMPLIED WARRANTY OF MERCHANTABILITY OR FITNESS FOR A PARTICULAR PURPOSE.

Specifically, McGraw-Hill makes no representation or warranty that the software is fit for any particular purpose and any implied warranty of merchantability is limited to the sixty-day duration of the Limited Warranty covering the physical diskette(s) only (and not the software) and is otherwise expressly and specifically disclaimed.

This limited warranty gives you specific legal rights; you may have others which may vary from state to state. Some states do not allow the exclusion of incidental or consequential damages, or the limitation on how long an implied warranty lasts, so some of the above may not apply to you.

IMPORTANT

Read the Disk Warranty terms on the previous page before opening the disk envelope. Opening the envelope constitutes acceptance of these terms and renders this entire book-disk package nonreturnable except for replacement in kind due to material defect.

If you need help with the enclosed disk

You might find it more convenient to keep these C, Pascal, and BASIC programs on your hard drive instead of having to access them from the floppy disk. To create a hard drive subdirectory to store these files in, type

MKDIR *directory-name*

at your hard drive prompt (most likely C:\), where *directory-name* is what you want to name the subdirectory.

To copy the files, place your disk in your floppy drive (probably drive B) and type

COPY B:*.*C:*directory-name*

The files on the disk will now be copied to your newly created subdirectory.